Review of Biological Research in Aging Volume 4

Review of Biological Research in Aging
Volume 4

EDITOR
Morton Rothstein
Department of Biological Sciences
State University of New York at Buffalo
Buffalo, New York

ASSOCIATE EDITORS

William H. Adler
Gerontology Research Center
Baltimore City Hospital
Baltimore, Maryland

Vincent J. Cristofalo
The Wistar Institute
Philadelphia, Pennsylvania

Caleb E. Finch
Gerontology Center
University of Southern California
Los Angeles, California

James R. Florini
Biology Department
Syracuse University
Syracuse, New York

George M. Martin
Pathology Department
University of Washington
Seattle, Washington

A JOHN WILEY & SONS, INC., PUBLICATION
New York • Chichester • Brisbane • Toronto • Singapore

Address all Inquiries to the Publisher
Wiley-Liss, Inc., 41 East 11th Street, New York, NY 10003

Copyright © 1990 Wiley-Liss, Inc.

Printed in United States of America

Under the conditions stated below the owner of copyright for this book hereby grants permission to users to make photocopy reproductions of any part or all of its contents for personal or internal organizational use, or for personal or internal use of specific clients. This consent is given on the condition that the copier pay the stated per-copy fee through the Copyright Clearance Center, Incorporated, 27 Congress Street, Salem, MA 01970, as listed in the most current issue of "Permissions to Photocopy" (Publisher's Fee List, distributed by CCC, Inc.), for copying beyond that permitted by sections 107 or 108 of the US Copyright Law. This consent does not extend to other kinds of copying, such as copying for general distribution, for advertising or promotional purposes, for creating new collective works, or for resale.

ISSN 0736-5055 • ISBN 0-471-56697-7

Contents

Contributors vii
Preface ix

SECTION I
EVOLUTION AND GENETICS OF AGING

Evolution of Aging
Michael R. Rose and Joseph L. Graves, Jr. 3

Aging in *Caenorhabditis elegans*: Update 1988
Thomas E. Johnson and Edward W. Hutchinson 15

Aging of Fungi
Kenneth D. Munkres 29

Protozoa
Joan R. Smith-Sonneborn 41

Aging and Life Span Correlates in Insects
Thomas Grigliatti 57

Genetic Aspects of Aging in *Mus musculus*
Richard L. Sprott and Carol A. Combs 73

SECTION II
IMMUNOLOGY OF AGING

Mechanisms of Impaired T-Cell Function in the Elderly
Rajesh K. Chopra 83

Humoral Immunosenescence: An Update
David L. Ennist 105

SECTION III
NEUROLOGY OF AGING

The Genetics of Alzheimer's Disease: A Review of Recent Work
Jeremy M. Silverman, Richard C. Mohs, Linda M. Bierer,
Richard S.E. Keefe, and Kenneth L. Davis 123

Transgenic Mouse Models of Amyloidosis in Alzheimer's Disease
Axel Unterbeck, Richard M. Bayney, George Scangos, and Dana O. Wirak 139

Contents

Amyloid Precursor Protein (APP) mRNA in Normal and Alzheimer's Disease Brain
Steven A. Johnson 163

Brain Amyloid in Aging and Alzheimer's Disease
Koji Ogomori, Tetsuyuki Kitamoto, and Jun Tateishi 171

The Striatum, A Microcosm for the Examination of Age-Related Alterations in the CNS: A Selected Review
J.A. Joseph, G.S. Roth, and R. Strong 181

Neurotoxic Effects of Estrogen, Glucose, and Glucocorticoids: Neurohormonal Hysteresis and Its Pathological Consequences During Aging
Charles V. Mobbs 201

SECTION IV
ENDOCRINOLOGY OF AGING

Problems in Design and Interpretation of Aging Studies: Illustration by Reports on Hormone Secretion and Actions
J.R. Florini and F.J. Mangiacapra 231

Hormone/Neurotransmitter Action During Aging: The Calcium Hypothesis of Impaired Signal Transduction
George S. Roth 243

Effects of Aging on the Hypothalamic–Pituitary Axis
Joseph Meites 253

SECTION V
CELL BIOLOGY OF AGING

Recent Advances in Cellular Aging Research: Understanding the Limited Life Span of Normal Human Fibroblasts
Paul D. Phillips and Vincent J. Cristofalo 265

In Vitro Studies of Aging Human Epidermis: 1975–1990
Barbara A. Gilchrest 281

SECTION VI
BIOCHEMISTRY OF AGING

The Molecular Pathology of Senescence: A Comparative Analysis
Robert J. Shmookler Reis 293

Age-Related Effects in Enzyme Metabolism and Catalysis
Ari Gafni 315

Alzheimer's Disease: A View Toward the Neurites
Kenneth S. Kosik 337

Food Restriction Research: Past and Present Status
Byung P. Yu 349

Index 373

Contributors

Richard M. Bayney, Molecular Therapeutics Inc., Miles Research Center, West Haven, CT 06516 [139]

Linda M. Bierer, Department of Psychiatry, Mount Sinai School of Medicine, New York, NY 10029, and Bronx VA Medical Center, Bronx, NY 10468 [123]

Rajesh K. Chopra, Clinical Immunology Section, Gerontology Research Center, National Institute on Aging, National Institutes of Health, Baltimore, MD 21224 [83]

Carol A. Combs, Department of Biomedical Research and Clinical Medicine, National Institute on Aging, National Institutes of Health, Bethesda, MD 20892 [73]

Vincent J. Cristofalo, The Wistar Institute, Philadelphia, PA 19104 [265]

Kenneth L. Davis, Department of Psychiatry, Mount Sinai School of Medicine, New York, NY 10029, and Bronx Medical Center, Bronx, NY 10468 [123]

David L. Ennist, Section on Molecular Genetics of Immunity, Laboratory of Developmental and Molecular Immunity, National Institute of Child Health and Human Development, National Institutes of Health, Bethesda, MD 20892 [105]

J.R. Florini, Biology Department, Syracuse University, Syracuse, NY 13244 [231]

Ari Gafni, Institute of Gerontology and Department of Biological Chemistry, The University of Michigan, Ann Arbor, MI 48109 [315]

Barbara A. Gilchrest, Cutaneous Gerontology Laboratory, USDA Human Nutrition Research Center on Aging, Tufts University, Boston, MA 02111 [281]

Joseph L. Graves, Jr., Department of Ecology and Evolutionary Biology, University of California, Irvine, CA 92717 [3]

Thomas Grigliatti, Department of Zoology, University of British Columbia, Vancouver, British Columbia, Canada V6T 2A9 [57]

Edward W. Hutchinson, Institute for Behavioral Genetics, University of Colorado, Boulder, CO 80309-0447 [15]

Steven A. Johnson, Andrus Gerontology Center and Department of Neurobiology, University of Southern California, Los Angeles, CA 90089-0191 [163]

Thomas E. Johnson, Institute for Behavioral Genetics, University of Colorado, Boulder, CO 80309-0447 [15]

J.A. Joseph, Gerontology Research Center, National Institute on Aging, Francis Scott Key Medical Center, Baltimore, MD 21224 [181]

Richard S.E. Keefe, Department of Psychiatry, Mount Sinai School of Medicine, New York, NY 10029, and Bronx VA Medical Center, Bronx, NY 10468 [123]

Contributors

Tetsuyuki Kitamoto, Department of Neuropathology, Neurological Institute, Faculty of Medicine, Kyushu University, Fukuoka 812, Japan [171]

Kenneth S. Kosik, Center for Neurologic Diseases, Brigham and Women's Hospital, Harvard Medical School, Boston, MA 02115 [337]

F.J. Mangiacapra, Biology Department, Syracuse University, Syracuse, NY 13244 [231]

Joseph Meites, Department of Physiology, Michigan State University, East Lansing, MI 48824 [253]

Charles V. Mobbs, Rockefeller University, New York, NY 10021 [201]

Richard C. Mohs, Department of Psychiatry, Mount Sinai School of Medicine, New York, NY 10029 and Bronx VA Medical Center, Bronx, NY 10468 [123]

Kenneth D. Munkres, Laboratory of Molecular Biology and Department of Genetics, University of Wisconsin, Madison, WI 53706 [29]

Koji Ogomori, Department of Neuropathology, Neurological Institute, Faculty of Medicine, Kyushu University 60, Fukuoka 812, Japan [171]

Paul D. Phillips, The Wistar Institute, Philadelphia, PA 19104 [265]

Michael R. Rose, Department of Ecology and Evolutionary Biology, University of California, Irvine, CA 92717 [3]

George S. Roth, Molecular Physiology and Genetics Section, Gerontology Research Center, National Institute on Aging, Francis Scott Key Medical Center, Baltimore, MD 21224 [181, 243]

George Scangos, Molecular Therapeutics Inc., Miles Research Center, West Haven, CT 06516 [139]

Robert J. Shmookler Reis, Departments of Medicine and Biochemistry and Molecular Biology, University of Arkansas for Medical Sciences, Little Rock, AR 72205 [293]

Jeremy M. Silverman, Department of Psychiatry, Mount Sinai School of Medicine, New York, NY 10029, and Bronx VA Medical Center, Bronx, NY 10468 [123]

Joan R. Smith-Sonneborn, Departments of Zoology and Physiology, University of Wyoming, Laramie, WY 82070 [41]

Richard L. Sprott, Department of Biomedical Research and Clinical Medicine, National Institute on Aging, National Institutes of Health, Bethesda, MD 20892 [73]

R. Strong, VA Medical Center, GRECC, St. Louis, MO 63125 [181]

Jun Tateishi, Department of Neuropathology, Neurological Institute, Faculty of Medicine, Kyushu University, Fukuoka 812, Japan [171]

Axel Unterbeck, Molecular Therapeutics Inc., Miles Research Center, West Haven, CT 06516 [139]

Dana O. Wirak, Molecular Therapeutics Inc., Miles Research Center, West Haven, CT 06516 [139]

Byung P. Yu, Department of Physiology, Health Science Center at San Antonio, University of Texas, San Antonio, TX 78284-7756 [349]

Preface

Volume 4 of *Review of Biological Research in Aging* follows the practice established by previous books in the series: It updates previously considered topics of aging research, and it brings new attention to areas not reviewed in earlier volumes.

The editors have been fortunate, once again, in the willingness of outstanding experts to contribute to this series. We are confident that the reviews they have provided are of the highest quality. They should serve the reader well in his or her efforts to keep abreast of recent progress in biological studies of aging. To the degree that *Review of Biological Research in Aging* can aid investigators toward achieving that goal, the Editors (and, we are sure, the authors) will feel that their efforts have been worthwhile.

The Editors

SECTION I
EVOLUTION AND GENETICS OF AGING

Evolution of Aging

Michael R. Rose and Joseph L. Graves, Jr.

INTRODUCTION

Evolutionary geneticists seem to have discovered the ultimate cause of senescence: the declining force of natural selection with the age of the adult soma. Evidence for this basic theory accumulated steadily throughout the 1980s, along with elaboration of some its corollaries for aging research. Aging concepts in evolutionary biology are far more developed now than was the case in the late 1970s. However, active debate remains between those who feel that the case for the evolutionary theory of aging is well established and those who feel that it is not. The present review will attempt to reflect some of this ongoing ferment.

Modern research on the evolution of senescence couched in terms of natural selection acting on gene frequencies begins with Haldane [1941] and Medawar [1946, 1952]. Both drew attention to the central point of all evolutionary analyses of senescence: in multicellular organisms, deleterious effects on the adult soma will have a strictly decreasing impact on fitness as the age at which these effects are expressed increases, all other things being equal, partly because such effects have a strictly decreasing probability of expression. Thus the force of natural selection must decline with the age of the adult soma, and with it the health of the soma. This theory has been further elaborated and clarified by Williams [1957], Hamilton [1966], Edney and Gill [1968], Emlen [1970], Charlesworth and Williamson [1975], and Charlesworth [1980], especially pp. 204–223].

Two cogent population genetic mechanisms for the evolution of senescence have been proposed, both based on the decline of the force of natural selection with adult age.

1. Mutation accumulation: maintenance of high-frequency deleterious alleles by mutation pressure, when they have such late effects that natural selection

Department of Ecology and Evolutionary Biology, University of California, Irvine, California 92717

has little impact on their frequency [Medawar, 1952; Edney and Gill, 1968; Charlesworth, 1980].

2. Antagonistic pleiotropy: natural selection favoring genes with early beneficial effects that outweigh later, pleiotropic, deleterious effects, because the latter have little impact on fitness [Medawar, 1952; Williams, 1957; Rose, 1985].

The general evolutionary theory of senescence is compatible with both the mutation-accumulation and antagonistic pleiotropy mechanisms, and can thus be tested independently of them. In particular, evidence in favor of the general theory may have no bearing on the validity of these two subsidiary mechanisms. On the other hand, the subsidiary mechanisms cannot hold if the general theory doesn't, as they both presume its validity.

The general evolutionary theory can be tested in two distinct ways, the first experimental and the second comparative. One important corollary of the general theory is that the reproductive schedule of a population should affect the evolution of senescence within it [Edney and Gill, 1968], in the absence of inbreeding, such that longevity should be *reduced* in populations with a relatively *earlier* average age of reproduction, and *increased* in populations with a relatively *later* average age of reproduction. Such dependence of senescence on a population's reproductive schedule has been demonstrated repeatedly. Some evidence for accelerated senescence with earlier reproduction in *Tribolium* has been provided by Sokal [1970] and Mertz [1975], although there were problems of consistency over replicates, and thus of statistical significance. Evidence for postponed senescence in *Drosophila* populations with delayed reproduction has been published by Wattiaux [1968a,b], Taylor and Condra [1980], Rose and Charlesworth [1980, 1981b], Rose [1984a], and Luckinbill et al. [1984], although not all of these authors discussed their results from the standpoint of natural selection molding senescence. Some have found anomalous results [e.g., Lints and Hoste, 1974, 1977; Lints et al., 1979; Flanagan, 1980], but these experiments have been subjected to thoroughgoing analyses by Clare and Luckinbill [1985] and Luckinbill and Clare [1985], who concluded that these and certain of their own experiments "would have obtained a positive outcome if only selection had been applied long enough to effect the derepression of the appropriate genes" [Luckinbill and Clare, 1985].

The second way of testing the evolutionary theory of senescence is based on the absolute correspondence between evolutionary age-structure and the evolution of senescence implicit in the theory. When reproduction is fissile, there should be no senescence, and vice versa. There are some broad comparative analyses of this prediction [e.g., Williams, 1957] and some intensive analyses of particular cases [e.g., Bell, 1984]. The comparative work has tended to corroborate, rather than falsify, the evolutionary theory.

The specific population genetic mechanisms for the evolution of aging can also be tested. Up to 1986, there were only two critical tests of the mutation–accumulation mechanism: Rose and Charlesworth [1980, 1981a] and Kosuda [1985]. In a sib analysis of adult female life history in an outbred *Drosophila melanogaster* laboratory population at evolutionary equilibrium, Rose and Charlesworth found that the additive genetic variance of 24-hour fecundity did not increase with age, contrary to expectation if mutations are accumulating with effects on fecundity confined to later ages. On the other hand, Kosuda [1985] used chromosomally manipulated *Drosophila* stocks to show that genetic variation had more impact on later male mating success compared with earlier male mating success, which fits the mutation-accumulation hypothesis.

A number of experiments support the antagonistic pleiotropy mechanism, one of the earliest being Gowen and Johnson's [1946] finding of a negative correlation between egg-laying rate and longevity over a group of wild-type *Drosophila* stocks. Wattiaux [1968b] demonstrated relatively reduced early male mating success in a longer-lived *Drosophila subobscura* population. Law et al. [1977] and Law [1979] provided evidence for antagonistic pleiotropy in a plant species, *Poa annua*. Rose and Charlesworth [1980, 1981a,b] obtained evidence of antagonistic pleiotropy between early fecundity and longevity in adult *Drosophila* females from a sib analysis and selection experiments. Comparable selection experiments by Rose [1984a], Luckinbill et al. [1984], and Luckinbill and Clare [1985] produced similar evidence for antagonistic pleiotropy between early reproduction and later survival in *Drosophila* females. Using methods similar to those of Kosuda [1985], Hiraizumi [1985] found evidence for negative correlation between early female reproduction and longevity. Evidence ostensibly against antagonistic pleiotropy [e.g., Giesel, 1979; Giesel and Zettler, 1980; Giesel et al., 1982] has been explained in terms of inbreeding artifacts and genotype–environment interaction by Rose [1984b] and Service and Rose [1985], respectively. All this work has utilized *Drosophila* species, the workhorse for the evolutionary biology of senescence, as it is for evolutionary biology generally.

REVIEW FROM MID-1986 TO LATE 1988

It used to be the case that virtually no efffort was made to study the evolutionary biology of aging as such. Alternative theories for the evolution of aging thus tended to use indirect evidence furnished by work in other fields. This is no longer the case. There is now an active group of experimenters who deliberately seek findings of relevance to alternative evolutionary theories. Fortunately or unfortunately, there is sharp disagreement between these workers, rather than a bland consensus. It is certainly healthy that the relationship between theory and experiment in this field is tight enough to permit rapid evaluation of the theoretical significance of particular experimental and comparative results.

It should be noted that we do not consider other review articles published during the review period. In addition, we do not discuss the extremely numerous evolutionary interpretations offered in experimental papers that do not directly test evolutionary theories. Essentially any finding in gerontology can be used to support or criticize evolutionary theories of aging, because these theories are so general.

Mathematical Research

Abugov [1986] generalizes the classic work of Charlesworth [1980] on the population genetics of populations with age-structure. Of particular relevance for evolutionary gerontology, he derives a more general form for the relationship between age and the force of natural selection, obtaining Hamilton's [1966] results on the decline in the force of natural selection as special cases. The most important features of the broadening of earlier results are that Abugov's treatment covers cases in which there is sex-specificity in fecundity and survival probability, in which both sexes determine the fertility of matings, and in which mating is nonrandom. Abugov [1988] follows a strategy analogous to that of his 1986 work, generalizing Lande's [1982] quantitative genetic model for the evolution of life history to cases in which the sexes differ in their selection patterns. Again, the Hamilton [1966] results are recovered as special cases.

Hirsch [1987] gives a nongenetic analysis of alternative demographic phenotypes in a population with age-structure, in order to address the question of why senescence should evolve in the first instance. The style of analysis is the same as that of Cole's [1954] classic paper on the evolution of patterns of reproduction and survival, using hypothetical constraints on an abstract phenotype, without any explicit genetic model. The puzzles solved are entertaining but essentially spurious beside the population genetics theory of Charlesworth [1980], Abugov [e.g., 1986], and others, given that evolution in age-structured populations normally proceeds on the basis of genetic variation.

A general "evolutionary reliability theory" for aging is proposed by Miller [1987]. However, "the theory of this paper is an abstract mathematical theory that assumes no physical, biological, or genetic mechanisms. It could apply to self-replicating machines that reproduce via means completely different from those of biological mechanisms." Miller's statement of this theory is in any event almost completely opaque, at least in the case under consideration. He asserts that he can fit life tables using his model, but the actual fitting procedures and results are not given. Therefore, the potential scientific value of this research for the evolution of aging is unclear, at least at present.

Experimental Research

In a series of related papers, Arking [1987a,b; Arking et al., 1988] gives further analyses of selected lines first reported in Luckinbill et al. [1984] and

Luckinbill and Clare [1985]. The Luckinbill publications, like others, show that culturing *Drosophila* at later ages over a number of generations leads to the laboratory evolution of increased longevity, as predicted by Edney and Gill [1968]. Arking goes over much of this same ground, somewhat awkwardly [see Luckinbill and Clare, 1987]. The original features of the Arking papers are: 1) the demonstration that the differentiation of Luckinbill's longer-lived lines is not dependent on temperature within the range 18–28°C; 2) evidence that there is no difference in metabolic rate between longer-lived lines and their controls; and 3) the proposal that the collapse in female fecundity just before death is a "biomarker" for aging. The first result is genuinely original, and supports the robustness of Luckinbill et al.'s findings with respect to the evolutionary interpretation of their results. The second result seems dubious, in light of Service's [1987] more complete analyses, discussed below. The third result seems reasonable, but of little value for hypothesis testing. Arking uses these findings to argue for a biphasic model of aging, which is stringently attacked by Luckinbill and Clare [1987], in a rejoinder to Arking [1987a]. Arking [1987c] responds with a plea for patience.

One of the more amusing demonstrations of the potential importance of antagonistic pleiotropy in the evolution of senescence is provided by Bellen and Kiger's [1987] study of the effects of the *dunce* allele on sexual activity and longevity in *D. melanogaster* females. The *dunce* allele interferes with memory in fruit flies by reducing or abolishing cAMP phosphodiesterase activity. Females homozygous for the allele mate about twice as frequently as normal females and undergo a mean life span reduction of about 50% when kept with males. However, the mechanistic basis of these joint effects is not made clear, beyond experiments showing that the reduced longevity is not a crowding effect.

Friedman and Johnson [1988a,b] describe the identification, characterization, and localization of a *Caenorhabditis elegans* mutant allele, *age-1*, which increases maximum life span by 60% at 20°C in this nematode species. Simultaneously, this allele decreases fertility by 75% at 20°C, at least in self-fertilizing hermaphrodites, a clear case of antagonistic pleiotropy [see Johnson and Hutchinson, this volume, for more details]. In addition, they appear to have found other mutant alleles from this same locus that exhibit antagonistic pleiotropy. The discovery of these alleles is perhaps the most direct support yet for either of the two principal population–genetic mechanisms for the evolution of aging.

Graves et al. [1988] and Luckinbill et al. [1988a] study the *D. melanogaster* lines with evolutionarily postponed senescence first reported in Luckinbill et al. [1984]. They find that body size has not increased in the longer-lived stocks, compared with controls, as reported by Rose et al. [1984]. They also find that the genetic differentiation of longer-lived stocks is not an artifact of altered sexual behavior, as suggested by Partridge and Andrews [1985]. The stocks reproduced at later ages had greater longevities when kept with or without mates, for both males and females. Tethered flight duration was greatly increased in the longer-lived stock. This increase in duration was not due to a reduction in flying speed. In addition, this enhancement of flight duration was sustained over the bulk of the life span.

Johnson [1986] reports molecular and quantitative genetic analyses of recombinant inbred (RI) lines of *C. elegans*. The main interest of this work for evolutionary purposes is that both molecular and quantitative genetic experiments suggest that polymorphic loci affecting life span are distributed throughout the six main chromosomes within the RI lines. This notion fits the evolutionary view that "aging" genes are fitness-component genes, so that many different loci could be polymorphic for alleles affecting aging.

Le Bourg et al. [1988] attempt to test for antagonistic pleiotropy in *D. melanogaster* using patterns of phenotypic correlation between life history characters. They assert that if genes with antagonistic "pleiotropic effects exist, such a negative correlation could possibly be observed *among individuals* within a single population" (their emphasis). The clear presumption of the paper is that such a negative correlation among individuals is a direct corollary of the hypothesis, not a "possible" prediction. This assertion is not logically correct, since population genetic hypotheses concern patterns of allele action, which determine patterns of *genetic*, not phenotypic, correlation. Such genetic correlation tests of the idea were published by Rose and Charlesworth [1981a]; it was also shown in that paper that phenotypic correlations and genetic correlations do not necessarily correspond for life history characters. The present study of LeBourg et al. therefore does not provide evidence concerning the validity of the antagonistic pleiotropy hypothesis.

Luckinbill et al. [1987, 1988b] study the number of loci giving rise to the longevity differences of the Luckinbill et al. [1984] *D. melanogaster* stocks. In the first study, quantitative–genetic methods are used to argue for a small number of loci affecting longevity, possibly only one. In the second, evidence is presented for the involvement of all three major chromosomes of the *D. melanogaster* genome. This latter result is like that found by Johnson [1986] in his study of RI lines in *C. elegans*. Luckinbill et al. [1988b] suggest that the contrast between their two studies arises from the preponderant contribution of one chromosome.

In yet another study of *D. melanogaster*, Mueller [1987] finds that stocks reproduced at early ages underwent a deterioration in later fecundity, relative to controls, apparently due to the accumulation of deleterious alleles with effects confined to later ages. This is one of the best studies published supporting the mutation-accumulation mechanism for the evolution of senescence.

Rose et al. [1987], Service [1987], and Service et al. [1988] report further studies of the stocks described in Rose [1984a]. This work builds on the finding of Service et al. [1985] that evolutionarily postponed senescence is associated with increased resistance to acute environmental stresses. In Service [1987], it is shown that the trade-off between starvation resistance and early fecundity shown in Service and Rose [1985] appears to be due to differential lipid allocation: longer-lived flies have more lipid with which to keep the soma alive. Service [1987] finds significantly decreased early metabolic rates in longer-lived stocks, unlike Arking et al. [1988]. This finding is probably due to increased experimental power, because Service used five stocks with postponed senescence and five matched controls, unlike the single pair of populations used by Arking et al. Rose et al. [1987] and Service et al. [1988] report that, when selection for increased longevity is relaxed, not all characters exhibit the reversed response expected on the antagonistic pleiotropy hypothesis. Rather, some differences between postponed senescence stocks and their controls appear to involve age-specific adaptations, in keeping with the mutation-accumulation mechanism for the evolution of aging. Other differences, however, continue to exhibit antagonistic pleiotropy. This finding, together with earlier work in *Drosophila* and Mueller's contemporaneous work, suggests that *both* of the basic population genetic mechanisms shape the evolution of aging in at least some species.

Comparative Research

Two populations of the leech *Erpobdella octoculata* are compared by Maltby and Calow [1986] in order to test for "trade-offs" (i.e., antagonistic pleiotropy) between reproduction and life span. One population lived for only 1 year but had high fecundity. The other had lower fecundity but lived for 2 or 3 years. These findings were obtained in both laboratory and field environments, suggesting that they are not invalidated by problems of genotype–environment interaction [see Rose and Service, 1985; Service and Rose, 1985]. As the differences between the populations assayed in the laboratory must be genetic, these results support the antagonistic pleiotropy hypothesis [cf. Law et al., 1977; Law, 1979].

Nesse [1988] analyzes life tables of various organisms in the wild in order to test whether or not these species undergo an increase in mortality due to senescence. The technique he uses is based on comparing mortality levels observed in the wild with those obtained by extrapolating from the mortality levels of wild animals that have just achieved reproductive maturity. While this procedure seems at least somewhat ad hoc, the effects he detects are so large in some cases that it seems indubitable that there is a genuine senescence effect being detected. However, the author argues that this finding constitutes evidence for antagonistic pleiotropy, since he supposes that mutation-accumulation could not generate the degree of senescent mortality that he detects. However, no apodictic argument for this supposition is provided.

Prothero and Jurgens [1987] provide an extensive analysis of the relationship between the maximal or mean life span and body or organ weight in homeotherms. This study draws on more recent and complete records than previous studies. Unlike Sacher [1959], they do not find that brain weight explains any more life span variation than body weight. They endeavor to explain their results in terms of molecular genetic models, but do not attempt to relate their findings to the evolutionary theory of senescence. One explanation of this type of comparative finding that *does* grow out of the evolutionary theory is that larger organisms are likely to have lower mortality rates, due to fewer potential predators among other mortality sources, and consequently the force of natural selection will be sustained at higher levels in later life. This sustained force of selection will then lead to the evolution of relatively greater longevity.

Schnebel and Grossfield [1988] compare the longevities of 12 *Drosophila* species and semispecies with a view to testing whether or not the between-species correlations were negative between early fitness-components and life span. The correlations found were not generally negative. Schnebel and Grossfield point out, however, that, "The hypothesis does not apply to life-history differences among species." It was originally framed, and is still evaluated today, in terms of allelic variation segregating within one outbred population evolving within one environment. Over a number of disparate species, it is unclear what prediction(s) the hypothesis can offer, particularly if epistasis modifies patterns of pleiotropy.

CONCLUSIONS

The later 1980s saw a solidification of interest in the evolutionary biology of aging. There are now a number of laboratories studying *Drosophila* stocks with evolutionarily postponed aging. The physiological and genetic analysis of these stocks has enabled us to evaluate evolutionary hypotheses with some power. Strictly genetic work in *C. elegans* has also proven highly relevant.

In terms of a scorecard comparing the theoretical alternatives, the basic evolutionary theory has received any number of corroborations. In fact, it has effectively disappeared from view as a point of debate. The specific population genetic mechanisms are still under consideration, especially in *Drosophila*. The antagonistic pleiotropy mechanism has turned up in so many different experimental tests that it would now seem to be merely churlish to doubt that it has some degree of importance in the evolution of aging [cf. Le Bourg et al., 1988]. On the other hand, there are now several different lines of evidence in favor of the mutation-accumulation mechanism, particularly in *Dro-*

sophila spp. [e.g., Mueller, 1987; Service et al., 1988]. For the time being, it would seem reasonable to conclude that *both* of these mechanisms act in the evolution of aging, perhaps simultaneously within a species. There is, of course, no logical difficulty with this conclusion; these mechanisms are not incompatible, merely different.

Perhaps the most important task for future research is the development of greater analytical refinement for the dissection of alternative genetic and physiological mechanisms of senescence. For example, the detection of multiple mechanisms for the postponement of senescence by Service et al. [1988] required the breakdown of the aging phenotype into multiple components, as reported by Service et al. [1985]. Another example of even greater power is that afforded by Friedman and Johnson's [1988a] study of the *age-1* mutant in *C. elegans*. Obtaining this mutant has provided unambiguous corroboration of the antagonistic pleiotropy mechanism for the evolution of aging. It is to be hoped that comparable breakthroughs will soon be achieved in other laboratories.

ACKNOWLEDGMENTS

We are grateful to L. MacPhee for comments on the manuscript. This work was supported in part by a University of California President's Fellowship to J.L.G. and NIA grant AG-06346 to M.R.R.

REFERENCES

Abugov R (1986): Genetics of Darwinian fitness. III. A generalized approach to age structured selection and life history. J Theor Biol 122:311–323.

Abugov R (1988): A sex specific quantitative genetic theory for life history and development. J Theor Biol 132:437–447.

Arking R (1987a): Successful selection for increased longevity in *Drosophila*: Analysis of the survival data and presentation of a hypothesis on the genetic regulation of longevity. Exp Gerontol 22:199–220.

Arking R (1987b): Genetic and environmental determinants of longevity in *Drosophila*. Basic Life Sci 42:1–22.

Arking R (1987c): Letters to the editor. Exp Gerontol 22:223–226.

Arking R, Buck S, Wells R, Pretzlaff R (1988): Metabolic rates in genetically based long lived strains of *Drosophila*. Exp Gerontol 23:59–76.

Bell G (1984): Evolutionary and nonevolutionary theories of senescence. Am Nat 124:600–603.

Bellen HJ, Kiger JA (1987): Sexual hyperactivity of dunce females of *Drosophila melanogaster*. Genetics 115:153–160.

Charlesworth B (1980): "Evolution in Age-Structured Populations." London: Cambridge University Press.

Charlesworth B, Williamson JA (1975): The probability of survival of a mutant gene in an age-structured population and implications for the evolution of life-histories. Genet Res 26:1–10.

Clare MJ, Luckinbill LS (1985): The effects of gene-environment on the expression of longevity. Heredity 55:19–29.

Cole LC (1954): The population consequences of life history phenomena. Q Rev Biol 29:103–137.
Edney EB, Gill RW (1968): Evolution of senescence and specific longevity. Nature 220:281–282.
Emlen JM (1970): Age-specificity and ecological theory. Ecology 51:588–601.
Friedman DB, Johnson TE (1988a): A mutation in the age-1 gene in *Caenorhabditis elegans* lengthens life and reduces hermaphrodite fertility. Genetics 118:75–86.
Friedman DB, Johnson TE (1988b): Three mutants that extend both mean and maximum life span of the nematode, *Caenorhabditis elegans*, define the age-1 gene. J Gerontol 43:B102–109.
Giesel JT (1979): Genetic co-variation of survivorship and other fitness indices in *Drosophila melanogaster*. Exp Gerontol 14:323–328.
Giesel JT, Zettler EE (1980): Genetic correlations of life historical parameters and certain fitness indices in *Drosophila melanogaster*: r_m, r_s, diet breadth. Oecologia 47:299–302.
Giesel JT, Murphy PA, Manlove MN (1982): The influence of temperature on genetic interrelationships of life history traits in a population of *Drosophila melanogaster:* What tangled data sets we weave. Am Nat 119:464–479.
Gowen JW, Johnson LE (1946): On the mechanism of heterosis. I. Metabolic capacity of different races of *Drosophila melanogaster* for egg production. Am Nat 80:149–179.
Graves JL, Luckinbill LS, Nichols A (1988): Flight duration and wing beat frequency in long- and short-lived *Drosophila melanogaster*. J Insect Physiol 34:1021–1026.
Haldane JBS (1941): "New Paths in Genetics." London: Allen and Unwin.
Hamilton WD (1966): The moulding of senescence by natural selection. J Theor Biol 12:12–45.
Hiraizumi Y (1985): Genetics of factors affecting the life history of *Drosophila melanogaster*. I. Female productivity. Genetics 110:453–464.
Hirsch H (1987): Why should senescence evolve? An answer based on a simple demographic model. Basic Life Sci 42:75–90.
Johnson TE (1986): Molecular and genetic analyses of a multivariate system specifying behavior and life span. Behav Genet 16:221–235.
Kosuda K (1985): The aging effect on male mating activity in *Drosophila melanogaster*. Behav Genet 15:297–303.
Lande R (1982): A quantitative genetic theory of life-history evolution. Ecology 63:607–615.
Law R (1979): The cost of reproduction in annual meadow grass. Am Nat 113:3–16.
Law R, Bradshaw AD, Putwain PD (1977): Life-history variation in *Poa annua*. Evolution 31:233–246.
Le Bourg E, Lints FA, Delince J, Lints CV (1988): Reproductive fitness and longevity in *Drosophila melanogaster*. Exp Gerontol 23:491–500.
Lints FA, Hoste C (1974): The Lansing effect revisited. I. Lifespan. Exp Gerontol 9:51–69.
Lints FA, Hoste C (1977): The Lansing effect revisited. II. Cumulative and spontaneously reversible parental age effects on fecundity in *Drosophila melanogaster*. Evolution 31:387–404.
Lints FA, Stoll J, Gruwez G, Lints CV (1979): An attempt to select for increased longevity in *Drosophila melanogaster*. Gerontology 25:192–204.
Luckinbill LS, Clare MJ (1985): Selection for life span in *Drosophila melanogaster*. Heredity 55:9–18.
Luckinbill LS, Clare MJ (1987): Letters to the editor. Exp Gerontol 22:221–222.
Luckinbill LS, Arking R, Clare MJ, Cirocco WC, Buck SA (1984): Selection for delayed senescence in *Drosophila melanogaster*. Evolution 38:996–1003.
Luckinbill LS, Clare MJ, Krell WL, Cirocco CC, Richards PA (1987): Estimating the number of genetic elements that defer senescence in *Drosophila*. Evol Ecol 1:37–46.
Luckinbill LS, Graves JL, Tomkiw A, Sowirka O (1988a): A qualitative analysis of some life history correlates of longevity in *Drosophila melanogaster*. Evol Ecol 2:85–94.

Luckinbill LS, Graves JL, Reed AH, Koetsawang S (1988b) Localizing genes that defer senescence in *Drosophila melanogaster*. Heredity 60:367–374.
Maltby L, Calow P (1986): Intraspecific life-history variation in *Erpobdella octocula* (Hirudinea: Erpobdellidae). II. Testing theory on the evolution of semelparity and iteroparity. J Anim Ecol 55:739–750.
Medawar PB (1946): Old age and natural death. Mod Quart 1:30–56.
Medawar PB (1952): "An Unsolved Problem of Biology." London: H.K. Lewis.
Mertz DB (1975): Senescent decline in flour beetles selected for early adult fitness. Physiol Zool 48:1–23.
Miller A (1987): Evolutionary reliability theory. Basic Life Sci 42:187–192.
Mueller LD (1987): Evolution of accelerated senescence in laboratory populations of *Drosophila*. Proc Natl Acad Sci USA 84:1974–1977.
Nesse RM (1988) Life table tests of evolutionary theories of theories of senescence. Exp Gerontol 23:445–453.
Partridge L, Andrews R (1985): The effect of reproductive activity on the longevity of male *Drosophila melanogaster* is not caused by an acceleration of aging. J Insect Physiol 31:393–395.
Prothero J, Jurgens KD (1987): Scaling of maximal life span in mammals: A review. Basic Life Sci 42:49–74.
Rose MR (1984a): Laboratory evolution of postponed senescence in *Drosophila melanogaster*. Evolution 38:1004–1010.
Rose MR (1984b): Genetic covariation in *Drosophila* life history: Untangling the data. Am Nat 123:565–569.
Rose MR (1985): Life history evolution with antagonistic pleiotropy and overlapping generations. Theor Popul Biol 28:342–358.
Rose M, Charlesworth B (1980): A test of evolutionary theories of senescence. Nature 287:141–142.
Rose MR, Charlesworth B (1981a): Genetics of life-history in *Drosophila melanogaster*. I. Sib analysis of adult females. Genetics 97:173–186.
Rose MR, Charlesworth B (1981b): Genetics of life-history in *Drosophila melanogaster*. II. Exploratory selection experiments. Genetics 97:187–196.
Rose MR, Service PM (1985): Evolution of aging. Rev Biol Res Aging 2:85–98.
Rose MR, Dorey ML, Coyle AM, Service PM (1984): The morphology of postponed senescence in *Drosophila melanogaster*. Can J Zool 62:1576–1580.
Rose MR, Service PM, Hutchinson EW (1987): Three approaches to trade-offs in life history evolution. In Loeschcke V (ed): "Genetic Constraints on Adaptive Evolution." Berlin: Springer Verlag, pp 91–97.
Sacher GA (1959): Relationship of lifespan to brain weight and weight in animals. In Wolstenholme GE, O'Connor M (eds) "The Life-Span of Animals." CIBA Foundation Colloquia on Ageing, Vol. 5. London: Churchill, pp 115–133.
Schnebel EM, Grossfield J (1988): Antagonistic pleiotropy: An interspecific *Drosophila* comparison. Evolution 42:306–311.
Service PM (1987): Physiological mechanisms of increased stress resistance in *Drosophila melanogaster* selected for postponed senescence. Physiol Zool 60:321–326.
Service PM, Rose MR (1985): Genetic covariation among life-history components: The effect of novel environments. Evolution 39:943–945.
Service PM, Hutchinson EW, MacKinley MD, Rose MR (1985): Resistance to environmental stress in *Drosophila melanogaster* selected for postponed senescence. Physiol Zool 58:380–389.

Service PM, Hutchinson EW, Rose MR (1988): Multiple genetic mechanisms for the evolution of senescence. Evolution 42:708–116.

Sokal RR (1970): Senescence and genetic load: Evidence from *Tribolium*. Science 167:1733–1734.

Taylor CE, Condra C (1980): r- and K-selection in *Drosophila pseudoobscura*. Evolution 34:1183–1193.

Wattiaux JM (1968a): Cumulative parental age effects in *Drosophila subobscura*. Evolution 22:406–421.

Wattiaux JM (1968b): Parental age effects in *Drosophila pseudoobscura*. Exp Gerontol 3:55–61.

Williams GC (1957): Pleiotropy, natural selection, and the evolution of senescence. Evolution 11:398–411.

Aging in *Caenorhabditis elegans:* Update 1988

Thomas E. Johnson and Edward W. Hutchinson

INTRODUCTION

The last few years have marked a transition for aging research in *Caenorhabditis elegans*. Several new lines of work have appeared, most notable of which is the derivation of long-lived strains obtained both from naturally occurring variation and by mutation. The loss of several workers in the field due to retirement or movement to other areas of research as well as the increasingly competitive nature of funding for fundamental, nonclinical research in aging has led to the loss of several labs that in the past have been among the most productive in the field. Other areas of research with *C. elegans* have continued to advance, and the physical map of the nematode is more than 95% complete. The background material for studying *C. elegans* has also become much more accessible as a result of the publication of a book detailing much of the nonaging background material for *C. elegans* [Wood, 1988].

The Physical Map

Coulson et al. [1986, 1988] have largely completed a massive study in which they sought to construct a physical map of the *C. elegans* genome. The map was constructed from an ordered overlapping collection of cloned DNA fragments; this allows easy access to a mapped genetic region without walking from previously cloned genes. The original map [Coulson et al., 1986] utilized both lambda and cosmid clones that were provisionally assigned an overlap on the basis of a shared "fingerprint" obtained from a digital characterization procedure following *Sau*3A1 digests of end-labeled *Hind*III-digested genomic clones. These techniques established a physical map consisting of some 860 contiguous overlapping arrays of cloned DNA fragments (contigs) containing about 60% of the genome and ranging in size from 35 to more than 350 kilobases [Coulson et al. 1986]. The contigs were subsequently correlated with the genetic

Institute for Behavioral Genetics, University of Colorado, Boulder, Colorado 80309

map using previously cloned and mapped genes or in situ hybridization [Albertson, 1985].

This approach reached its practical limit at about 700 contigs containing 90–95% of the genome when few new sequences were being detected; it was then supplemented with an independent cloning strategy [Coulson et al., 1988], the use of yeast artificial chromosome vectors (YACs). YACs [Burke et al., 1987] can carry 50–1,000 kilobase inserts, involve a novel host, and use an independent cloning strategy; they were thus found highly useful in linking previously unlinked contigs. This resource has proven invaluable in numerous cloning strategies for the isolation and characterization of novel genes and is a model for the eventual cloning of the human genome.

AGING RESEARCH
Nuclear Proteins as Aging Markers

Meheus et al. [1987], in conjunction with studies on nematode histones [Vanfleteren et al., 1986, 1987], examined age-specific changes in the pattern of nuclear proteins using two-dimensional gel electrophoresis. Several age-specific proteins were detected, including several glycoproteins, and one subsequently called S-28 was partially sequenced. Although no sequence homology was detected with existing proteins, there was some indication of similarity with a high-mobility-group protein from calf thymus. Subsequent studies showed that the glycoproteins, but not S-28, probably were not age-specific per se but, instead, were induced by the treatment with fluorodeoxyuridine (FUdR) used to sterilize adult hermaphrodites. These studies provide an age-specific marker protein (S-28) that had not been previously found in two-dimensional gels on pulse-labeled, whole-worm extracts [Johnson and McCaffrey, 1985] and show the usefulness of a organism in which different methods for maintaining age-synchrony are available [see Johnson, 1984 for a summary of alternate methods of culture].

Studies on Lysosomes

Clokey and Jacobson [1986] showed that fluorescently labeled proteins are endocytosed and can be found within the gut cells of *C. elegans,* localized to autofluorescent granules. These autofluorescent granules resemble both by spectral and solubility properties the lipofuscin granules seen in aged vertebrate cells. Rhodamine-B-isothiocyanate (RITC)-labeled bovine serum albumin (BSA) accumulates in the same organelle that autofluoresces (Fig. 1). These studies and those with other probes suggest that the autofluorescent granules are the ultimate site of deposition of these proteins. Together with the fact that the granules could be loaded with acridine orange, a lysosomotropic agent, these observations suggest that the autofluorescent granules are secondary lysosomes and that this body is the site of lipofuscin accumulation.

Aging in *Caenorhabditis elegans* 17

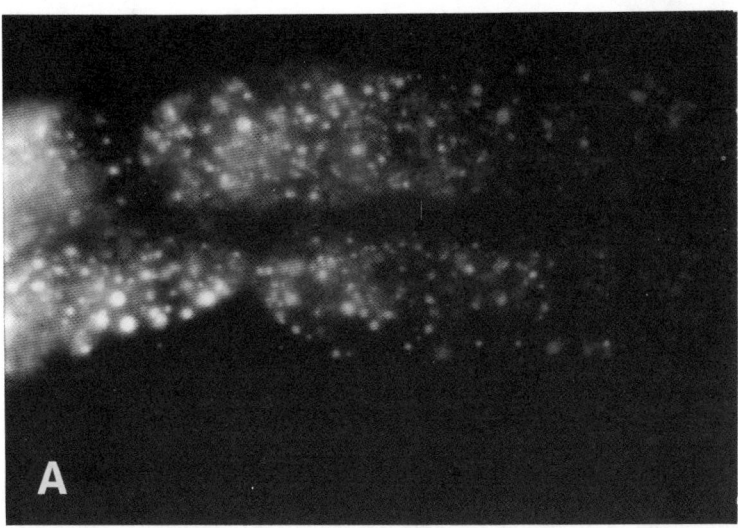

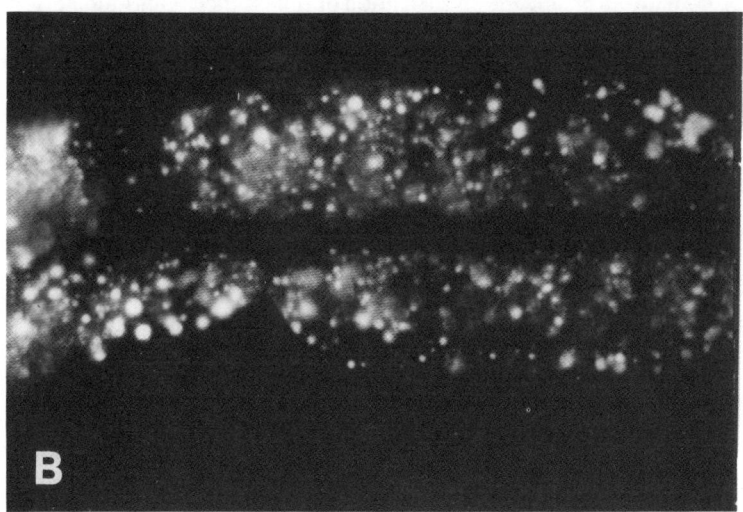

Fig. 1. Coincidence of autofluorescence and ingested RITC-BSA in intestinal granules. Animals raised at 16°C to 90 hours of age were fed RITC-BSA for 8 hours and then "chased" for 4 hours on lawns of *E. coli* OP50. **A:** Cell immediately posterior to the pharynx, autofluoresence (blue) viewed with 365–420 epiillumination × 167. **B:** Same view, but RITC-BSA fluorescence (red) viewed with 546–590 epiillumination × 167. (From Clokey and Jacobson, 1986, with permission.)

Sarkis et al. [1988] show that three lysosomal proteases decline with chronological age, in contrast to lysosomal hydrolases [Bolanowski et al., 1983]. The largest change in activity is associated with cathepsin D, which shows a tenfold drop between days 3 and 11, a period representing the transition from young adulthood to the mean life span. Two other proteases show declines: 2.5-fold for cathepsin Ce1 and eightfold for cathepsin Ce2. A nonlysosomal protease, cathepsin CeX, shows no age-related decline in the rate of protein turnover.

Karey and Rothstein [1986] reported a 30% increase in total proteolytic activity in crude lysosomal preparations of *Turbatrix aceti*. The increase in proteolytic activity was observed even though half-lives of total TCA-insoluble protein increased from 67 to 170 hours. This estimate of half-life is affected in an unknown manner by the reported decrease in leucine transport in aged worms but is consistent with earlier observations using other techniques [Reznik and Gershon, 1979]. Karey and Rothstein further showed that loading with lysosomotropic agents did not alter protein breakdown rates, thus suggesting a nonlysosomal pathway for the endogenous breakdown of proteins.

Further work in *C. elegans* on protein half-lives in older worms and in mutants with altered life spans could resolve the relationship between lysosomal function, protein turnover, and the specification of senescence. Unfortunately, none of the "players" mentioned above seems to be pursuing these goals.

DNA Repair

DNA repair has been suggested to be causally involved in the specification of aging and senescence. Two studies took advantage of either mutations that decreased DNA repair or long-lived strains to ask whether there was any evidence for the involvement of DNA repair in the specification of life span in *C. elegans*.

Hartman et al. [1988] examined four recombinant inbred (RI; see below) strains for their sensitivities with regard to three DNA damaging agents: UV, γ-radiation, and methyl methanesulfonate (MMS). There were no significant correlations between sensitivity of embryos to UV- or γ-radiation and mean or maximum life span. The ability of these strains to repair UV photoproducts was also tested directly. In contrast to the predictions should UV repair be involved in the specification of life span (i.e., that strains with longer life spans have increased UV repair capacity), the longest-lived RI strain had a significant reduction in UV repair at 24 hours of development but not at 48 hours. It should be noted that this study examined repair only in larval stages and leaves open the possibility that the RIs could show altered repair only later in life.

In a parallel study, Johnson and Hartman [1988] examined several radiation-sensitive (Rad) mutants to see if these strains showed corresponding deficits in length of life. Even at acute doses ranging up to 300 krads, the Rad mutants were no more sensitive to loss of life than were the wild type; all stages showed shorter life expectancies at doses near 2×10^5 rads. In contrast to a poorly controlled earlier study [Yeargers, 1981], there was no indication that dauer larvae were any more radiation resistant than were other larval stages; nor was there any difference in sensitivity when adults were irradiated at 8 days of life.

In several experiments low levels of irradiation resulted in a slight, but statistically significant, extension of life expectancy, a phenomenon referred to as hormesis [see Hickey et al., 1983 and Luckey, 1982 for reviews]. Such slight extensions have been used to suggest that DNA repair is involved in limiting life. If this were true, the Rad mutants would be expected to fail to show hormetic responses. Not only did several of the Rad mutants have normal life expectancies and normal maximum life spans, but they also showed slight increases in life span after low levels of irradiation, suggesting that hormesis does not result from induction of DNA repair.

Vitamin E Hormesis

Harrington and Harley [1988] examine the effect of vitamin E on life span. They find that at intermediate concentrations (200 μg/ml) vitamin E has a slight but replicable and statistically significant effect on life expectancy. Similar effects in lowering total fertility and delaying the age of maximum fertility led the authors to conclude that much of the effect was due to slowed development and that "these data do not support the free-radical scavenging mechanisms for vitamin E-induced life span extension in *C. elegans*. Instead they support a general mechanism in which, within a narrow range of treatment, otherwise deleterious agents can increase longevity by slowing growth and development," another example of hormesis.

Aging in Dauer Larvae

An interesting alternative approach has been touched upon in work by Albert and Riddle [1988], who examined mutants in either of two loci *(daf-9* or *daf-15)*. Normally the dauer larvae represents a time out from aging, and even those dauers surviving for periods of several months have a normal life span after return to normal growth conditions [Klass and Hirsh, 1976]. Albert and Riddle discovered that mutants in either of these loci failed to complete development and, nevertheless, had life spans that were typical of normal adult hermaphrodites; they argue that increased survival of dauers necessitates the completion of the entire dauer developmental cycle. However, Albert and Riddle did not rule out the possibility that the shorter life span of the Daf mutants is mere chance or an indirect effect of the Daf phenotype and that death is due to a cause other than normal adult aging. This observation is interesting in light of studies [Johnson et al., 1984] showing that prolongation of any larval stage can prolong life, consistent with a model in which aging does not begin until devel-

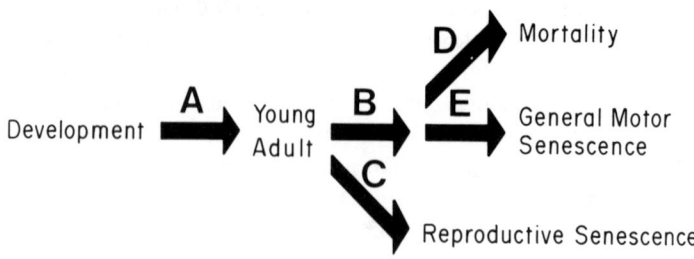

Fig. 2. Diagram describing the order of dependency of events in senescence of *C. elegans*. Arrows indicate dependency relationships. (From Johnson, 1987a, with permission).

opment is completed (Fig. 2). If the interpretation of Albert and Riddle is correct, then aging and senescence would have already begun in these abnormally arrested partial-dauer larvae.

Derivation of Long-Lived Genetic Strains

Three studies focused on the derivation and characterization of long-lived genotypes of *C. elegans* [Johnson, 1987a; Friedman and Johnson, 1988a,b]. Our prejudice is that this is the direction in which future research into the fundamental mechanisms of aging should be going. Three presentations at the Symposium on Modern Biological Theories of Aging, held June 3–6, 1986, in New York, focused on the nature of genetic specification of aging processes and pointed to potential future directions in aging research [Hayflick, 1987; Russell, 1987; Johnson, 1987b].

Genetic dissection of aging processes. Johnson [1988] argued for a systematic approach to the dissection of aging using extant genetic techniques and pursuing directions similar to the successful approach to the dissection of developmental processes. Such an approach could entail the isolation of mutants with altered (particularly slowed) aging, the genetic and phenotypic characterization of these mutants, and, ultimately, the determination of the flow of control from the level of the gene to that of the whole organism. This is a difficult, but perhaps not impossible, objective; the major constraint is the length of time required for assays on life span.

The most challenging aspect of this approach is the identification of genetic variants that have truly altered the aging process(es). We have made the assumption that "mutations" or other genetic alterations "that extend life span must involve altering the rate-limiting steps specifying length of life" [Johnson, 1988], and thus we should seek strains with longer life expectancies and increased maximum life spans.

Recombinant inbred strains. The first approach was to derive long-lived RI strains by mating two inbred laboratory strains followed by inbreeding by self-fertilization [Johnson and Wood, 1982; Johnson, 1987a; Johnson et al., 1988]. These RI strains displayed life expectancies and maximum life spans both much shorter and up to 70% longer than the wild type; this life extension resulted from a slower rate of increase in the mortality rate with chronological age (Fig. 3) [Johnson, 1987a; Johnson et al., 1988]. It was shown that there was no corresponding change in length of reproduction or rate of development in the RI strains, showing that these three life history components are independently specified in this genetic background. The rate of movement decline in this genetic background was slowed in the long-lived and more rapid in the short-lived strains, consistent with joint genetic specification of one or

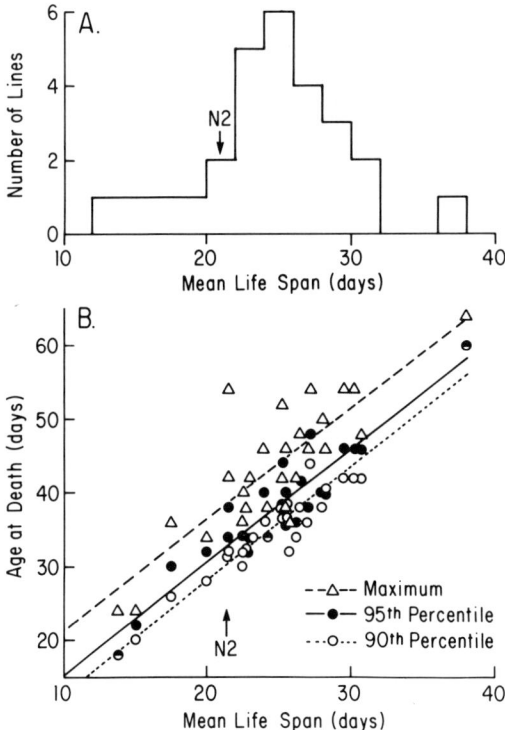

Fig. 3. Life span of hermaphrodites from RI strains. **A:** Mean life spans of 27 RI strains. Data are the average of two survival experiments, each containing 50 nematodes. The entire experiment involved the assay of 2,950 nematodes; 2,206 died of natural causes. **B:** Regression of mean life span (same nematodes as described in Fig. 1A) on either maximum life span, the 95th percentile of life span, or the 90th percentile of life span. Mean life span is highly correlated ($P \ll 0.001$) with maximum life span ($r = 0.83$), the 95th percentile of life span ($r = 0.93$), and the 90th percentile of life span ($r = 0.96$) (From Johnson, 1987a, with permission).

more processes limiting life and general motor function. These findings are summarized in Figure 2.

***age-1*, a long-lived mutant.** Friedman and Johnson [1988a,b] studied four long-lived mutants, originally identified by Klass [1983]. We were able to replicate Klass's observations that the strains were indeed longer-lived. Two mutants (MK542 and MK546) were particularly long-lived, showing extensions in life expectancy at 25°C of 65–85% (Table I). Three mutants (MK31, MK542, and MK546) displayed a reduction in fertility of almost fivefold when compared with the N2 wild type or the DH26 parental strain [Friedman and Johnson, 1988b]. These same three strains displayed partial paralysis that was mapped to the *unc-31* locus on linkage group V. Subsequent complementation tests among these three mutants used both life expectancy and fertility [Friedman and Johnson, 1988b] and showed that all three strains (MK31, MK542, and MK546) carry noncomplementing mutations, in a gene we called *age-1*.

We mapped *age-1* by taking advantage of the fact that *C. elegans* is a self-fertilizing hermaphrodite and shows no inbreeding depression for life span [Johnson and Wood, 1982]. The longest-lived strains were mated to N2, and F2 progeny were allowed to self-fertilize for 5 to 15 generations, leading to a series of reisolate lines (Fig. 4A) each of which is almost completely homozy-

TABLE I. Mean Life Spans and Self-Fertilities of *age-1(hx546)* and Wild-Type Strains

Strain	Genotype	Temp (°C)	Life span[a] Mean ± SEM	N	Fertility[b,c] Mean ± SEM	N
N2	*age-1+ fer-15+*	20	22.6 ± 0.7	5	243.7 ± 28.3	7
		25	15.0 ± 0.8	2		
DH26	*age-1+ fer-15(b26)*	20			248.9 ± 27.4	4
		25	15.7 ± 1.3	3		
MK31	*age-1(hx31)fer-15(b26) unc-31(z3)*	20			56.1 ± 10.1	2
		25	18.4 ± 0.4	4		
MK542	*age-1(hx542)fer-15(b26) unc-31(z2)*	20			16.2 ± 6.2	2
		25	28.1 ± 6.3	2		
MK546	*age-1(hx546)fer-15(b26) unc-31(z1)*	20	32.2 ± 2.2	5	30.7 ± 4.1	6
		25	27.0 ± 3.7	4		
TJ401	*age-1(hx546)fer-15(b26)*	20	36.4 ± 2.0	3	59.9 ± 10.4	7
		25	25.9 ± 1.0	3		
TJ411	*age-1(hx542)fer-15(b26)*	20			46.0	1
		25	30.1 ± 4.1	2		

[a]Mean ± standard error of the life span (days) in several (N) replicate experiments.
[b]Mean ± standard error of the hermaphrodite self-fertilities in several (N) replicate experiments.
[c]At 25°C the *fer-15(b26)* genotypes are sterile; therefore there are no data for fertility at this temperature.

gous. When the lines were assayed for life span and fertility, it was found that the two traits cosegregated (Fig. 4B–D). It was quite surprising that the long-life trait failed to recombine with the *fer-15* gene, which had been used by Klass to facilitate the mutant isolation. The cosegregation of long life, reduced fertility, and *fer-15* suggests two things: first, both long life and reduced fertility result from a mutation at a gene, *age-1*, which has pleiotropic effects; and second, *age-1* maps less than a map unit from *fer-15* in the middle of linkage group II.

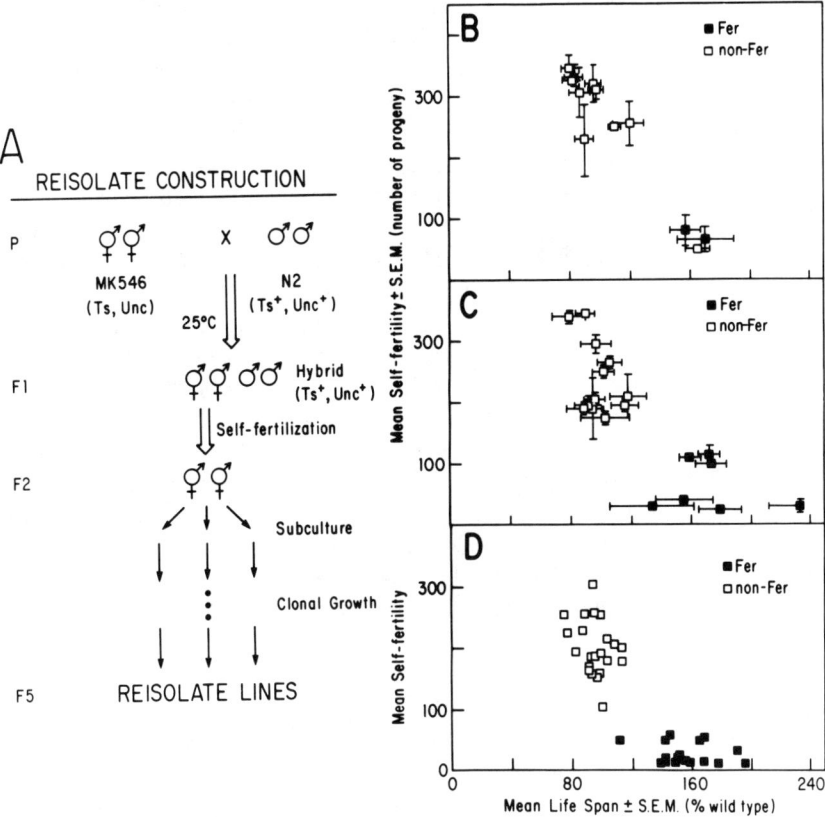

Fig. 4. **A:** Method for constructing homozygous populations from crosses between N2 and MK546. **B and C:** Life expectancy at 20°C of reisolates from the cross of MK546 [*age-1(hx546) fer-15(b26ts)* II; *unc-31(zl)* IV] to N2 is plotted relative to hermaphrodite self-fertility. F_5 reisolates from experiment 1 (B) and F_{15} reisolates from experiment 2 (C). **D:** Life expectancy at 25°C of F_{10} reisolates from crosses of MK542 [*age-1 (hx542) fer-15(b26ts)* II; *unc-31 (z2)* IV] to N2. Fer and non-Fer stocks are indicated; because of the large number of points, standard errors are not shown in Figure 4D but ranged from 5% to 15% of the mean life span while self-fertility is the average of three to five hermaphrodites whose progeny were counted collectively rather than individually. (From Friedman and Johnson, 1988a, with permission).

Caveats. Several problems with these results remain. First, the fact that all three strains contained mutations in both *unc-31* and *age-1* suggested that the original mutants may not have been independently derived; this possibility was subsequently confirmed by Klass (personal communication). However, the fact that these screens yielded only one mutant does not imply that there are no additional genes that, when mutated, lead to longer life in *C. elegans*. Indeed, Russell [1987], as well as ourselves, has isolated mutants that extend life, and Russell (personal communication) has shown that some alleles of *cha-1* [Rand and Russell, 1984, 1985] extend life. There is also some evidence that the long-life genes associated with TJ143, the longest-lived RI strain [Johnson, 1987a], are additive to *age-1* in their effect on life span [Johnson, unpublished data]. All of these lines of evidence suggest that more genes may still be identified by mutation. It would be surprising indeed if the effects of dietary restriction could not be mimicked by behavioral or metabolic mutations that limit dietary intake. Klass [1983] suggested that the mutants just described might result from reduced food intake. However, Johnson [1986] and Friedman and Johnson [1988b and unpublished observations] have obtained data suggesting that this simple explanation does not seem to be true.

Another possibility is that the *age-1* mutation is a double mutant resulting perhaps from a rearrangment flanking or near to *fer-15* and that it suppresses recombination. No depression of recombination in this region has been detected in three point crosses with flanking markers nor has it been possible to separate *age-1* and *fer-15* in these same crosses [Fitzpatrick and Johnson, unpublished data]. Deficiency mapping has assigned *fer-15* and the reduced fertility phenotype of *age-1* to the same nominal region that is flanked by eight deficiency breakpoints on one side and three on the other [Shoemaker, Friedman, and Johnson, unpublished data]. These facts are consistent with the proposed model that *age-1* results in reduced fertility and longer life [Friedman and Johnson, 1988a,b] but are also consistent with *age-1* being a further mutational event in *fer-15*.

Implications of longer-lived mutants. Evolutionary theory argues that senescence is a result of the indifference of natural selection to deleterious genetic effects expressed some time after the onset of reproduction: the declining force of natural selection with the age of the adult soma [Charlesworth, 1980]. Two population genetic mechanisms for the evolution of senescence have been proposed, both based on the decline of the force of natural selection with age: 1) mutation accumulation—maintenance of high-frequency deleterious alleles by mutation pressure, when they have such late effects that natural selection has little impact on their frequencies [Medawar, 1952]; and 2) antagonistic pleiotropy—natural selection favoring genes with beneficial effects early in life that outweigh later, pleiotropic, deleterious effects, because the latter have little importance on fitness [Medawar, 1952; Williams, 1957; Rose, 1985].

age-1 is the first clear example of a genetic locus showing antagonistic pleiotropy. The wild-type allele increases reproductive fitness fivefold while shortening life some 70%.

There are possible scientific constraints on the ability to generalize the findings on *age-1* to humans. There is no reason to suspect evolutionary conservation of gene function for such genes; consequently, even if humans carry a homologous gene, it may play a different role. Nevertheless, the discovery of what seems to be a single gene that, when eliminated by mutation, lengthens life, is an important discovery. Moreover, it may be that the aging processes can be further dissected by means of additional mutational analyses.

ACKNOWLEDGMENTS

This publication has been made possible, in part, by grants from the USPHS (R01-AG08322 and K04-AG00369).

REFERENCES

Albertson DG (1985): Mapping muscle protein genes by *in situ* hybridization using biotin-labeled probes. EMBO J 4:2493–2498.

Albert PS, Riddle DL (1988): Mutants of *Caenorhabditis elegans* that form dauer-like larvae. Dev Biol 126:270–293.

Bolanowski MA, Jacobson LA, Russell RL (1983): Quantitative measures of aging in the nematode *Caenorhabditis elegans:* II. Lysosomal hydrolases as markers of senescence. Mech Ageing Dev 21:295–319.

Burke DT, Carle FG, Olson MV (1987): Cloning of large segments of exogenous DNA into yeast by means of yeast artificial chromosome vectors. Science 236:806–812.

Charlesworth B (1980): "Evolution in Age-Structured Populations." London: Cambridge University Press.

Clokey GV, Jacobson LA (1986): The autofluorescent "lipofuscin granules" in the intestinal cells of *Caenorhabditis elegans* are secondary lysosomes. Mech Ageing Dev 35:79–84.

Coulson A, Sulston J, Brenner S, Karn J (1986): Toward a physical map of the genome of the nematode *Caenorhabditis elegans*. Proc Natl Acad Sci USA 83:7821–7825.

Coulson A, Waterston R, Kiff J, Sulston J, Kohara Y (1988): Genome linking with yeast artificial chromosomes. Nature 335:184–186.

Friedman DB, Johnson TE (1988a): A mutation in the *age-1* gene in *Caenorhabditis elegans* lengthens life and reduces hermaphrodite fertility. Genetics 118:75–86.

Friedman DB, Johnson TE (1988b): Three mutants that extend both mean and maximum life span of the worm, *Caenorhabditis elegans,* define the *age-1* gene. J Gerontol Biol Sci 43:B102–B109.

Harrington LA, Harley CB (1988): Effect of vitamin E on life span and reproduction in *Caenorhabditis elegans*. Mech Ageing Dev 43:71–78.

Hartman PS, Simpson VJ, Mitchell D, Johnson TE (1988): Radiation sensitivity and DNA repair in *Caenorhabditis elegans* strains with different life spans. Mutat Res 208:77–82.

Hayflick L (1987): Origins of longevity. In Warner HR, Butler RM, Sprott RL, Schneider EL (eds): "Modern Biological Theories of Aging." New York: Raven Press, pp 21–34.

Hickey RJ, Bowers EJ, Clelland RC (1983): Radiation hormesis, public health, and public policy: A commentary. Health Phys 44:207–219.
Johnson TE (1984): Analysis of the biological basis of aging in the nematode, with special emphasis on *Caenorhabditis elegans*. In Johnson TE, Mitchell DH (eds): "Invertebrate Models in Aging Research." Boca Raton, FL: CRC, pp 59–93.
Johnson TE (1986): Molecular and genetic analyses of a multivariate system specifying behavior and life span. Behav Genet 16:221–235.
Johnson TE (1987a): Aging can be genetically dissected into component processes using long-lived lines of *Caenorhabditis elegans*. Proc Natl Acad Sci USA 84:3777–3781.
Johnson TE (1987b): Developmentally programmed aging: Future directions. In Warner HR, Butler RM, Sprott RL, Schneider EL (eds): "Modern Biological Theories of Aging". New York: Raven Press, pp 63–76.
Johnson TE (1988): Genetic specification of life span: processes, problems and potentials. J Gerontol Biol Sci 43:B87–B92.
Johnson TE, Wood WB (1982): Genetic analysis of lifespan in *Caenorhabditis elegans*. Proc Natl Acad Sci USA 79:6603–6607.
Johnson TE, McCaffrey G (1985): Programmed aging or error catastrophe? An examination by two-dimensional polyacrylamide gel electrophoresis. Mech Ageing Dev 30:285–297.
Johnson TE, Hartman PH (1988): Radiation effects on life span in *Caenorhabditis elegans*. J Gerontol Biol Sci 43:B137–B141.
Johnson TE, Mitchell DH, Kline S, Kemal R, Foy J (1984): Arresting development arrests aging in the nematode *Caenorhabditis elegans*. Mech Ageing Dev 28:23–40.
Johnson TE, Conley WL, Keller ML (1988): Long-lived lines of *Caenorhabditis elegans* can be used to establish predictive biomarkers of aging. Exp Gerontol 23:281–295.
Karey KP, Rothstein MR (1986): Evidence for the lack of lysosomal involvement in the age-related slowing of protein breakdown in *Turbatrix aceti*. Mech Ageing Dev 35:169–178.
Klass MR (1983): A method for the isolation of longevity mutants in the nematode *Caenorhabditis elegans* and initial results. Mech Ageing Dev 22:679–286.
Klass MR, Hirsh D (1976): Non-aging developmental variant of *Caenorhabditis elegans*. Nature 260:523–525.
Luckey TD (1982): Physiological benefits from low levels of ionizing radiation. Healthy Phys 43:771–789.
Medawar PB (1952): "An Unsolved Problem in Biology." London: HK Lewis.
Meheus LA, Van Beeumen JJ, Coomans AV, Vanfleteren JR (1987): Age-specific nuclear proteins in the nematode worm *Caenorhabditis elegans*. Biochem J 245:257–261.
Rand JB, Russell RL (1984): Choline acetyltransferase-deficient mutants of the nematode *Caenorhabditis elegans*. Genetics 106:227–248.
Rand JB, Russell RL (1985): Properties and partial purification of choline acetyltransferase from the nematode *Caenorhabdities elegans*. J Neurochem 44:189–200.
Reznik AZ, Gershon D (1979): The effect of age on the protein degradation system in the nematode *Turbatrix aceti*. Mech Ageing Dev 11:403–415.
Rose MR (1985): Life history evolution with antagonistic pleiotropy and overlapping generations. Theor Popul Biol 28:342–358.
Russell RL (1987): Evidence for and against the theory of developmentally programmed aging. In Warner HR, Butler RM, Sprott RL, Schneider EL (eds): "Modern Biological Theories of Aging." New York: Raven Press, pp 35–61.
Sarkis GJ, Ashcom JD, Hawdon JM, Jacobson LA (1988): Decline in protease activities with age in the nematode *Caenorhabditis elegans*. Mech Ageing Dev, 45:191–201.

Vanfleteren JR, Van Bun SM, Delcambe LL, Van Beeumen JJ (1986): Multiple forms of histone H2B from the nematode *Caenorhabditis elegans*. Biochem J 235:769–773.

Vanfleteren JR, Van Bun SM, Van Beeumen JJ (1987): The primary structure of histone H2A from the nematode *Caenorhabditis elegans*. Biochem J 243:297–300.

Williams GC (1957): Pleiotropy, natural selection, and the evolution of senescence. Evolution 11:398–411.

Wood WB (1988): ''The Biology of *Caenorhabditis elegans*.'' Cold Spring Harbor, NY: Cold Spring Harbor Press.

Yeargers E (1981): Effect of γ-radiation on dauer larvae of *Caenorhabditis elegans*. J Nemat 13:235–237.

Aging of Fungi

Kenneth D. Munkres

INTRODUCTION

This review summarizes recent advances in research on aging of fungi using the Myxomycete slime mold *Physarum* and the Ascomycetes *Podospora* and *Neurospora*. Additionally, unlike previous reviews of this series, the Ascomycete yeast *Saccharomyces cerevisiae* is also included. During the past 2 years, 5 of a total of 13 papers on the aging of yeast have been written; therefore it is timely and appropriate to review that subject now.

PHYSARUM

Sohal et al. [1986] continue to develop their unified theory about the relationship between metabolic rate, free radicals, differentiation, and aging. Evidence from a variety of sources supports the view that oxygen free radicals play a role in cellular differentiation. It is postulated that cellular differentiation is accompanied by changes in the redox state of cells. Differentiated cells have a relatively more prooxidizing or less reducing intracellular environment than undifferentiated or dedifferentiated cells. Changes in the redox balance during differentiation appear to be due to an increase in the rate of superoxide radical generation. Differentiated cells, in general, exhibit higher rates of cyanide-resistant respiration, cyanide-insensitive superoxide dismutase (SOD) activity, and peroxide concentration and lower levels of reduced glutathione (GSH) as compared with undifferentiated cells. The effects of free radicals on cellular differentiation may be mediated by the consequent changes in ionic composition. Changes in the rates of free radical generation in developing systems may alter cellular charge distribution and ion composition in ways that affect chromatin activity. The existence of such a mechanism would permit oxygen free radicals to influence chromatin activity without directly reacting with it.

Laboratory of Molecular Biology and Department of Genetics, University of Wisconsin, Madison, Wisconsin 53706

To further test the unified theory, Nations et al. [1987] examined the effect of calcium on differentiation, SOD activity, and GSH concentration of *Physarum polycephalum* microplasmodia. The microplasmodia differentiate into spherules when the $CaCl_2$ concentration of their nutrient medium is increased to 54 mM as compared with no added calcium. The salts starvation medium routinely used to induce differentiation contains 8 mM $CaCl_2$. Calcium-free salts medium will not induce spherulation; no other metal is essential. High-calcium (54 mM) also induces the spherulation of a strain that had not been previously observed to spherulate in either nutrient or salts medium. The striking increase in SOD activity and decrease of GSH in the normal strain that are characteristic of salts-induced spherulation with 8 mM calcium do not occur in salts medium containing high calcium. In calcium-free salts medium, no significant change in SOD is observed, and very little change in GSH occurs in the normal strain. The immediate effect of the oxidative stress associated with spherulation may be the release of calcium from mitochondrial stores into the cytosol. The parameters modulating this stress are, in turn, sensitive to exogenous calcium concentrations.

SOD activity increases 46-fold during differentiation of *Physarum*, whereas no change occurs in a strain that does not differentiate in 8 mM calcium salts medium. Addition of SOD, via liposomes, to the nondifferentiating strain induces differentiation; this effect is enhanced by an inhibitor of GSH synthesis. Water- or lipid-soluble antioxidants failed to induce differentiation. Conversely, oxidative treatments including introduction of D-amino acid oxidase, via liposomes, induced differentiation. Cellular oxidation is the probable cause of the SOD effect [Allen et al., 1988]. The results of this paper and previous ones were reviewed in the context of the unified theory [Sohal et al., 1988]. It is hypothesized that the mechanism by which oxidative stress influences gene expression is by its effect on the concentration of cytosolic free calcium, which acts as the specific or obligatory signal for differentiation. A survey of the literature suggests that cells of disparate organisms employ elevation of oxidative stress to induce differentiation and to mobilize calcium.

PODOSPORA

The biochemical genetics of quiescence, the capacity to remain viable and differentiate after nutrient starvation, was reviewed in the preceding volume of this series [Munkres, 1987a]. *Incoloris* mutants, initially defined in terms of pigmentation, are quiescence-defective; their hyphal tips do not survive nutrient starvation [Bernet, 1986].

The senescence phenomenon has been reviewed [Bockelman et al., 1986; Cummings et al., 1986; Esser et al., 1986]. The major mitochondrial (mt) plasmid DNA (p1 DNA), also called α-sen DNA, is thought to be the causal factor in senescence.

The CoI gene of p1 DNA contains group I introns coding for polypeptides homologous to maturases (already identified in yeast) whose function is splicing and maturation of mRNA [Karsch et al., 1987].

Two senescence escape mutants *(mex)* with molecules rearranged in the same mt-DNA region, called β-sen, contain foreign DNA sequences at each recombination border consisting of short stretches of A and T residues. The origin of the foreign DNA may involve reverse transcriptase activity [Koll et al., 1987].

The well-documented senescence that occurs on solid medium does not occur when mycelia are serially passaged in liquid medium. The apparent immortality was not dependent upon either unique mt-DNA rearrangments or the presence of α-sen DNA. The authors speculate that growth in liquid cultures exerts selective pressure to maintain the wild-type mt-genome [Turker and Cummings, 1987].

Three novel plasmids, derived from a common region of the mt-genome, and other previously described plasmids are associated with an 11 base pair consensus sequence. The sequence may be an integral part of plasmid formation and the senescence process [Turker et al. 1987a].

Longevity mutants, defined as strains capable of growth renewal after one or more senescence crises, contain a novel family of plasmids with very short monomeric units derived from a highly ordered 368 base pair region of the mt-genome. It was hypothesized that those small plasmids may act as antagonists of the α-sen DNA function [Turker et al., 1987b].

A longevity mutant (TS-1) with a temperature-sensitive phenotype for senescence was identified. At 34°C, TS-1 becomes senescent and contains α-sen DNA, whereas at 27°C neither senescence nor the plasmid are detected. Other plasmids from specific regions of the mt-genome are also present in senescent cultures. The results provide additional evidence that α-sen DNA is involved in senescence and suggest that it may play a role in the formation of other mt-plasmids. This apparently represents the first mutant of *Podospora* or any other organism with a temperature-sensitive senescence phenotype. The mutant may contain more than one genetic alteration for the continuous growth and temperature-sensitive phenotypes. A formal genetic analysis will be required to determine the genomic location of the gene(s) underlying those phenotypes [Turker et al., 1987c].

Integration of α-sen DNA in the nuclear genome during senescence could not be detected [Koll, 1986], contrary to a previous report [Wright and Cummings, 1983]. Koll's observation is consistent with the well-documented maternal inheritance of the senescence phenotype.

Both juvenile and senescent states of wild-type strain *s* exhibit two transcripts homologous to the first intron of CoI and p1 DNA, which exist in lariat and linear conformations. The latter probably results from breakage of the former. The findings suggest that the transcripts arise from processing of CoI pre-mRNA rather than from transcription of the excised plasmid. The significance of the lariat RNA concerning p1 DNA amplification via a postulated reverse transcriptase mechanism is discussed [Schmidt et al., 1987].

In addition to the class II intron of α-sen DNA, another class II intron fragment linked to it also exhibits amino acid homology with known reverse transcriptases. This new intron is present in race *A* but not race *s*, a fact that may account for their twofold difference in senescence rate. It is not clear whether the reverse transcriptase previously observed in race *A* is derived from those introns [Matsura et al., 1986].

Two extrachromosomal long-life mutants (*ex*-1, *ex*-2) showing indefinite growth are characterized by an almost complete deletion of the CoI gene and its first intron (p1 DNA), an absence of active gene product, cytochrome oxidase, and multiple rearrangements of their mt-DNA. The rearrangements produce novel gene arrangements that are expressed as indicated by transcript analysis. By analogy with *Neurospora*, the authors suggest that a nuclear-encoded mitochondrial alternate oxidase provides oxidative production of ATP, although less efficiently than normal, thereby allowing growth, albeit at a reduced rate, in the absence of cytochrome oxidase. The authors discuss a model in which p1 DNA-directed gene product catalyzes recombination processes leading to a heterogeneous population of rearranged circular submolecules in senescent mitochondria that no longer encode functional polypeptides. The sorting out of free-replicating p1 DNA or of defective mt-DNA lacking recombination sites may result in hyphal segments in which no further recombination occurs; hence, such segments are protected from complete disintegration of the mt-genome and give rise to long-life mutant clones [Schulte et al., 1988].

Two sen DNAs, α and β, were tested for their ability to drive autonomous replication in yeast and *Podpospora*. The α but not the β sequence has autoreplicative (ARS) properties in yeast; the ARS sequences of β are not included in the region common to all the β sen-DNAs. Neither the α nor the β sequences can confer ARS properties in *Podospora*. These sequences inserted into a hybrid vector carrying the suppressor tRNA *su*4-1 do not change the mode of transformation of a suppressible *leu*-1-1 strain of *Podospora:* the transformation is by integration whether or not the plasmid carries a sen-DNA sequence. The *su*4-1 gene integrates at its homologous site in a minority of cases. It is possible to reisolate free plasmids at a low frequency from some transformants. The presence of the sen-DNA on the transforming vector has no effect upon the longevity of the transformants [Sainsard-Chanet and Begel, 1986].

SACCHAROMYCES CEREVISIAE

In previous reviews of this series, this reviewer thought that the limited information on aging of this yeast was not sufficiently advanced to warrant review;

however, within the past 2 years, 5 of a total of 13 papers on the subject have been written; therefore, it is timely and appropriate to present a complete review of the subject now.

Insofar as possible, the review will be divided into Methods and Results sections. The methods are summarized in a general manner to provide the general reader with their conceptual and biological basis. The methods are not cited in detail. The serious student of the subject will undoubtedly wish to read the original papers [Barton, 1950, Bartholomew and Mittwer, 1953; Chen et al., 1989; Egilmez et al., 1990; Egilmez and Jazwinski, 1989; Hough, 1961; Jazwinski et al., 1989; Johnston, 1966; Mortimer and Johnston, 1959; Müller, 1971, 1985; Müller et al., 1980; Pohley, 1987].

Budding yeasts are convenient for studying the aging of single mitotic cells since the budding process permits easy identification of mother and daughter cells at each division. The mother's cell wall is maintained after each division with a scar remaining after bud abscission.

Methods

Aging is measured microscopically in terms of the time-dependent increases of cell size, number of bud scars, number of cell divisions, generation time, and cytological changes acompanying death after division ceases.

Cell volume and number of bud scars of mother cells increase linearly with age. Although mother cells typically bud 20–30 times, no more than ten scars can be seen by conventional light microscopy; however, a microspectrofluorimetric method permits measurement of the total number of scars of individual mothers. The method depends upon a fluorescent dye with affinity for chitin, the major scar component.

The number of buds formed, i.e., the reproductive capacity (RC), and generation times of individual mother and daughter cells are determined by micromanipulation technique.

RC is typically 20–30, ranging from 1 to never more than 50. RC declines with time in a sigmoidal manner. Such survival data are compatible with the Gompertz theorem [Pohley, 1987; Jazwinski et al., 1989]; hence, populations of yeast cells exhibit a feature of the aging of populations of multicellular organisms: the age-specific exponential increase of mortality rate.

Generation time increases linearly with time from 60–100 min initially. During the terminal one to three divisions, generation time increases sharply to as long as 6 hours.

Generation time is a measure of physiological age (senescence), whereas the other parameters are measures of nominal age. Thus, even an age-synchronized population in terms of the number of divisions exhibits a statistically normal distribution of RC or remaining life spans [Egilmez et al., 1990].

In addition to the sharp increase in generation time, old mothers in the final one to three divisions are characterized by a wrinkled surface, loss of turgor, and inability to mate with young cells [Müller, 1985].

After cessation of budding, the cells die, exhibiting granular, vacuolate cytoplasm and, in some cases, lysis. Apparently no one has attempted to use vital stains to detect senescent or dead cells.

To develop yeast as a useful model system for studies of cellular aging at the molecular level, Egilmez et al. [1990] devised a method for the preparation of large quantities of pure age-synchronized populations involving a combination of methods of growth synchronization and cell separation by rate-zonal sedimentation in density gradients.

Results

The ideas that finite RC is a consequence of accumulation of bud scars or change of cell surface–volume ratio [Mortimer and Johnston, 1959; Johnston, 1966] appear to be disproven or discounted [Müller, 1971, 1985; Müller et al., 1980; Egilmez and Jazwinski, 1989].

Aging could not be attributed to the accumulation of nuclear or mitochondrial mutations, chromosome irregularities, or "error catastrophe" in transcription or translation [Müller, 1971].

The RC of daughters was not correlated with their mothers' age; thus, Müller [1985] concluded that aging could not be attributed to infectious plasmids analogous to the situation in *Podospora anserina*; however, since no data were presented, those results cannot be critically evaluated.

Since the α-sen plasmid of *P. anserina* has autoreplicative properties in yeast [Sainsard-Chanet and Begel, 1986], a test of its influence on yeast life span may be feasible.

Biological life span (RC) is unrelated to chronological age [Müller et al., 1980].

Genotypically different strains are characterized by their average life spans [Müller, 1971, 1985; Jazwinski et al., 1989]. The mean and maximum life spans are characteristic features of a given strain and are always correlated [Jazwinski et al., 1989]. In strains with different life spans, the senescence phenotype, with lengthening of generation times, appears at different ages, the longer the life span the later senescence occurs. That observation suggests that the senescence phenotype possesses a genetic component [Jazwinski et al., 1989].

In crosses of strains with different life spans, short life span is generally dominant over long life span in the zygotes [Müller, 1985]. Conversely, Jazwinski et al. [1989] concluded that the genotype of the longer-lived parent is dominant. Both the mean and maximum life spans of the diploid progeny of the zygotes were very similar to those of the longer-lived parent. Those two types of observations are not necessarily contradictory because the division histories of zygotes and their diploid progeny differs [Jazwinski, personal communication].

Heterosis of life span, the superiority of sexual progenies' lifespan over that of either parent, is the rule for inbred crossbreeding multicellular organisms, but has not been observed in yeast.

Both the senescent and RC phenotypes exhibit dominance. The slow division rate of very old mothers is transmitted to their daughters and granddaughters, but not to their great granddaughters—a phenomenon that may represent transmission of a dominant cytoplasmic factor that is subsequently diluted or inactivated [Egilmez and Jazwinski, 1989]. Dominance of the RC phenotype is observed in matings. In crosses between haploid *virgin* cells, the zygotes exhibit the RC of the short-lived parent; however, if one of the parent cells had budded several times before fusion, the zygotes' lifespan is reduced correspondingly, i.e., there is no "rescue hybridization." "The damage which occurs in a cell during aging is transferred to the zygote and cannot be repaired by the young partner. It is a dominant feature" [Müller, 1985].

Senescence in yeast and human diploid fibroblasts in culture is similar in that both exhibit lengthening in generation time with age and dominance of the senescent phenotype. By further analogy, the yeast senescence factor may be a membrane protein that elicits a DNA-synthesis inhibitor [Jazwinski et al., 1989]. Conversely, human cells apparently differ from yeast in that the former do not necessarily die after they have ceased to divide and can be kept alive for long periods [Bell et al., 1978]. If, indeed, mortality is a general feature of yeast cells when they cease to divide, a point that in the opinion of this reviewer has not been critically tested, than it appears that experimental selection of mutants with superior life spans (RC) may be feasible because of the possibility of exerting intense selection pressure.

To estimate the number of different genetic loci involved in aging, c-DNA prepared from m-RNA of old and young cells was used to probe a yeast genomic DNA library. Six nuclear genes were identified, five young-specific and one old-specific. The expression of at least two of those genes (one young-specific and the other old-specific) is not cell cycle-dependent, indicating that their expression is truly age-specific. Of course, it is not clear whether the expression of those genes is the cause or the effect of aging [Chen et al., 1989; Jazwinski, et al. 1989].

Although yeast is championed as the ideal eucaryotic organism for biochemical and molecular genetics in general, a serious methodological shortcoming is the lack of mutants in isogenic lines that exhibit life spans superior or inferior to their wild-type parent. According to theories that DNA repair or free radicals are important in the aging process, short-lived mutants may already exist, namely, those known to be deficient in DNA repair or antioxienzymes such as superoxide dismutase or catalase.

NEUROSPORA

Bertrand [1986] reviewed the properties of the *kalilo* senescence factor of *Neurospora intermedia*, a mitochondriala insertional (IS) element initially thought to be derived from a nuclear plasmid.

In the *kalilo* strains, senescence is initiated by insertion of a 9.0 kb foreign nucleotide sequence *kal* DNA into mt-DNA. A 9.0 kb linear DNA plasmid homologous to *kal* apparently occurred in high copy number in close association with nuclei of presenescent and senescent cells, but not in cells of long-lived normal strains. The "nuclear" plasmid, like the mt-IS element, is maternally inherited. The authors surmise that the plasmid is the etiological prescursor of the *kal* DNA insertion sequences that appear in the mt-DNAs of senescent cell lines and conclude that the *kalilo* element induces senescence because it is a mutator of mitochondrial genes [Bertrand et al., 1986].

The close association of the plasmid with nuclei was unexpected because it as well as the mt-IS element are maternally inherited. Subsequently, it was concluded that the plasmid is also a mitochondrial element. The observed nuclear association may be an artifact produced by the unusual affinity of the element for nuclear membranes or chromatin after mitochondria are dissolved with detergents during the extraction of nuclei [Bertrand and Griffiths, 1989].

Kalilo cytoplasms express senescence in two ways: death in a subculture series or growth cessation in a race tube. In the latter case, resumption of growth usually occurs, whereas in the former, growth resumption is rare. Growth resumption is associated with a resurgence of normal mt-DNA and a decline of abnormal genomes. Those observations and others led to the conclusion that some feature of the development or germination of quiescent cells enhances the expression of senescence [Griffiths et al., 1986].

In some field-isolates from Hawaii and India, integration of their respective linear plasmids *kalilo* and *maranhar* into mt-DNA induces senescence. Although the two plasmids show little homology at the DNA level, both have inverted long terminal repeats, and each potentially encodes a DNA polymerase and a RNA polymerase. Both plasmids generate very long inverted repeats of mt-DNA at their ends upon integration. Hence, they appear to integrate by a mechanism that involves pairing of both ends of the plasmid with short stretches of homologous nucleotide sequences in mt-DNA. This recombinogenic association apparently generates an origin for an unscheduled round of mt-DNA replication. In the process, the resulting two copies of the mitochondrial chromosome are joined to opposite ends of the plasmid. A model for the senescence-associated accumulation of mt-DNAs with plasmid insertion sequences is proposed on the basis of common features that characterize senescence in a variety of filamentous fungi [Bertrand and Griffiths, 1989].

Senescence in the filamentous fungi is usually associated with deteriorations in mitochondrial function and progressive accumulations of a great diversity of mutant forms of mt-DNA. The diversity of molecular defects in the mitochondrial chromosome that can result in senescence-like processes indicates that certain kinds of mitochondrial mutations, rather than specific genetic elements, cause the accumulation of defective mitochondrial chromosomes. Several arguments indicate it is unlikely that senescence plasmids contain information, such as a mitochondrial origin of replication, that conveys a replicative advantage to mitochondrial chromosomes containing plasmid-derived insertion sequences. "It is more likely that the senescence-inducing activity of *kalilo* and *maranhar* is inherent in their activity as mutators of mitochondrial genes" [Bertrand and Griffiths, 1989].

Survey of 36 wild-type isolates revealed 7, all from different geographic locations, that contain dsRNAs ranging from 500 base pairs to 18 kb. A total of seven distinct dsRNA species was identified. Only one of the seven dsRNA species was homologous to genomic DNA prepared from all strains in the survey. That dsRNA may be of cellular origin; whereas the others may be of viral origin. A review indicates that most fungal viruses contain dsRNA genomes or dsRNA replicative intermediates and reveals a few well-documented cases in which they alter the host's phenotype; however, in those cases or in *Neurospora,* senescence has not been associated with such viruses. The biological function of the *Neurospora* dsRNAs is unknown [Myers et al., 1988].

The age-dependent accumulation of fluorescent pigment, lipofuscin, is a universal indicator of cell aging, including senescence in *Neurospora* and *Podospora* [Munkres, 1981]. Analysis of fluorescence pigments from normal and senescent *Neurospora* mycelia by two-dimensional thin-layer chromatography revealed eight relatively nonpolar components, five being senescence-specific. Those five fluorogens were identical or similar to synthetic fluorescent malondialydehyde polymers by eight physiochemical criteria and different in several of those criteria from synthetic fluorescent compounds with Schiff-base linkages derived from reaction of the aldehyde with amino or amine groups [Munkres, 1987b].

Dietary free radical scavengers delay senescence in *Neurospora* and *Podospora* and inhibit accumulation of the fluorescent pigments [Munkres, 1981]; therefore, they probably arise from malonaldehyde formed during free radical-mediated peroxidation reactions of unsaturated lipids or deoxyribose. In aerobic aqueous solution, malondialdehyde rapidly polymerizes upon the addition of ferrous ions, but it is not known whether the in vivo polymerization occurs by metal or enzymatic catalysis. It was postulated that cells may have enzymes that catalyze the polymerization since the free aldehyde is known to be toxic and mutagenic [Munkres, 1981; 1987b]. *Neurospora* may also have enzymes that "repair" malonaldehyde-linked Schiff-base molecular damage. [Munkres, 1981] The proposed enzymes may constitute part of the elaborate system of defense against free radical-mediated oxidative stress and senescence.

ACKNOWLEDGMENTS

The preparation of this review was supported by the College of Agriculture and Life Sciences, the Graduate School, and the Laboratory of Molecular Biology of the University of Wisconsin, Madison. Drs. Griffiths, Jazwinski, and Sohal provided reprints and manuscripts of their research. This is contribution 3053 from the Department of Genetics.

REFERENCES

Allen RG, Balin AK, Reimer RJ, Sohal RS, Nations C (1988): Superoxide dismutase induces differentiation in the slime mold, *Physarum polycephalum*. Arch Biochem Biophys 261:205–211.
Bartholomew JW, Mittwer T (1953): Demonstration of yeast bud scars with the electron microscope. J Bacteriol 65:272–275.
Barton AA (1950): Some aspects of cell division in *Saccharomyces cerevisiae*. J Gen Microbiol 4:84–86.
Bell E, Marek LF, Levinstone DS, Merrill C, Sher S, Young IT, Eden M (1978): Loss of division potential in vitro: Aging or differentiation. Science 202:1158–1163.
Bernet J (1986): *Podospora* mutations reducing cell survival following nutrient exhaustion. Curr Microbiol 14:133–136.
Bertrand H (1986): The *kalilo* senescence factor of *Neurospora intermedia:* A mitochondrial IS-element derived from a nuclear plasmid. In Wickner RB, Hinnebusch AM, Lambowitz AM, Gunsalus IC, Hollaender A (eds): "Extrachromosomal Elements in Lower Eucaryotes." New York: Plenum Press, pp 93–103.
Bertrand H, Griffiths AJF (1989): Linear plasmids that integrate into mitochondrial DNA in *Neurospora*. Genome, in press.
Bertrand H, Griffiths AJF, Court DA, Cheng CK (1986): An extrachromosomal plasmid is the etiological precursor of kal DNA insertion sequences in the mitochondrial chromosome of senescent *Neurospora*. Cell 47:829–837.
Bockelman B, Osiewacz HD, Schmidt FR, Shulte E (1986): Extrachromosomal DNA in fungi-organization and function. In Buck WE (ed): "Fungal Virology." Boca Raton, FL: CRC Press, pp 237–283.
Chen JB, Egilmez NK, Jazwinski SM (1989): Differential gene expression during aging of the yeast *Saccharomyces cerevisiae* [Abstract]. Fed Am Soc Exp Biol 3:A570.
Cummings DJ, Turker MS, Domenico JM (1986): Mitochondrial excision-amplification plasmids in senescent and long-lived cultures of *Podospora anserina*. In Wickner RB, Hinnebusch A, Lambowitz AM, Gunsalus JC, Hollaender A (eds): "Extrachromosomal Elements in Lower Eucaryotes." New York: Plenum Press, pp 129–146.
Egilmez NK, Jazwinski SM (1989): Evidence for the involvement of a cytoplasmic factor in the aging of the yeast *Saccharomyces cerevisiae*. J Bacteriol 171:37–42.
Egilmez NK, Chen JB, Jazwinski SM (1990): Preparation and partial characterization of old yeast cells. J Gerontol, in press.

Esser K, Küch U, Lang-Hinrich C, Lemke P, Osiewacz HD, Stahl U, Tudzynski P (1986): "Plasmids of Eukaryotes." Berlin: Springer-Verlag.

Griffiths AJF, Kraus S, Bertrand H (1986): Expression of senescence in *Neurospora intermedia*. Can J Genet Cytol 28:459–467.

Hough JS (1961): Estimation of the age of cells in a population of yeast. J Inst Brewing 67:494–495.

Jazwinski SM, Egilmez NK, Chen JB (1989): Replication control and cellular lifespan. Exp Gerontol, in press.

Johnston JR (1966): Reproductive capacity and mode of death of yeast cells. Antonie van Leeuwenhoek J Microbiol Serol 32:94–98.

Karsch T, Kück U, Esser K (1987): Mitochondrial group I introns from the filamentous fungus *Podospora anserina* code for polypeptides related to maturases. Nucleic Acids Res 15:6743–6744.

Koll F (1986): Does nuclear integration of mitochondrial sequences occur during senescence in *Podospora*? Nature 324:597–599.

Koll F, Begel O, Belcour L (1987): Insertion of short poly d(A)d(T) sequences at recombination junctions in mitochondrial DNA of *Podospora*. Mol Gen Genet 209:630–632.

Matsura ET, Domenico JM, Cummings DJ (1986): An additional class II intron with homology to reverse transciptase in rapidly senescing *Podospora anserina*. Curr Genet 10:915–922.

Mortimer RK, Johnston JR (1959): Lifespan of individual yeast cells. Nature 183:1751–1752.

Müller I (1971): Experiments on aging in single cells of *Saccharomyces cerevisiae*. Arch Mikrobiol 77:20–25.

Müller I (1985) Parental age and the life-span of zygotes of *Saccharomyces cerevisiae*. Antonie van Leeuwenhoek J Microbiol Serol 51:1–10.

Müller I, Zimmerman M, Becker D, Flomer M (1980) Calender lifespan versus budding lifespan of *Saccharomyces cerevisiae*. Mech Ageing Dev 12:47–52.

Munkres KD (1981) Biochemical genetics of aging of *Neurospora crassa* and *Podospora anserina*: A review. In Sohal RS (ed): "Age Pigments." Amsterdam: Elsevier/North Holland Biomedical Press, pp 83–100.

Munkres KD (1987a): Aging of Fungi. In Rothstein M (ed): "Review of Biological Research in Aging," Vol. 3. New York: Alan R. Liss, pp 41–50.

Munkres KD (1987b): *Neurospora* age pigments resemble malonaldehyde polymers. In Totaro EA, Glees P, Pisanti FA (eds): "Advances in the Biosciences," Vol. 64, "Advances in Age Pigments Research." Oxford: Pergammon Press, pp 165–184.

Myers CJ, Griffiths AJF, Kraus SR, Martin RR (1988): Double stranded RNA in natural isolates of *Neurospora*. Curr Genet 13:495–501.

Nations C, Allen RG, Balin AK, Reimer RJ, Sohal RS (1987): Superoxide dismutase activity and glutathione concentration during the calcium-induced differentiation of *Physarum polycephalum* microplasmodia. J Cell Physiol 133:181–186.

Pohley H-J (1987): A formal mortality analysis for populations of unicellular organisms *(Saccharomyces cerevisiae)*. Mech Ageing Dev 38:231–243.

Sainsard-Chanet A, Begel O (1986): Transformation of yeast and *Podospora*: Innocuity of senescence-specific DNAs. Mol Gen Genet 204:443–451.

Schmidt U, Kosach M, Stahl U (1987): Lariat RNA of a group II intron in a filamentous fungus. Curr Genet 12:291–295.

Schulte E, Kück U, Esser K 91988): Extrachromosomal mutants from *Podospora anserina*: Permanent vegetative growth in spite of multiple recombination events in the mitochondrial genome. Mol Gen Genet 211:342–349.

Sohal RS, Allen RG, Nations C (1986): Oxygen free radicals play a role in cellular differentiation: An hypothesis. J Free Radicals Biol Med 2:175–181.

Sohal RS, Allen RG, Nations C (1988): Oxidative stress and cellular differentiation. Ann NY Acad Sci 551:59–74.

Turker MS, Cummings DJ (1987): *Podospora anserina* does not senesce when serially passaged in liquid culture. J Bacteriol 169:454–460.

Turker MS, Domenico JM, Cummings DJ (1987a): Excision-amplification of mitochondrial DNA during senescence in *Podospora anserina*. J Mol Biol 198:171–185.

Turker MS, DOmenico JM, Cummings DJ (1987b): A novel family of mitochondrial plasmids associated with longevity mutants of *Podospora anserina*. J Biol Chem 262:2250–2255.

Turker MS, Domenico JM, Cummings DJ (1987c): A *Podospora anserina* longevity mutant with a temperature-sensitive phenotype for senescence. Mol Cell Biol 7:3199–3204.

Wright RM, Cummings DJ (1983): Integration of mitochondrial gene sequences within the nuclear genome during senescence in a fungus. Nature 302:86–88.

Protozoa

Joan R. Smith-Sonneborn

HISTORICAL BACKGROUND

Molecular diversity and evolutionary time separate the myriads of different single-celled organisms, the Protozoa. Reproduction in protozoans can include an asexual vegetative cycle (when binary fission occurs) and a sexual cycle (when fertilization occurs).

The life span of cells ranges from a few divisions to apparent immortality. Immortal strains show no loss in ability to divide and do not require a fertilization to "reset" their calendar. Survival in many ciliates is achieved by successive episodes of fertilization to establish the next generation with its species-specific limited cell division potential. Life span begins with the fertilized cell, the founder of the clone. The number of cell divisions after fertilization marks the fission age of a cell. Senescence is defined as the reduced probability that a viable cell will be produced at the next division. Clonal life span is defined as the number of cell divisions or days elapsed since the origin of the clone at fertilization until the death of all representatives derived from that clone. The species survive by fertilization and production of successive generations and, like humans, must reproduce before cell senescence impairs production of viable offspring.

In the last decade, and especially within the last few years, an explosion of information on the molecular and cell biology of the ciliates has occurred. The ciliates are those protozoans with hair-like organelles (cilia) used for locomotion and feeding in their watery home. Typically, these ciliates possess two different kinds of nuclei, the relatively transcriptionally inert germ line micronuclei and the active macronucleus. The ciliates *Paramecium, Oxytricha, Stylonychia, Tetrahymena, Euplotes,* and *Tokophyra* as well as the colonial flagellate *Volvox,* have been favored as experimental organisms to study protozoan aging [see previous reviews in Smith-Sonneborn, 1981, 1983a,b, 1984, 1985a,b,c,d, 1987a, 1990a,b; Aufderheide, 1984a; Takagi, 1987]. With the

Departments of Zoology and Physiology, University of Wyoming, Laramie, Wyoming 82070

41 © 1990 Wiley-Liss, Inc.

exception of *Tokophyra,* all other ciliates have enjoyed popularity for molecular studies, especially *Tetrahymena. Tetrahymena,* with both mortal and apparently immortal strains [Nanney, 1974], has been adopted by molecular biologists as a generalized eukaryotic cell. The ciliates have been used to probe mechanisms of gene activation, gene and gene product rearrangements, and evolutionary relations. Ribosomal genes have been used both as models of an active gene and as chronometers to measure evolutionary change [Sogin and Elwood, 1986; Van Bell, 1985; Kumazaki et al., 1982; Kuntzel et al., 1983; see review in Blackburn, 1982].

Successful DNA transformations both in *Tetrahymena* and *Paramecium* [Tondravi and Yao, 1986; Godiska et al., 1987] and the new molecular data greatly enhance the value of ciliates as research organisms.

Ciliates As Model Eukaryotes

Ciliates may occupy a strategic evolutionary position between the prokaryotes and higher eukaryotes and as such offer a unique perspective on primitive differences in gene regulation between kingdoms [see reviews in Smith-Sonneborn, 1987b, 1990a,b; Nanney, 1986]. Like higher eukaryotes, ciliates have: 1) a nucleosome chromatin structure [see review in Gorovsky, 1986]; 2) a typical ciliary structure; 3) an environmentally responsive phagocytic membrane; 4) an electrically excitable surface membrane like nerve and muscle cells [Kung and Saimi, 1982; Kung, 1985]; 5) a functional response to hormones like insulin [Kohida et al., 1986] and neuropeptides [Quinones-Madonado and Renaud, 1987; Castrodadt et al., 1988]; 6) a common DNA sequence recognition site for topoisomerase, important for regulation of gene function [Andersen et al., 1987]; 7) chemical homologies with histones and histone variant genes associated with active genes [White et al., 1988], as well as actin [Yasawa et al., 1981], tubulin [Adoutte et al., 1985], and telomeres [Blackburn and Szostak, 1984; see review in Blackburn, 1986]; and 8) post-transcriptional modification of gene products [Zaug and Cech, 1986; Andreadis et al., 1987].

In contrast to the similarities between molecules in ciliates and higher eukaryotes, certain molecules, like cytochrome c in *Tetrahymena,* mimic the prokaryotic kingdom [Tarr and Fitch, 1976; Baba et al., 1981]. Conservation of DNA sequence in ribosomal DNA has been found in higher eukaryotes, but one family of repeat sequences is similar to a cAMP-dependent catabolite regulatory protein site in *E. coli* [Niles, 1985]. The regulatory protein is capable of binding with several genes to regulate their expression.

Also, unlike eukaryotes, the ochre and amber stop words UAA and UAG are used to code for glutamine rather than as termination codons [Horowitz and Gorovsky, 1985; Caron and Meyer, 1985; Preer et al., 1985; Helftenbein, 1985].

MICRO- VERSUS MACRONUCLEI
Role in Aging

The role of the macronucleus in longevity was highlighted by the macronuclear transplant experiments of Aufderheide [1987]. Whereas young macronuclear transplants could significantly extend the life of old recipients, old macronuclear transplants did not alter longevity. The importance of the nucleus was implied by the fact that the transplanted macronucleus "remembered" its age. Since the recipient was a short-lived mutant with an inability to properly divide its nucleus as cell division [Aufderheide and Schneller, 1987] there is a possibility that the extended life span only corrected that defect. The old macronucleus also has difficulty with equal macronuclear division. Therefore, failure to extend life span with the old macronucleus may reflect its inability to complement the mutational defect. Since, as clonal age increases in *Paramecium,* there is a dramatic loss of total macronuclear DNA [Schwartz and Meiseter, 1973; Takagi and Kanazawa, 1982; Klass and Smith-Sonneborn, 1976], probably nucleosomal DNA [Heifetz and Smith-Sonneborn, 1981], only the young, not old, donor macronuclear transplants may be able to replenish the necessary levels of rDNA.

The macronucleus is considered the determinant of the protozoan life span since: 1) only the macronucleus is strongly active in transcription during the vegetative life cycle; 2) young, not old, macronuclear transplants extend life span duration in recipients [Aufderheide, 1987]; and 3) rarely, amicronucleated cells are known to survive. However, the importance of the micronucleus for oral development and cell survival has been recently reviewed [Chau and Ng, 1988; Ng, 1986, 1988], and amicronucleates were found to possess micronuclear-specific DNA sequences in their macronucleus [Karrer et al., 1984; Ammermann, 1987; Stein-Gavens et al., 1987]. An important role for the micronucleus, and/or micronucleus-specific DNA sequences has become evident.

Nuclear Chemical and Functional Differences

The micronuclei are realtively transcriptionally inactive, while the macronuclei are active during vegetative growth. The micronuclei differ from the macronuclei in chromatin composition as well as in DNA sequences [see review in Gorovsky, 1986]. The histone composition differences are as follows. Micronuclei have three peptides (alpha, beta, and gamma) instead of the typical H1 in the linker region. They lack hv1 and hv2 (H2 histone variants found associated with active genes) and have little acetylation or phosphorylation (in contrast to the high acetylation and phosphorylation found in macronuclei). Micronuclei have two H3 forms (there is only one form in macronuclei). The concentration of H2A.1 is greater than H2A.2 (the concentrations are equal in macronuclei). In addition, the micronucleus has telomere ends (C4A2) internal to chromosomes rather than at the ends, is insensitive to DNase I (the macronucleus is sensitive), does not exhibit the Z conformation of DNA found in the macronucleus, and does not exhibit the N6 methylation of adenine found in the macronucleus. Finally, the multiple DNA breakage sites found in micronuclei are not observed in macronuclear DNA.

NUCLEOCYTOPLASMIC INTERACTIONS
The Fertilization Cell Cycle

Relevant background. In *Paramecium*, fertilization is a rejuvenation process allowing a cell destined to die to reset the calendar for its 200-fission life span [Sonneborn, 1954]. Fertilization is by conjugation (cross-fertilization) or autogamy (self-fertilization). During fertilization, the old macronucleus disintegrates, the micronuclei undergo meiosis, and one haploid gamete is preserved in the paroral cone region, which duplicates. One nucleus is either exchanged with a mate during conjugation or the duplicate nuclei fuse during autogamy, the self-fertilization process. The fusion nucleus undergoes division to differentiate two new macronuclei and micronuclei for the daughter cells. The macronucleus undergoes multiple endoreduplication. In *Paramecium tetraurelia*, after one more micronuclear division, the cells divide and begin their clonal life span of about 200 possible binary fissions before death.

During the "sensitive" period of nuclear development in *Paramecium*, the interval between formation of the fertilized nucleus and the second postfertilization division of the cell, both nuclear differentiation and determination of macronuclear gene function occurs, some early and some late in the period. During nuclear differentiation and determination, the cytoplasm can critically alter the phenotype of cells [Sonneborn, 1977; Mikami, 1987; Koizumi et al., 1986].

In the cytoplasm of the fertilized cell, the ancestral macronucleus continues to persist during the development of the new macronucleus. New data indicate that the old macronucleus can direct the processing of the micronuclear DNA for the new macronucleus [Forney and Blackburn, 1988].

The cytoplasm has been shown to affect both the survival of offspring of cells of different ages and the resultant phenotypes of the cells during the sexual cycle. Dramatic age-related effects of cytoplasm of cell survival have been reported [Sonneborn and Schneller, 1960]. A young micronucleus transferred to an old cytoplasm during mating usually fails to survive, though occasional survivors show a complete youthful phenotype and longevity. Nuclear transplants of an old micronucleus to a young cytoplasm in *Paramecium caudatum* can lead to a more youthful progeny at the next fertilization, though there is an old age limit at which transplantation to the young cytoplasm can no longer rescue the micronucleus [Karino and Hiwatashi, 1984a,b] and at which fertilization can no longer completely rejuvenate the offspring [Smith-Sonneborn et al., 1974]. Also, the old member of a mating pair, despite genetic identity with its young partner after nuclear exchange, shows more lethality than its mate during a subsequent self-fertilization [Sonneborn and Schneller, 1960; WIlliams, 1980].

In ciliates, not only the cytoplasm, but also the localization within the cytoplasm and environmental and physiological conditions can influence cell phenotype. Positional effects influence differentiation of micro- and macronuclei [Mikami, 1980; Grandchamp and Beisson, 1981] and the survival of micronuclear meiotic products [Yanagi, 1987]. Environmental and physiological states can lead to alternative states of gene expression [Sonneborn, 1977].

Molecular changes during fertilization. The changes in nuclear chromatin structure during differentiation have been followed in *Tetrahymena* [see the work of Allis and Wiggins, 1984a,b; Wenkert and Allis, 1984; Allis et al., 1984; summarized by Gorovsky, 1986].

During starvation, which induces fertilization, the micronuclei show selective loss of their micronuclear-specific histone proteins (alpha, beta, and gamma histones on the linker DNA and a processed H3 histone). After nuclear fusion and mitotic division, a newly synthesized alpha protein appears in the nuclei. Further nuclear differentiation appears to be related to nuclear position within the cell cytoplasm and is associated with the loss of the alpha protein. The presumptive macronucleus initially has no detectable histones; by the eighth endoreduplication, it contains the typical macronuclear proteins, including those histone proteins specific for the macronucleus, the H2 variants hv1 and hv2 (which are associated with transcriptional activity). The appearance of beta and gamma protein in the micronucleus occurs after differentiation of the macronucleus has begun.

Nucleic acid processing. Developmentally regulated DNA processing and sequence elimination occurs during the differentiation of the macronucleus from the micronuclear zygote [Preer et al., 1987; Howard and Blackburn, 1985; Forney and Blackburn, 1988]. Fragmentation of the micronuclear genome occurs in a sequence-specific manner in *Tetrahymena* [Austerberry and Yao, 1987] and involves specific micronuclear breakage sites [Yao et al., 1987]. Using in situ hybridization of micronuclear-specific DNA sequences to developing macronuclei, amplification of both micro- and macronuclear-specific DNA sequences was found in early macronuclear development; later, specific micronuclear DNA sequences are eliminated [Preer et al., 1987].

The molecular basis for cytoplasmic inheritance of antigens in ciliates has been discovered. In *Tetrahymena,* the mutant phenotype has been rescued by wild-type cytoplasm [Doerder and Berkowitz, 1987]; this phenotype is believed to be related to gene processing. In *Paramecium,* micronuclear processing errors were found to be the basis of a cytoplasmically regulated A antigen mutation [Preer et al., 1987; Forney and Blackburn, 1988]. DNA rearrangements between the different antigens was not found during antigenic expression and therefore cannot explain exclusive expression of one of an array of antigens in vegetative cells [Forney et al., 1983; Godiska, 1987].

Incorrect splicing of the mRNA can occur [Mannella et al., 1979] and has been suggested as a inducer of clonal aging in fungi [Kuck et al., 1981; Esser, 1985]. Incorrect splicing can also cause mutation in *Paramecium*.

The Vegetative Cell Cycle

During the vegetative cell cycle, the macronucleus directs the cell phenotype. In contrast to the significant events of macronuclear transplants, repeated cytoplasmic injection using 5–30% of the cytoplasm did not alter life span [Aufderheide, 1984b]. Although cytoplasm could not alter longevity, an injected cytoplasmic fraction from immature cells could restore youthful sexual activity to inactive senescent cells [Haga and Karino, 1986]. However, the cell division rate was not restored in such cells. Thus, one of the "biomarkers" of senescence could be altered by cytoplasmic influences, while others were unchanged.

Mating type determinations were found to be predetermined by micronuclear cytoplasmic interactions prior to fertilizations according to studies by Brygoo [1977] and Brygoo et al. [1980]. Initiation of macronuclear DNA synthesis is set during the previous cell interfission interval [Berger and Ching, 1988], as is the onset of the self-fertilization process of autogamy [Mikami and Koizumi, 1983; Berger, 1986]. Cell cycle regulation has been suggested as a mechanism for targeting specific histones to their proper nuclei [Wu et al., 1988]. The gene coding for micronuclear linker histone is expressed only in association with micronuclear DNA replication, whereas the gene for macronuclei histone H1 is expressed during macronuclear, not micronuclear S phase.

RIBOSOMAL DNA AS A MODEL ACTIVE GENE

In contrast to the 17S an 25S rRNA genes, the 5S and tRNA genes are not greatly amplified during macronuclear development in *Tetrahymena*. The 5S RNA hybridizes to only 0.01% of the total genome (equivalent to about 300–350 copies of the gene). The 25S and 17S rDNA comprise about 2% of the genome, with 18,000 copies [see reviews by Karrer, 1986; Yao, 1986]. Fragments of rDNA of *Tetrahymena* can replicate in yeast [Kiss et al., 1981] and serve as an autonomous replication sequence. In *Tetrahymena*, the majority of the fragments that contain 5S DNA are similar in size in the micro- and macronucleus, though some bands were found to be specific to the micronucleus [Allen et al., 1984]. Some macronuclear and micronuclear differences in 5S genes may be due to differences in genomic processing and sensitivity of methyl adenine to EcoRI [Dugaiczyk et al., 1974], since only the macronucleus contains N6 methylated adenine [Gorovsky et al., 1973]. During aging, deletions of micronuclear chromosomes and 5S rDNA have been found [Allen et al., 1984].

During differentiation of the micro- and macronuclei of *Tetrahymena*, the 30 clusters of 5S genes show structural changes. Exonuclease digestion shows that DNA fragmentation does not account for the differences. Rearrangements appear random with respect to the 5S genes. Neither DNA sequence alterations or gene rearrangements appear to account for the transcriptional activation of 5S genes during macronuclear development [Pederson et al., 1984].

In *Paramecium*, rDNA is a short extrachromosomal DNA with a repeat length of 8.2 kilobases (11 repeats on a single rDNA molecule). Restriction analysis has shown heterogeneity of spacer lengths in tandem repeats implying multiple copies of rDNA in the micronuclei [Findly and Gall, 1978, 1980].

ENVIRONMENTALLY ALTERED LONGEVITY

Antioxidants (vitamin C) were found to prolong the life span of *Paramecium*. In vitro lipid peroxidation assays adapted for *Paramecium* [Leibovitz, 1986] from the calorimetric thiobarbituric aldehyde assay of Tappel and Zalkin [1959] indicated that both vitamin E and C reduced lipid peroxidation. Extracts of *Tetrahymena*, a long-lived ciliate, and *Paramecium*, could protect *Paramecium* from the initial oxygen free radical attack, and the short-lived species was also more sensitive to iron-induced peroxidation [George and Leibovitz, 1988]. These results favor a role for antioxidants in the longevity of *Paramecium*. However, studies of endogenous levels of antioxidants (glutathione and the activities of catalase, peroxidase, and superoxide dismutase) did not shown alterations with age in *Paramecium*.

Growth of *Paramecium* in a dilute culture medium shortened the life span [Takagi et al., 1987], but this result may well be attributable to malnutrition and, therefore, may not be comparable to animal experiments showing enhanced life span associated with restriction of caloric content in otherwise sufficient diets.

The induction of beneficial effects by low doses of an otherwise harmful physical or chemical agent is called hormesis [see *Health Physics*, 1987]. Beneficial effects have been reported in *Paramecium* using low-frequency electromagnetic fields and cosmic, gamma and ultraviolet irradiation. A selective sensitivity to extremely low-frequency electromagnetic fields using electromagnetic coils with different signal parameters was found in *Paramecium* [Darnell, 1988]. Waveforms found to extend longevity in wild-type cells [Smith-Sonneborn and Darnell, 1984] were harmful to mutant cells with altered membrane characteristics. Also, the millimolar concentration of sodium required to induce backward swimming in *Paramecium* was used as an index of electromagnetic field effects on cells. The studies indicated that the cell surface was the likely target of electromagnetic effects [Dihel and Smith-Sonneborn, 1985] and that variation of frequency could alter the behavioral response of cells to ionic environment [Darnell, 1988].

Stimulation of cell division was found in response to zero gravity or cosmic radiation [Richoilley et al., 1986] and to ionizing radiation [Luckey, 1986]. Cells, shielded from cosmic or gamma radiation, did not grow as rapidly as those exposed to low levels of radiation or high altitude [Tixador et al., 1981; Planel et al., 1987].

Increased longevity in *Paramecium* after treatment with ultraviolet and photoreactivation [Smith-Sonneborn, 1979] may also be an example of beneficial effects induced by low levels of DNA damage. In *Tetrahymena*, a dose-dependent induction of DNA polymerase by radiation was found [Keiding and Westergaard, 1971; Westergaard and Marcker, 1976; Ostergaard et al., 1987], and mitochondrial DNA polymerase was induced by thymine starvation [Ostergaard et al., 1987]. In *Paramecium*, crude preparations of irradiated cells showed radiation-induced DNA polymerase [Williams, 1980]. Long-lived survivors of radiation treatment may be examples of those cells capable of expression of a stress-related enhancement of life span induced by hormesis. The existence of rare long-lived *Paramecia* has been reported under natural conditions [Takagi et al., 1987]; these organisms have been unknowingly subjected to stress. The possible existence of two physiologically distinct regulators of potential life span, one stress-induced, utilizing a reserve capacity to cope with environmental stress and the other, giving a shorter life span, associated with normal homeostatic controls [Stebbing, 1987] could explain the apparent paradox that DNA polymerase does not decline during normal clonal aging, i.e., that level of polymerase is sufficient to support a normal life span, while increased repair is required to support an extended longevity.

In the ciliate *Stylonychia*, a comparison of DNA repair capacity in the micro- and macronucleus during vegetative growth was made. Using unscheduled DNA synthesis as an index of DNA repair capacity, no difference in repair was found between micro- and macronuclear DNA [Ammermann, 1988]. The difference in DNA copy number in the micronucleus versus the macronucleus was considered a possible explanation for the increased sensitivity of the micronuclei to DNA damage.

EFFECTS OF NEW WORK OF AGING IN PROTOZOA

Recent evolutionary data indicate that the ciliates may be organisms that span characteristics of prokaryotes and higher eukaryotes and as such provide a unique evolutionary perspective of the origin of aging in higher eukaryotic cells. The explosion of information on the molecular biology of ciliates within the last 5 years also greatly enhances their value as research tools for studies on aging. The protozoa are both an in vitro and in vivo model of cell senescence since a protozoan is a single cell and an organism.

The new information indicates that alternative pathways for processing both DNA and RNA within a single cell before and after transcription can generate phenotypic variation in "genotypically identical" cells and provide a possible explanation for phenotypic variation within clones. Changes in phenotype with age could theoretically be expressed without any change in the genome. Instead, changes could occur in the processing of gene products. DNA repair enzymes, which may overlap in function with those involved in nucleic acid processing, could have a critical role in cell phenotype and longevity—rather than only a role in the correction of DNA damage with age.

The interaction of the nucleus and cytoplasm at different times in the life cycle indicates that although the newly developing macronucleus is responsive to the cytoplasm (containing the ancestral macronucleus), cytoplasm injected into vegetative cells can influence some, but not all, of the biomarkers of senescence. The molecular basis for "cytoplasmic" inheritance of mutation of the A antigen gene was found to be a nucleic acid processing mutation under the domination of the ancestral macronucleus. Such studies raise the possibility that if the ancestral macronucleus is from an old cell, then defects in its ability to properly process nucleic acid could impact on the survival and longevity of progeny even with a young micronucleus in the old cytoplasm.

Differences in the relatively transcriptionally inert micronucleus and the active macronucleus were found associated with chromatin structure differences. A greater role for the micronucleus in cell survival has been found both with respect to the development of the oral structure and from the finding that micronuclear-specific sequences have been found in the macronucleus of amicronucleated strains.

Using the technology developed to compare starved and well-fed cells, comparison of aged and young chromatin could be made to examine any age-related changes in chromatin.

Aged cells show abnormalities in cell cycle timing events. New data reveal that both the targeting of proper chromatin molecules for normal cell function and determination of commitments to cell division, autogamy, and mating type are cell cycle-dependent. Thus, an alteration in cell cycle could have pleiotropic effects on cell function and possibly explain some age-related changes in gene function.

There is a possibility that organisms possess two types of life span potentials, one the normal homeostatically controlled life span, and the other a stress-induced life span enhancement. Under this view, the unusually long-lived individuals observed in life table data would reflect a stress-induced enhancement of life span potential. Dietary restriction and low-level radiation might be examples of such modulation of life span.

REFERENCES

Adoutte A, Claisse M, Maunoury R, Beisson J (1985): Tubulin evolution: Ciliate specific epitopes are conserved in the ciliary tubulin of metazoa. J Mol Evol 22:220–229.
Allen SL, Ervin PR, McLaren NC, Brand RE (1984): The 5S ribosomal RNA gene cluster in *Tetrahymena thermophila:* Strain differences, chromosomal location, and loss during micronuclear aging. Mol Gen Genet 197:244–253.
Allis CD, Wiggins JC (1984a): Proteolytic processing of micronuclear H3 and histone phosphorylation during conjugation in *Tetrahymena thermophila.* Exp Cell Res 153:287–298.
Allis CD, Wiggins JC (1984b): Histone rearrangements accompany nuclear differentiation and dedifferentiation in *Tetrahymena.* Dev Biol 101:282–294.
Allis CD, Allen RL, Wiggins JC, Chicoine LG, Richman R (1984): Proteolytic processing of H1-like histones in chromatin: A physiologically and developmentally regulated event in *Tetrahymena* micronuclei. J Biol Chem 99:1669–1677.
Ammermann D (1987): Germ line specific DNA and chromosomes of the ciliate *Stylonychia lemnae.* Chromosoma 95:37–43.
Ammermann D (1988): DNA damage and repair in *Stylonychia lemnae* (Ciliata, Protozoa). J Protozol 35:264–267.
Andersen A, Christiansen K, Sorensen B, Westergaard O (1987): Sequence specific action of eukaryotic topoisomerases I and II. Second International Ciliate Molecular Genetics Conference, August 2–6, Berkeley, California.
Andreadis A, Gallego ME, Nadal-Ginard B (1987): Generation of protein isoform diversity by alternative splicing. Annu Rev Cell Biol 3:207–242.
Aufderheide KJ (1984a): Cellular ageing: An overview. In Sauer HW (ed): "Cellular Ageing." Basel: Karger, pp 2–8.
Aufderheide KJ (1984b): Clonal ageing in *Paramecium tetraurelia.* Absence of evidence for a cytoplasmic factor. Mech Ageing Dev 28:57–66.
Aufderheide KJ (1987): Clonal aging in *Paramecium tetraurelia.* II. Evidence of functional changes in the macronucleus with age. Mech Ageing Dev 37:265–279.
Aufderheide KJ, Schneller MV (1985): Phenotypes associated with early clonal death in *Paramecium tetraurelia.* Mech Ageing Dev 32:299–309.
Austerberry CF, Yao M-C (1987): Nucleotide sequence structure and consistency of a developmentally regulated DNA deletion in *Tetrahymena thermophila.* Mol Dev Biol 7:435–443.
Baba ML, Darga LL, Goodman M, Czelusniak J (1981): Evolution of cytochrome C investigated by the maximum parsimony method. J Mol Evol 17:197–213.
Berger JD (1986): Autogamy in *Paramecium:* Cell cycle state-specific commitment to meiosis. Exp Cell Res 166:475–485.
Berger JD, Ching AS-L (1988): The timing of initiation of DNA synthesis in *Paramecium tetraurelia* is established during the preceding cell cycle as cells become committed to cell division. Exp Cell Res 174:355–366.
Blackburn EH (1982): Characterization and species differences of rDNA: Protozoans. In Busch H, Rothblum L (eds): "The Cell Nucleus: rDNA," Vol. 10A. New York: Academic Press, pp 145–170.
Blackburn EH (1986): Telomeres. In Gall JG (ed): "The Molecular Biology of Ciliated Protozoa." New York: Academic Press, pp 155–178.
Blackburn EH, Szostak JW (1984): The molecular structure of centromere and telomeres. Annu Rev Biochem 78:2263–2267.
Brygoo Y (1977): Genetic analysis of mating type differentiation in *Paramecium tetraurelia.* Genetics 87:633–653.
Brygoo Y, Sonneborn TM, Keller AM, Dippell RV, Schneller MV (1980): Genetic analysis of mating type differentiation in *Paramecium tetraurelia.* Genetics 94:951–959.

Caron F, Meyer E (1985): Does *Paramecium primaurelia* use a different genetic code in its macronucleus? Nature 314:185–188.

Castrodad FA, Renaud FL, Ortiz J, Phillips DM (1988): Biogenic amines stimulate regeneration of cilia in *Tetrahymena thermophila*. J Protozool 35:260–264.

Chau MF, Ng SF (1988): The somatic function of the micronucleus in sexual reproduction of *Paramecium tetraureliaa:* Initiation of oral membranelle assembly. J Cell Sci 89:157–166.

Croute F, Vidal S, Dupouy D, Soleilhavoup JP, Serre G (1985): Studies on catalase, glutathione peroxidase and superoxide dismutase activities in aging cells of *Paramecium tetraurelia*. Mech Ageing Dev 29:53–62.

Darnell CM (1988): Effects of extremely low frequency electromagnetic radiation on *Paramecium* lifespan and ion conductance. M.S. Thesis, Laramie, Wyoming.

Dihel LE, Smith-Sonneborn J (1985): Effects of low frequency electromagnetic field on cell division and the plasma membrane. Bioelectromagnetics 6:61–71.

Doerder FP, Berkowitz MS (1987): Nucleo-cytoplasmic interaction during macronuclear differentiation in ciliate protists: Genetic basis for cytoplasmic control of Ser H expression during macronuclear development in *Tetrahymena thermophila*. Genetics 117:13–23.

Dugaiczyk A, Hedgpeth J, Boyer HW, Goodman HM (1974): Physical identity of the SV40 deoxyribonucleic acid sequence recognized by the EcoRI restriction endonuclease and modification methylase. Biochemistry 13:503–512.

Esser K (1985): Genetic control of aging: The mobile intron model. In Bergener M, Ermini M, Stahelin HB (eds): "The Thresholds in Aging." London: Academic Press, pp 4–20.

Findly RC, Gall JG (1978): Free ribosomal RNA genes in *Paramecium* are tandemly repeated. Proc Natl Acad Sci USA 75:3312–3316.

Findly RC, Gall JG (1980): Organization of ribosomal genes in *Paramecium tetraurelia*. J Cell Biol 84:547–559.

Forney JD, Blackburn EH (1988): Developmentally controlled telomere addition in wild-type and mutant paramecia. Mol Cell Biol 8:251–258.

Forney JD, Epstein LM, Preer LB, Rudman BM, Widmayer DJ, Klein WH, Preer JR Jr (1983): Structure and expression of genes for surface proteins in *Paramecium*. Mol Cell Biol 3:466–474.

George RP, Leibovitz BE (1988): *Tetrahymena* extract protects *Paramecium* from oxygen radicals. Age 11:66–69.

Godiska R (1987): Structure and sequence of the H surface protein gene of *Paramecium* and comparison with related genes. Mol Gen Genet 208:529–536.

Godiska R, Aufderheide KJ, Gilley D, Hendrie P, Fitzwater T, Preer LB, Polisky B, Preer JR (1987): Transformation of *Paramecium* by microinjection of a cloned serotype gene. Proc Natl Acad Sci USA 84:7590–7594.

Gorovsky MA (1986): Ciliate chromatin and histones. In Gall JG (ed): "The Molecular Biology of Ciliated Protozoa." New York: Academic Press, pp 227–261.

Gorovsky MA, Hattman S, Pleger GL (1973): 6N methyl adenine in the nuclear DNA of a eukaryote, *Tetrahymena pyriformis*. J Cell Biol 56:697–791.

Grandchampe S, Beisson J (1981): Positional control of nuclear differentiation in *Paramecium*. Dev Biol 81:336–341.

Haga N, Karino S (1986): Microinjection of immaturin rejuvenates sexual activity of old *Paramecium*. J Cell Sci 86:263–271.

Health Physics (1987): 52:250–688.

Heifetz S, Smith-Sonneborn J (1981): Nucleolar changes in aging and autogamous *Paramecium tetraurelia*. Mech Ageing Dev 16:255–263.

Helftenbein E (1985): Nucleotide sequence of a macronuclear DNA molecule coding for a-tubulin from the ciliate *Stylonychia lemnae*. Special codon usage: TAA is not a termination codon. Nucleic Acids Res 13:415–433.

Horowitz S, Gorovsky MA (1985): An unusual genetic code in nuclear genes of *Tetrahymena*. Proc Natl Acad Sci USA 82:2452–2455.

Howard EA, Blackburn EH (1985): Reproducible and variable genomic rearrangements occur in the developing somatic nucleus of the ciliate *Tetrahymena thermophila*. Mol Cell Biol 5:2039–2050.

Karino S, Hiwatashi K (1984a): Analysis of germinal aging in *Paramecium caudatum* by micronuclear transplantation. Exp Cell Res 136:407–415.

Karino S, Hiwatashi K (1984b): Resistance of germinal nucleus to aging in *Paramecium:* Evidence obtained by micronuclear transplantation. Mech Ageing Dev 26:51–66.

Karrer KM (1986): The nuclear DNAs of Holotrichous ciliates. In Gall J (ed): "The Molecular Biology of Ciliates." New York: Academic Press, pp 85–110.

Karrer K, Stein-Gavens S, Allitto BA (1984): Micronucleus specific sequences are present in an amicronucleate mutant of *Tetrahymena*. Dev Biol 105:121–129.

Keiding J, Westergaard O (1971): Induction of DNA polymerase activity in irradiated *Tetrahymena* cells. Exp Cell Res 64:317–322.

Kiss GB, Amin AA, Pearlman RE (1981): Two separated regions of the extrachromosomal rDNA *Tetrahymena thermophila* enable autonomous replication of plasmids in yeast. Mol Cell Biol 1:535–543.

Klass M, Smith-Sonneborn J (1976): Studies in DNA content, RNA synthesis, and DNA template activity in aging cell of *Paramecium aurelia*. Exp Cell Res 98:63–72.

Kohida K, Thomka M, Csaba G (1986): Age of the cell culture: A factor influencing hormonal imprinting in *Tetrahymena*. Acta Microbiol Hung 33:295–300.

Koizumi S, Kobayashi S, Mikami K (1986): Analysis of mating type determination by transplantation of O macronuclear karyoplasm in *Paramecium tetraurelia*. Dev Genet 7:187–195.

Kuck U, Stahl U, Esser K (1981): Plasmid like DNA is part of mitochondrial DNA in *Podospora anserina*. Curr Genet 3:151–156.

Kumazaki T, Hori H, Osawa S, Mita T, Higashinakagawa T (1982): The nucleotide sequences of 5S RNA from three ciliated protozoans. Nucleic Acids Res 10:4409–4412.

Kung C (1985): Calcium channels of *Paramecium:* A multidisciplinary study. Curr Top Membr Trans 23:45–66.

Kung C, Saimi Y (1982): The physiological basis of taxis in *Paramecium*. Annu Rev Physiol 44:519–534.

Kuntzel H, Piechulla B, Hahn U (1983): Consensus structure and evolution of 5S rRNA. Nucleic Acids Res 11:893–900.

Leibovitz B (1986): Lipid peroxidation antioxidants and aging in *Paramecium tetraurelia*. Ph.D. Thesis, University of Wyoming, Laramie, Wyoming.

Luckey TD (1986): Ionizing radiation promotes protozoan reproduction. Radiat Res 108:215–221.

Mannella CA, Collins RA, Green MR, Lambowitz AM (1979): Defective splicing of mitochondrial rRNA in cytochrome deficient nuclear mutants of *Neurospora crassa*. Proc Natl Acad Sci USA 6:2635–2639.

Mikami K (1980): Differentiation of somatic and germinal nuclei correlated with intracellular localization of in *Paramecium caudatum* exconjugants. Dev Biol 80:46–55.

Mikami K (1987): Macronuclear development and gene expression in exconjugants of *Paramecium caudatum*. Dev Biol 123:161–168.

Mikami K, Koizumi S (1983): Microsurgical analysis of the clonal age and cell-cycle stage required for the onset of autogamy in *Paramecium tetraurelia*. Dev Biol 100:127–132.

Nanney DL 91974): Aging and long term temporal regulation in ciliated protozoa. A critical review. Mech Ageing Dev 3:81–105.

Nanney DL (1986): Introduction. In Gall JG (ed): "The Molecular Biology of Ciliated Protozoa." New York: Academic Press, pp 1–26.

Niles EG (1985): Identification of multiple sites in the promoter region of the *Tetrahymena* rRNA gene which binds the *E. coli* catabolite regulatory binding protein. J Biol Chem 256:12849–12856.

Ng SF (1986): The somatic function of the micronucleus of ciliated protozoa. Prog Protistol 1:215–286.

Ng SF (1988): Persistence of oral structures in the sexual process of amicronucleate *Paramecium tetraurelia*. J Protozool 35:321–326.

Ostergaard E, Brams P, Westergaard O, Nielsen OF (1987): Purification and characterization of an inducible mitochondrial DNA polymerase from *Tetrahymena thermophila*. Biochem Biophys Acta 908:150–157.

Pederson DS, Shupe K, Gorovsky MA (1984): Changes in chromatin structure accompany modulation of the rate of transcription of 5S ribosomal genes in *Tetrahymena*. Nucleic Acids Res 12:3003–3121.

Planel H, Soleilhavoup JP, Tixador R, Richoilley G, Conter A, Croute F, Caratero C, Gaubin Y (1987): Influence on cell proliferation of background radiation or exposure to very low chronic gamma radiation. Health Phys 52:571–578.

Preer JR, Preer LB, Rudman BM, Barnett AJ (1985): Deviation from the universal code shown by the gene for surface protein 51A in *Paramecium*. Nature 314:188–190.

Preer JR Jr, Preer LB, Rudman B, Barnett A (1987): Molecular biology of the genes for immobilization antigens in *Paramecium*. J Protozool 34:418–423.

Quinones-Maldonado V, Renaud FL (1987): Effect of biogenic amines on phagocytosis in *Tetrahymena thermophila*. J Protozool 34:435–438.

Richoilley G, Tixador R, Gasset G, Templier J, Planel H (1986): Preliminary results of the "Paramecium" Experiment. Naturwissenschaften 73:404.

Schwartz V, Meister H (1973): Eine Alterveranderung des makronucleus von *Paramecium*. Z Naturforsch Sec[C] 38c:232.

Smith-Sonneborn J (1979): DNA repair and longevity assurance in *Paramecium tetraurelia*. Science 203:1115–1117.

Smith-Sonneborn (1981): Genetics and aging in protozoa. Int Rev Cytol 73:319–354.

Smith-Sonneborn J (1983a): Aging in protozoa. In Rothstein M (ed): "Review of Biological Research in Aging," Vol. 1. New York: Alan R. Liss, pp 29–35.

Smith-Sonneborn J (1983b): Programmed increased longevity by weak pulsating current in *Paramecium*. Bioelectrochem Bioeng 11:373–382.

Smith-Sonneborn J (1984): Protozoan aging. In Mitchell D, Johnson T (eds): "Selected Invertebrate Models for Aging Research." Boca Raton: CRC Press, pp 79–99.

Smith-Sonneborn J (1985a): Aging in unicellular organisms. In Finch CE, Schneider EL (eds): "Handbook of the Biology of Aging." New York: Van Nostrand Press, pp 79–104.

Smith-Sonneborn J (1985b): Aging in protozoa. In Rothstein M (ed): "Review of Biological Research in Aging," Vol. 2. New York: Alan R. Liss, pp 13–28.

Smith-Sonneborn J (1985c): Genome interaction in the pathology of aging. In Cristofalo V (ed): "Handbook of the Cell Biology of Aging." Boca Raton: CRC Press, pp 453–479.

Smith-Sonneborn J (1985d): Protozoa. In Lints FA (ed): "Nonmammalian Models for Research on Aging: Interdisciplinary Topics in Gerontology," Vol. 21. Basel: Karger, pp 201–230.

Smith-Sonneborn J (1987a): Aging in Protozoa. In Rothstein M (ed): "Review of Biological Research in Aging," Vol. 3. New York: Alan R. Liss, pp 33–40.

Smith-Sonneborn J (1987b): Longevity in the Protozoa. In Woodhead A, Thompson KH (eds): "Evolution of Longevity in Animals, A Comparative Approach." Basic Life Sciences, Vol. 42. New York: Plenum Press, pp 101–109.

Smith-Sonneborn J (1990a): Protozoa. In Schneider EL, Row J (eds): "Handbook on the Biology of Aging," Vol. 3. New York: Van Nostrand Press (in press).

Smith-Sonneborn J (1990b): Protozoa. In Woodhead A (ed): "Invertebrate Models of Aging." Boca Raton: CRC Press (in press).
Smith-Sonneborn J, Darnell C (1984): Difference in the longevity of wild type ion transport mutant in normal and altered environment. Gerontologist 24:110.
Smith-Sonneborn J, Klass M, Cotton D (1974): Parental age and life span versus progeny life span in *Paramecium* J Cell Sci 14:691–699.
Sogin ML, Elwood HJ (1986): Primary structure of the *Paramecium tetraurelia* small subunit rRNA coding region: Phylogenetic relationships within the ciliophora. J Mol Evol 23:53–60.
Sonneborn TM (1954): The relation of autogamy to senescence and rejuvenescence in *P. aurelia*. J Protozool 1:36–53.
Sonneborn TM (1977): Genetics of cellular differentiation: Stable nuclear differentiation in eukaryote unicells. Annu Rev Genet 11:349–367.
Sonneborn TM, Schneller MV (1960): Age induced mutations in *Paramecium*. In Strehler BL (ed): "The Biology of Aging," American Institute of Biological Sciences Symposium 6. Baltimore: Waverly, pp 286–287.
Stebbing ARD (1987): Growth hormesis. Health Phys 52:543–547.
Stein-Gavens S, Wells JM, Karrer KM (1987): A germ line specific DNA sequence is transcribed in *Tetrahymena*. Dev Biol 120:259–269.
Takagi Y (1987): Aging. In Gortz H-D (ed): *"Paramecium."* Berlin: Springer-Verlag, pp 131–140.
Takagi Y, Kanazawa N (1982): Age associated change in macronuclear DNA content in *Paramecium caudatum*. J Cell Sci 54:137–147.
Takagi Y, Nobuoka T, Doi M (1987): Clonal lifespan of *Paramecium tetraurelia:* Effect of selection on its extension and uses of fission for its determination. J Cell Sci 88:129–138.
Tappel AL, Zalkin H (1959): Inhibition of lipid peroxidation in mitochondria by vitamin E. Arch Biochem Biophys 80:333–336.
Tarr GE, Fitch WM (1976): Amino acid sequence of cytochrome c from *Tetrahymena pyriformis* phenoset A. Biochem J 159:193–197.
Tixador R, Richoilley G, Monrozies E, Planel H, Tap G (1981): Effects of very low doses of ionizing radiation on the clonal life-span in *Paramecium tetraurelia*. Int J Radiat Biol 39:47–54.
Tondravi MM, Yao M-C (1986): Transformation of *Tetrahymena thermophila* by microinjection of ribosomal RNA genes. Proc Natl Acad Sci USA 83:4369–4373.
Van Bell CT (1985): 5S and 5.8S ribosomal RNA evolution in the suborder *Tetrahymena* (Coliophora: Hymenostomatida). J Mol Evol 22:231–236.
Wenkert D, Allis CD (1984): Timing of the appearance of macronuclear specific histone variant hv1 and gene expression in developing new macronuclei of *Tetrahymena thermophila*. J Cell Biol 98:2107–2117.
Westergaard O, Marcker KA (1976): Accumulation of replicative DNA intermediates in respone to damage of DNA in *Tetrahymena pyriformis*. In Keifer J (ed): "Radiation and Cellular Control Processes." Heidelberg: Springer-Verlag P, pp 162–169.
White EM, Shapiro DL, Allis DC, Gorovsky MA (1988): Sequence and properties of the message encoding *Tetrahymena* hv1, a highly evolutionarily conserved histone H2A variant that is associated with active genes. Nucleic Acids Res 16:179–198.
Williams TJ (1980): Determination of clonal lifespan in *Paramecium tetraurelia*. Ph.D. Thesis. University of Wyoming, Laramie, Wyoming.
Wu M, Allis D, Gorovsky M (1988): Cell-cycle regulation as a mechanism for targeting proteins to specific DNA sequences in *Tetrahymena thermophila*. Proc Natl Acad Sci USA 85:2205–2208.
Yanagi A (1987): Positional control of the fates of nuclei produced after meiosis in *Paramecium caudatum:* Analysis of nuclear transplantation. Dev Biol 122:535–539.

Yao M-C (1986): Amplification of ribosomal RNA genes. In Gall JG (ed): "The Molecular Biology of Ciliated Protozoa." New York: Academic Press, pp 179–198.

Yao M-C, Zheng K, Yao C-H (1987): A conserved nucleotide sequence at the sites of developmentally regulated chromosomal breakage in *Tetrahymena*. Cell 48:779–788.

Yasawa M, Yagi K, Toda H, Kondo K, Narita K, Yamazaki C, Sobue K, Yakiuchi S, Nosawa Y (1981): The amino acid sequence of *Tetrahymena* calmodulin which specifically interacts with guanylate cyclase. Biochem Biophys Res Commun 99:1951–1957.

Zaug AJ, Cech TR (1986): The intervening sequence RNA of *Tetrahymena* is an enzyme. Science 231:470–475.

Aging and Life Span Correlates in Insects

Thomas Grigliatti

INTRODUCTION

Insects, especially *Drosophila*, have been used in the study of aging for many decades. In addition to their relatively small size and short life span, there is little, or virtually no, cell division in the adult soma. Thus the problem of tissue renewal or rejuvenation through cell replacement is eliminated. Some view the absence of this variable regenerative capacity, a property characteristic of vertebrates, as a research advantage. Others, however, would conclude that insects are therefore poor models of vertebrate aging. However, since we know so little about the comparative biology of senescence and are ignorant of its principal causes, it would seem prudent to continue to utilize insects as one of a number of multicellular models for aging studies. They have certainly contributed immensely to our understanding of basic physiology and development. Insects, especially *Drosophila*, are probably one of the more useful organisms for the study of the evolution of senescence and will probably be helpful in deciphering the molecular basis of at least some aspects of aging and senescence.

No review of the recent studies on aging in insects would be complete without mentioning current monographs on the subject. *Insect Aging: Strategies and Mechanisms* [Collatz and Sohal, 1986] focuses on phenotypic changes associated with aging in insects. The book has a wide scope. It covers many different levels of biological organization, from basic biochemical and physiological processes through life history and behavioral correlates with aging. Drosophila *as a Model Organism for Ageing Studies* [Lints and Soliman, 1988], as is evident, reviews the literature on *Drosophila*.

This review makes no pretense at being comprehensive; the reader should, therefore, consult the two recent volumes cited above. Rather, I have attempted to review the most recent literature in a very factual manner. I have tried to

---

Department of Zoology, University of British Columbia, Vancouver, British Columbia, Canada V6T 2A9.

provide detailed abstracts of the authors' approach to a particular question, their data, and their conclusions, and I have tried to avoid the imposition of personal viewpoints or inferences.

FREE RADICALS AND FREE RADICAL DEFENSES

Free radicals are generated as a natural by-product of aerobic metabolism. They are highly reactive, and biological systems have evolved defense mechanisms to reduce their harmful effects. These include enzymes such as the superoxide dismutases (SOD), catalase, and peroxidase, as well as other reducing agents such as glutathione. About 30 years ago Harman [1956] suggested that free radicals may be a major factor in aging. Gershman [1964] modified this hypothesis by suggesting that degeneration of the antioxidant defense mechanisms might be a principal factor in the aging process. Since that time considerable effort has been expended to determine the role free radicals and free radical defenses have on aging and longevity.

Previous studies from the Sohal laboratory have indicated that the levels of antioxidants such as SOD and catalase decline in older houseflies [Sohal et al., 1984b]. Similarly, indicators of free radical action, lipofuscin-like compounds, increase in older houseflies [Sohal and Donato, 1979; Sohal, 1981]. Recently, the Sohal laboratory looked for relationships between loss of antioxidant defenses, accumulation of peroxides, and longevity in houseflies. Houseflies lose the ability to fly prior to death. The Sohal lab used this phenotype as a biomarker of aging and longevity. They separated a group of similarly aged flies into two cohorts: those that were unable to fly (crawlers) and those that were still capable of flight (fliers). They then measured the remaining longevity of each group of adults as well as total SOD activity, cyanide-resistant SOD activity, catalase activity, and amount of glutathione, all of which act as free radical scavengers [Sohal et al., 1986]. In addition, they measured the amount of peroxide and the concentration of thiobarbituric acid reactants in each group as an estimate of the amount of potential free radical formation or action. These measurements were made approximately 1 week after crawlers were separated from fliers. The remaining life span of crawlers was considerably shorter than fliers (8 vs. 18 days, respectively). The activity of SOD, cyanide-resistant SOD, and catalase was lower among the crawlers than the fliers. Similarly, the amount of glutathione among crawlers was lower than that in fliers. In contrast, the amounts of inorganic peroxide and thiobarbituric acid-reactants were much higher in crawlers than fliers. Hence, shorter life span was correlated with lower antioxidant levels and higher products of free radical formation or action. The Sohal group imply that the crawlers have a faster rate of aging than the fliers, accounting for their reduced life expectancy. Nonetheless, we do not know whether the loss of free radical protection is a consequence or a cause of physiological decline.

The Sohal group attempted to address this issue. They altered the metabolic rate in houseflies by varying temperature and then determined its effect on antioxidant defenses and longevity [Farmer and Sohal, 1987]. Upon eclosion, they separated a group of houseflies into two populations, one maintained at 20°C and the other at 28°C. They measured the life span of each group, and on days 3, 8, and 15 they measured SOD, cyanide-insensitive SOD, and catalase activity and amount of glutathione. The life spans of the 20°C and 28°C populations were 45 and 22 days, respectively. Total respiration and amount of cyanide-resistant respiration were both increased at elevated temperatures. Curiously, the SOD activity increased with age in both populations, and there was significant difference in activity between populations. In contrast, the catalase activity decreased with age. The amount of glutathione also tended to decrease with age, and there was less glutathione/individual in the 28°C populations than in the 20°C populations. Finally, the amount of hydrogen peroxide in 28°C populations was less than in 20°C populations. These results are somewhat confounding. First, as might be expected, higher temperature appears to increase metabolic activity as determined by respiration rate. However, the amount of peroxide is lower at the higher temperature. They offer the possibility that the peroxide level may appear reduced simply because it reacts faster at the higher temperature. Second, the amount of SOD activity increases with age while the catalase activity decreases. This finding differs from the previous paper, and one might expect that the synthesis of the two enzymes would be more tightly coupled (even though they measured activity, not synthesis). Third, in a previous paper [Sohal et al., 1984a] they show that if the metabolic rate is increased via exercise rather than elevated temperature, then peroxide and glutathione levels increase rather than decrease. Nonetheless, they argue that elevated temperature may shorten life span in houseflies by increasing the rate of free radical generation and decreasing the level of antioxidant defenses.

Finally, Sohal [1988] examined the effects of hydrogen peroxide on life span and antioxidant levels. Hydrogen peroxide was administered in the drinking water from day 1 posteclosion, and the activities of SOD and catalase, the concentration of glutathione, and the amount of peroxide were measured on day 15. The hydrogen peroxide had small, and quite variable, effects on longevity when administered at concentrations of 50 mM or less. The average life span of flies given 5 mM peroxide was about 85% of controls; those given to 10 mM H_2O_2 lived about 10% longer than controls, while those given 25 mM were about the same as controls. There were no significant differences in SOD and catalase activity between controls and flies given H_2O_2. The reduced form of glutathione (GSH) increased markedly (over 200%) in flies fed 10 mM H_2O_2 and decreased slightly (to about 75% of the control level) in flies fed 100 mM H_2O_2. In contrast, the oxidized form of glutathione (GSSG) was reduced by about 33% and increased by 350% in flies fed 10 mM and 100 mM H_2O_2, respectively. The endogenous level of H_2O_2 was elevated (nearly 100%) in flies given 5 mM or 100 mM H_2O_2 but was nearer to control levels in flies given 10 mM. Hence, there is an interesting correlation between large increases in GSH levels and a slight increase in the average life span in flies fed 10 mM H_2O_2. The only dramatic decline in life span occurred in the groups fed 75 and 100 mM H_2O_2, in which the life span was 78 and 73% of controls. Unfortunately, only the 100 mM group was examined for GSH and GSSG levels; both were decreased.

The results of these three studies are somewhat variable. Perhaps the only consistent trend is the decrease in glutathione levels that was seen with an increase in age, an increase in metabolic rate (temperature), and administration of high doses of peroxide. This correlation is intriguing and demands further study. However, there is still no compelling evidence to suggest that a decline in glutathione level leads to a reduction in longevity.

The effect of SOD activity on longevity is unclear. For example, a shorter-lived strain of *Drosophila* was found that had reduced SOD activity [Bartosz et al., 1979]. However, a SOD null mutant strain that produces less than 3.5% of the normal amount of SOD protein has a life span that is only 30% shorter than its wild-type counterpart, and a mutant strain with only 50% of the SOD protein has normal life span [Seto and Tener, manuscript in preparation]. The *Drosophila* SOD gene has been cloned and sequenced [Seto et al., 1989]. More importantly, this SOD wild-type gene has been reintroduced into a SOD^+ isogenic strain by transformation. It will be interesting to see what effect additional SOD activity has on longevity and how it affects other mitochondrial and physiological factors. The effects of antioxidants on longevity have been examined in a number of insects with variable results [Miquel et al., 1982]. When antioxidants had a positive effect on life span, it was often assumed that the antioxidants provided protection against the ravages of free radicals. Miquel et al. [1982] put forth an alternative hypothesis. They suggested that, at least in insects, antioxidants may actually reduce the metabolic rate and thereby increase longevity. In an attempt to distinguish between these two possibilities, Ruddle et al. [1988] administered propyl gallate to adult *Drosophila* for different periods of time and at different intervals. Adult flies were maintained on media with propyl gallate from day 7 posteclosion onward, from day 7 to day 28, or from day 28 onward. The authors suggest that if the life-extending property of propyl gallate results from protection against free radicals, then the group that was treated late in life should derive the most benefit, for it is at this time that the endogenous protective systems decline. The mean life span of the cohorts given propyl gallate either early (days 7–28) or late in life (day 28 onward) increased very slightly, from 35 ± 6.5 days (controls) to approximately 40 ± 9 days for the treated groups. The authors use an extensive statistical analysis to establish that these differences, which seem rather small with respect to the standard deviations, are indeed significant. The mean life span of adults maintained on propyl gallate continuously from day 7 posteclosion was 45.5 ± 13.1 days. Hence the duration of treatment seems more important than the interval during which they were treated. Finally, the authors refer to unpublished data demonstrating that flies given propyl gallate actually have a reduced oxygen consumption. Therefore, they suggest that the life-extending property of propyl gallate, at least in part, results from a reduction in metabolic rate.

Massie et al. [1985] examined the effect of a large number of anti-inflammatory agents on the life span of *Drosophila*. None of these agents had any positive effect on life span, with the exception of very low doses of 2,3-dihydroxybenzoic acid (10^{-5} M), which increased the life span by about 10%. High doses of some compounds seemed to have a negative effect on life span. The authors concluded that anti-inflammatory agents generally have no effect on longevity and that low doses are certainly not deleterious. They also suggest that the inhibition of hydroxyl radical formation or protection from the action of such agents is not a useful approach for altering senescence in *Drosophila*.

PHENOTYPIC AND LIFE HISTORY CORRELATES OF AGING

A variety of laboratories have studied phenotypic changes that may be associated with aging. Massie and Kogut [1987] looked for changes in mitochondrial enzyme levels in *Drosophila*. It has been proposed that the mitochondria and especially the inner membrane might be particularly susceptible targets for free radical damage. Hence the loss of mitochondrial DNA in older flies, which they had reported on previously, might have been a consequence of free radical damage. To examine the susceptibility of mitochondrial membranes to damage, they measured the activity of NADH cytochrone-c reductase (located on the outer membrane), adenylate kinase (compartment between outer and inner membrane), and malate dehydrogenase (matrix) in flies aged 1–5, 6–9, 10–19, 20–29, 30–39, and 40–50 days. They reported no significant change in activity of these enzymes regardless of whether the activity was expressed on a per fly or per mg protein basis. Previous studies have reported changes in the activity of some mitochondrial enzymes with age. Hence it is clear that the levels of both mitochondrial and cytoplasmic enzymes can change with age. However, no clear pattern has emerged. Whether the alterations in enzyme level result from the accumulation of damage, or are "programmed" remains unclear.

Massie and Williams [1987] examined the relationship between metabolic activity, longevity, and amount of mitochondrial DNA. Metabolic rate was altered in two ways. First, flight was eliminated by removing the wings from newly eclosed flies and comparing them with their winged siblings. Second, winged and dewinged flies were placed at four different temperatures: 11°, 20°, 25°, and 30°C. In general, dewinged flies lived slightly longer than the winged individuals. Nevertheless, the magnitude of the difference was slight and only occasionally reached statistical significance. There was significant loss of mitochondrial DNA with increasing age among winged populations maintained at 25°C and 30°C. Among the dewinged populations (maintained at the same temperature), the age-associated mitochondrial loss was very gradual and significantly less than the loss observed among their winged siblings. Hence the authors conclude that the large decline in mitochondrial DNA is associated with flight. There was only a very gradual decline in mitochondrial DNA with increasing age in populations maintained at 11°C, and there was no difference between the dewinged and winged populations. At 11°C no flights were observed among the winged population. Finally, there is virtually no loss of mitochondrial DNA with age in head tissue but a significant loss from the remainder of the body in populations maintained at 25°C. This finding suggests that the dramatic decline in mitochondrial DNA in populations maintained within the normal physiological range is either tissue-specific or is associated with flight activity, and hence the loss is associated principally with the indirect flight muscles. The rather gradual decline in mitochondrial DNA in populations maintained at 11°C or in populations of dewinged flies maintained at higher temperatures may be age-related.

Oxygen consumption was consistently higher in winged flies than dewinged flies (25°C), suggesting a slightly higher metabolic rate. However, the difference was not significant. The measurements for the winged population were more variable than those from the dewinged population. The variation was greatest among young flies and declined with age of the population. This variability in oxygen consumption may be a cosequence of variation in flight activity among younger flies.

In general, these authors concluded that the rather dramatic loss in mitochondrial DNA normally observed in aging populations of *Drosophila* is not a consequence of the aging process per se, but is more likely a consequence of flight activity.

Fleming et al. [1986] examined alterations in protein synthesis with age in *Drosophila*. They used two-dimensional gel electrophoresis and a sophisticated computer analysis to detect both qualitative and quantitative changes in a large number of proteins. This procedure examines the most abundant proteins in the animal at the time of protein labeling or extraction. They found no qualitative changes in the pattern of protein synthesis with age. They suggested that this observation is not consistent with theories of aging that posit stochastic loss of function and is equally inconsistent with those suggesting that aging is genetically programmed. They observed quantitative alterations in synthesis of seven proteins with increasing age of *Drosophila*. They suggest that changes in the amounts of specific proteins may reflect changes in homeostasis, age-related changes in synthesis from one tissue to another, or alterations in isotope content of these particular proteins. The authors should be congratulated for their conservative interpretation of a very nice set of data.

Le Bourg [1987] examined the relationship between aging and spontaneous locomotor activity. Populations of 50 adult males and 50 females were placed in individual vials at 25°C. To assess spontaneous locomotion the flies were observed in residence (without handling) at three different times during the 12 hour light period, and any movement (other than preening behavior) during the 6 second observation period was scored positive. Spontaneous locomotor activity measurements were made once per week from 4–67 days posteclosion. Male flies were slightly more active during dawn and dusk; the spontaneous locomotor activity of females was more uniform. There were no significant changes in the spontaneous motion of males or females with age. Hence this character is not a useful bio-marker for aging.

Partridge et al. [1987] examined the effects of egg production and exposure to males on the longevity of female *Drosophila*. The rate of egg production per day was altered by varying the nutrient content of the medium. Flies on nutrient medium produced more eggs/day, on average, than flies on nutrient-deprived medium. There was a correlation between a high rate of egg production and decrease in female survival. Second, they compared the longevity of females that cohabitated with males continuously with that of females exposed to males intermittently (1 day out of 3). They concluded that continuous exposure to males reduces female longevity with no decrease in egg production.

POPULATION GENETICS

A number of people have addressed the issue of the evolution of senescence. For the last decade or so two schools of thought regarding the evolution of senescence have prevailed. The mutation-accumulation theory, first advanced by Medawar [1957], suggests that senescence is due to the accumulation of deleterious alleles whose effects are expressed in the latter portion of life. Since these alleles have a minor effect on fitness, they escape selection. The antagonistic pleiotropy theory suggests that alleles that promote fecundity early in adult life will have deleterious effects later in life [Williams, 1957]. The positive effects of such alleles favor their selection. Since the mutation-accumulation theory invokes no negative selection, while the antagonistic pleiotropy theory invokes a positive selection and increased fitness, one might intuitively conclude that the antagonistic pleiotropy theory is more reasonable. Indeed, it has received a great deal of support [see Rose and Graves, this volume). Nonetheless, there is some evidence for the mutation-accumulation theory [Kosuda, 1985].

Mueller [1987] examined these alternatives by establishing two populations of *Drosophila*. One replicate set of populations was maintained under r selection conditions and another set was kept under K selection for 120 generations. The populations in the r environment were maintained at low density with abundant food for both larvae and adults, and the females used to produce the next generation were 3–6 days old. In contrast, the K populations were maintained at very high density, with high larval and adult mortality. Females in this population were allowed to reproduce late in life. The r group environment should favor antagonistic pleiotropy, while the K group environment should favor mutation-accumulation. After 120 generations under r or K environments, F1 hybrids between each of the r populations and between each of the K populations were produced. Each of the r-F1 and K-F1 females were tested for fecundity early and late in life. If senescence resulted from mutation-accumulation, then the F1-K hybrid populations should escape the deleterious effects of recessive alleles at different loci (they would now be heterozygous at these loci). In contrast, if senescence resulted from antagonistic pleiotropy, the alleles favoring early fecundity in the three r populations should be fixed after 120 generations, and the F1 hybrids should remain homozygous for these alleles. Indeed the r population hybrids showed improved late fecundity, and Mueller concludes that the decline in late fecundity of the r populations reflects the accumulation of deleterious recessive alleles. He argues that this result does not conflict with the earlier experiments from Rose's lab. Mueller simply states he has identified environmental conditions permitting the accumulation of alleles that promote senescence.

A great deal of effort has been expended during the last decade, especially in the labs of Arking, Lints, Luckinbill, and Rose, to select long- and short-lived lines from randomly breeding populations. During the last 2 years the life history parameters of these selected lines have been examined extensively. The Luckinbill and Arking labs have collaborated on a series of selection experiments in *Drosophila* that covered more than 20 generations [Luckinbill et al., 1985]. They selected for long- and short-lived strains under uncrowded (density-controlled) or crowded (density not controlled) conditions. Under each environment (crowded or uncrowded), two lines were established. One line was selected for early reproduction (2–6 days posteclosion); a second line was selected for late reproduction (during the last quarter of the life span). Theoretically, the latter regime selects for extended life span, while the former will accommodate a shorter life span but does not necessarily select for it. Arking [1987a] re-examined this data as individual populations instead of pooled data from each replicate. The results are, of course, much the same as in the original paper. In non-density-controlled, late-reproducing population, life span extension becomes apparent in generations 21 and 25. In the density-controlled (uncrowded), late-reproducing lines, life span extension appears in generation 21. Unfortunately, later generations were not examined, or at least the data were not presented. Nonetheless, it appears that selection for late reproduction in high-density cultures leads to life span extension. Arking presents a long and interesting discussion of the data. Based on the observation that life span extension was clearly and dramatically increased only in the crowded lines, Arking speculates on the nature of the mechanism controlling longevity in *Drosophila*. He posits that genes influencing life span are regulatory genes that control homeostatic mechanisms and that senescence itself is not controlled genetically but is a stochastic process beginning when the homeostatic mechanism deteriorates. This hypothesis unites, or draws from, both the environmental–stochastic loss and genetically programmed theories of aging.

A printed discussion ensues from this article of both the data and the hypothesis—Luckinbill and Clare [1987] are rebutted by Arking [1987b]. It is apparent that the very powerful collaboration between these two excellent labs has tattered. Nonetheless, each group continues to make highly useful, strong, and interesting contributions.

In an attempt to test the rate of living hypothesis [Pearl, 1928], the Arking laboratory has measured the metabolic rate of the long-lived and short-lived strains [Arking et al., 1988]. The two selected lines from the Luckinbill and Arking collaboration were kept at 18°, 22°, 25°, and 28°C as adults. The fecundity and metabolic rate of each strain was measured over the adult life span. They observed an inverse effect of temperature on life span, which is typical of poikilotherms. Temperature had no affect on the difference in longevity between the two strains, that is, the life span of the long-lived strain was consistently longer than that of the short-lived strain. There was no significant difference between the metabolic rates of the two strains. The females of the long-lived strain, if anything, had a slightly higher metabolic rate than the short-lived females; however, the difference was not significant. The total lifetime metabolic potential (O_2/mg/lifetime) of the long-lived strains was higher than the short-lived strain (about 135%). The difference in metabolic potential appears to result from the life span extension, since there is no difference in the daily metabolic rate. Females of the long-lived strain generally deposited far more eggs/lifetime than females from the short-lived strain, yet the daily fecundity of the two strains was similar. Using these criteria, the long-lived strain appears to have an extended presenecent period. There seems to be a correlation between fecundity and lifetime metabolic rate. Indeed, if one divides the lifetime metabolic potential by the lifetime egg production there is no difference between the long- and short-lived strains. Finally, Arking et al. [1988] show that there is no difference between the mean daily metabolic rate of the long- and short-lived strains and four other strains including, among others, Oregon-R (a standard laboratory control strain) and a wild-caught strain. They conclude that strains selected for long and short life spans clearly violate the predictions made by the rate of living hypothesis, as originally stated. They speculate whether the life histories of these selected strains conform to the rate of living hypothesis as redefined by Sohal [1986].

Luckinbill and colleagues [1987] attempted to estimate the number of genes responsible for the life span extension in the selected strains. Using a protocol described by Lande [1981], they estimated that one or two genes controlled the extended life span phenotype. While exciting, this result was intuitively unlikely. Indeed, the F2 individuals from crosses between the long-lived and short-lived strains did not yield the appropriate 1:2:1 phenotypic ratio. The Luckinbill lab reapproached the problem by assessing the contribution of each chromosome to the extended life span phenotype [Luckinbill et al., 1988a]. They substituted chromosomes from the long-lived strain for those of the short-lived strain. While laborious, this approach is simple in *Drosophila melanogaster* since there are only four chromosomes. In fact, the X chromosome, and chromosomes 2 and 3 represent approximately 20%, 40% and 40% of the genome respectively; the fourth chromosome is very small by comparison (hence the genome is distributed essentially among the three major chromosomes). In addition, in *D. melanogaster* the randomizing effects of recombination can be effectively eliminated using multiply inverted chromosomes. Hence, the Luckinbill lab created eight different strains representing all possible combinations of chromosomes X, 2, and 3 from the long- and short-lived strains (LLL, LLS, LSL, SLL, LSS, SLS, SSL, SSS). They then measured the life span of males and females from each group. The extended life span phenotype in the originally selected line was a polygenic character. In females the third chromosome and X chromosome contributed most heavily to this phenotype. The influence of the second chromosome, while not absent in females, was more striking in males. Hence much of the variability between lines is located on 60% of the genome.

The Luckinbill lab then reconstructed the parental lines from the lines with chromosome substitutions and showed that their life span was quite similar to the original lines. This result, again, suggested that much of the variation between lines was genetic and that recombination was effectively eliminated during stock construction.

In general, it appears that genes influencing the longevity of the short- and long-lived strains are distributed to all three major chromosomes, with the third chromosome having the most powerful effect (it represents 40% of the genome), the X chromosome having a distinct influence, and second chromosome loci having the least effect. This very solid analysis strongly infers that the genetic basis of the life span differences between two selected lines is polygenic. The total number of genes, and the relative effect of each gene, as well as the precise distribution of these genes on the three linkage groups remains undetermined at this time. Presumably, the Luckinbill group will attempt to measure the effect that specific segments of the third chromosome, from these long- and short-lived strains, have on life span.

The Luckinbill group [1988b] went on to examine traits that might be associated with increased life span in these lines. They compared the effect of presence or absence of sexual partners on the longevity. Flies were kept as either single individuals or as pairs in shell vials. The presence of a sexual partner had no effect on the longevity of the long-lived line. However, survival of females from the short-lived line was considerably longer (32%) among single than among mated individuals. There was an influence of the presence of females on the longevity of males, but it was not nearly as dramatic. Finally, pairing of flies of the same sex had no effect on adult survival. Hence, much of the increase in longevity that was established by selection for late reproduction may be attributed to increased tolerance of females to males.

Body weight was used to estimate the size of the flies. There was no significant difference in body weight between long-lived and short-lived strains. Therefore, selection for late reproduction or increased longevity need not influence body size.

Finally, the stamina of flies was tested by examining the frequency and duration of tethered flights [Graves et al., 1988]. Both the duration and number of flights were measured in very young (4-day-old), young (5–15-days-old), and older adults (30 or more days) from the long- and short-lived lines. There was little difference in the flight endurance of very young flies from these two lines. However, as the populations aged, the flight endurance of flies from the long-lived line exceeded that of the short-lived line. There was no effect of humidity (68 vs. 100%) on flight endurance of either line. Reciprocal crosses were made between long- and short-lived lines, and the flight endurance of the F1 individuals was always near the mean of the two parental strains. They suggest that the difference in flight endurance of the two parental lines could be attributed to a single gene located on the third chromosome. However, it is just as likely that it could be associated with multiple factors with additive effects. Since they have the 8 chromosome substitution lines (see above), one presumes that these will be analyzed for flight endurance and will be the subject of another paper.

To further test the antagonistic pleiotropy and mutation-accumulation theories of the evolution of senescence, Service et al. [1988] carried out a reversed selection experiment. Lines that had been selected for extended life span and had also acquired resistance to dessication and starvation and tolerance to alcohol were subjected to selection for early reproduction for more than 21 generations. They found that when the larvae were raised under high-density conditions there was an increase in early fecundity, which is probably expected under the selection regime they used. Under these same conditions resistance to starvation declined. The latter result is consistent with the antagonistic pleiotropy theory. In constrast, under these same conditions, there was no consistent, significant change in either resistance to dessication (it remained high) or alcohol tolerance. This is more consistent with the mutation accumulation theory for the evolution of senescence. Curiously, using the same strains and selection protocol but in lines where the larval density was controlled and low (about 30 larvae/vial), the selection was unsuccessful. Hence selection for extended life span, using early and late reproduction as the selection force, seems to require an environmental stress factor [Lints et al., 1979].

SINGLE GENE MUTATIONS

Several labs have isolated single gene mutations that influence the life span of either the larva or the adult of holometabolous insects. Nagata et al. [1987] have isolated a third mutant, called nonmolting glossy (*nm-g*), in *Bombyx mori* that fails to molt and eventually dies as a larva. This mutation is recessive and is located on chromosome 17. *Bombyx mori* goes through four larval molts (five instar stages), with the transition from one instar stage to another occuring on days 4, 7, 11, and 17 after egg hatch. In the *nm-g* strain the molt between first and second instar is delayed about twofold (from 3–4 days to 6–9 days). Depending on the genetic background, about 35–70% of the first instar larvae actually molt to form second instars. Those that fail to molt die between days 9 and 12 (approximately a threefold extension of the first instar stage). During this period they continue to grow, and they attain a body weight about twice that of normal first instar larvae. Individuals that do form second instar larvae never ecdyse to third instar. They die 16–20 days after hatching, attaining a body weight about 2.6 times that of normal second instars. Hence, the *nm-g* mutant larvae live about two to three times longer than their wild-type counterparts. This particular mutant appears to be ecdysterone-deficient. Injection with 20-hydroxy ecdysone at appropriate intervals allows some *nm-g* larvae to develop to the third instar stage and a few to survive to the fourth instar stage.

Soliman [1987] examined the effects of the visible markers *ebony* (black body) and *yellow* (yellow body instead of tan) and the double mutant *ebony–yellow* on longevity of *D. melanogaster* on two types of media: standard yeast–sucrose–agar medium and this standard medium plus 10% ethanol. All three mutant strains and the wild-type control lived longer on the ethanol-containing medium. The wild-type showed the biggest difference in longevity on the two media followed by the *ebony*, and *yellow* strains, with the double mutant showing the least difference. Soliman discusses the role of alcohol on life extension and the possible pleiotropic effects of *ebony* and *yellow* on other fitness characters.

Koivisto and Portin [1987] attempted to estimate the number of loci that, if mutant, might have a negative effect on life span. They treated a *yellow* strain of *D. melanogaster* (X chromosome marker) with ethyl methanesulfonate (EMS) to induce mutations. A total of 252 lines each representing a different EMS-exposed X chromosome was recovered. The sex-linked lethal mutation rate was quite high, 50% (127/152), suggesting that a large number of lines would have two or more lethal mutations and many sublethal mutations. The recessive lethals were discarded. From the remaining 125 lines, they recovered 51 lines in which goodly numbers of both *yellow* males and females survived to adulthood (the remainder were strong semilethals). The longevity of all 51 lines was examined at 25°C. In 21 lines the mean life span of both sexes was significantly decreased compared with the control (*yellow*) flies. In six lines the mean life span of the males but not females was reduced; in six other lines the reciprocal was observed. However, the survival curves of many of these 33 lines was somewhat linear, that is, there was an immediate and continuous decline in viability, with only slight shortening of the maximum life span. Only 13 strains fit their final criteria of "mortality" mutants. They estimate that the frequency of mortality mutants induced by EMS mutagenesis was 25% (13/51). Since the sex-linked lethal frequency was 50%, they estimate that the ratio of lethal to mortality mutants was about 2:1. With x-rays, this ratio was 4:1 [Gould and Clark, 1977]. Hence they conclude that one-fourth to one-half of the genes can affect mortality. What this says of course is that loss of function or hypomorphy at any number of loci can have a negative effect on life span, which is probably not surprising.

ACKNOWLEDGMENTS

This work was supported by grant 3005 from the Natural Sciences and Engineering Research Council of Canada.

REFERENCES

Arking R (1987a): Successful selection for increased longevity in *Drosophila*: Analysis of the survival data and presentation of a hypothesis on the genetic regulation of longevity. Exp Gerontol 22:199–220.

Arking R (1987b): Letter to the editor. Exp Gerontol 22:223–226.

Arking R, Buck S, Wells RA, Pretzlaff R (1988): Metabolic rates in genetically based long lived strains of *Drosophila*. Exp Gerontol 23:59–76.

Bartosz G, Leyko W, Fried W (1979): Superoxide dismutase activity and life span of *Drosophila melanogaster*. Experientia 35:1193.

Collatz KG, Sohal RS (1986): "Insect Aging: Strategies and Mechanisms." Berlin: Springer-Verlag.

Farmer KJ, Sohal RS (1987): Effects of ambient temperature on free radical generation, antioxidant defenses and lifespan in the adult housefly, *Musca domestica*. Exp Gerontol 22:59–65.

Fleming JE, Quattrocki E, Latter G, Miquel J, Marcuson R, Zuckerkandl E, Bensch KG (1986): Age-dependent changes in proteins of *Drosophila melanogaster*. Science 231:1157–1159.

Gershman R (1964): Biological effects of oxygen in the animal organism. In Dickens F, Neil E (ed): "Oxygen in the animal organism." New York: McMillan, pp 475–492.

Gould AB, Clark AM (1977): X-ray induced mutations causing adult life-shortening in *Drosophila melanogaster*. Exp Gerontol 12:107–112.

Graves JL, Luckinbill LS, Nichols A (1988): Flight duration and wing-beat frequency in long- and short-lived *Drosophila melanogaster*. J Insect Physiol 34:1021–1026.

Harman D (1956): Aging: A theory based on free radical and radiation chemistry. J Gerontol 11:298–300.

Koivisto K, Portin P (1987): Induction of mutations which shorten the adult life span in *Drosophila melanogaster*. Hereditas 106:83–87.

Lande R (1981): The minimum number of genes contributing to quantitative variation between and within populations. Genetics 99:541–553.

Le Bourg E (1987): The rate of living theory, spontaneous locomotor activity, aging and longevity in *Drosophila melanogaster*. Exp Gerontol 22:359–369.

Lints FA, Soliman MH (1988): "Drosophila as a Model System for Ageing Studies." Glasgow: Blackie and Son Ltd.

Lints FA, Stull J, Gruwez G, Lints CV (1979): An attempt to select for increased longevity in *Drosophila melanogaster*. Gerontology 25:192–204.

Luckinbill LS, Clare MJ (1987): Letter to the editor. Exp Gerontol 22:221–222.

Luckinbill LS, Arking R, Clare MJ, Cirocco, Buck SA (1985): Selection for delayed senescence in *Drosophila melanogaster*. Evolution 38:996–1003.

Luckinbill LS, Clare MJ, Krell WL, Cirocco WC, Richards PA (1987): Estimating the number of genetic elements that defer senescence in *Drosophila*. Evol Ecol 1:37–46.

Luckinbill LS, Graves JL, Reed AH, Koetsuwang S (1988a): Localizing genes that defer senescence in *Drosophila melanogaster*. Heredity 60:367–374.

Luckinbill LS, Graves JL, Tomkin A, Sowirka O (1988b): A qualitative analysis of some life-history correlates of longevity in *Drosophila melanogaster*. Evol Ecol 2:85–94.

Massie HR, Kogut KA (1987): Influence of age on mitochondrial enzyme levels in *Drosophila*. Mech Ageing Dev 38:119–126.

Massie HR, Williams TR (1987): Mitochondrial DNA and life span changes in normal and dewinged *Drosophila* at different temperatures. Exp Gerontol 22:139–153.

Massie HR, Williams TR, Iodice AA (1985): Influence of anti-inflammatory agents on the survival of *Drosophila*. J Gerontol 40:257–260.

Medawar PB (1957): "The Uniqueness of the Individual." London: Metheun and Co.

Miquel J, Fleming JE, Economos AC (1982): Antioxidants, mitochondrial respiration and aging in *Drosophila*. Arch Gerontol Geriatr 1:349–363.

Mueller LD (1987): Evolution of accumulated senescence in laboratory strains of *Drosophila*. Proc Natl Acad Sci USA 84:1974–1977.

Nagata M, Tsuchida K, Schimizu K, Yoshitake N (1987): Physiological aspects of *nm-g* mutant: An ecdysteroid-deficient mutant of the silkworm, *Bombyx mori*. J Insect Physiol 33:723–727.

Partridge L, Green A, Fowler K (1988): Effects of egg-production and exposure to males on female survival in *Drosophila melanogaster*. J Insect Physiol 33:745–749.

Pearl R (1928): "The Rate of Living." London: University of London Press.

Ruddle DL, Yengoyan LS, Miquel J, Marcuson R, Fleming JE (1988): Propyl gallate delays senescence in *Drosophila melanogaster*. Age 11:54–58.

Service PM, Hutchinson EW, Rose MR (1988): Multiple genetic mechanisms for the evolution of senescence in *Drosophila melanogaster*. Evolution 42:708–716.

Seto NOL, Hayashi S, Tener GM (1989): Cloning, sequence analysis and chromosomal location of the Cu-Zn superoxide dismutase gene of *Drosophila melanogaster*. Gene 75:85–92.

Sohal RS, Donato H (1979): Effect of experimental prolongation of life span on lipofuschin

content and lysosomal activity in the brain of the housefly, *Musca domestica*. J Gerontol 34:489–496.
Sohal RS (1981): Relationship between metabolic rate, lipofuscin accumulation and lysosomal enzyme activity during aging in the adult housefly, *Musca domestica*. Exp Gerontol 6:347–355.
Sohal RS (1986): The rate of living theory: A contemporary interpretation. In Collatz KG, Sohal RS (ed): "Insect Aging." New York: Springer-Verlag, pp 23–44.
Sohal RS (1988): Effect of hydrogen peroxide administration on the life span, superoxide dismutase, catalase, and glutathione in the adult housefly, *Musca domestica*. Exp Gerontol 23:211–216.
Sohal RS, Allen RG, Farmer KJ, Procter J (1984a): Effect of physical activity on superoxide dismutase, catalase, inorganic peroxides and glutathione in the housefly, *Musca domestica*. Mech Ageing Dev 26:75–81.
Sohal RS, Farmer KJ, Allen RG, Cohen NR (1984b): Effect of age on oxygen consumption, superoxide dismutase, catalase, glutathione, inorganic peroxides and chloroform-soluble antioxidants in the adult male housefly, *Musca domestica*. Mech Ageing Dev 24:185–195.
Sohal RS, Toy PL, Allen RG (1986): Relationship between life expectancy, endogenous antioxidants and products of oxygen free radical reactions in the housefly, *Musca domestica*. Mech Ageing Dev 36:71–77.
Soliman MH (1987): Effect of *ebony* and *yellow* mutants of *Drosophila melanogaster* on adult survival and alcohol dehydrogenase activity. Gerontology 33:57–63.
Williams GC (1957): Pleiotropy, natural selection, and the evolution of senescence. Evolution 11:398–411.

Genetic Aspects of Aging in *Mus musculus*

Richard L. Sprott and Carol A. Combs

INTRODUCTION

The number of mouse studies primarily concerned with the genetics of aging is still very limited, undoubtedly reflecting the relatively long times needed to conduct selection studies in this organism compared with lower organisms like *Drosophila* or *Caenorhabditis elegans*. The enormous expense of such a study also makes it difficult to contemplate for many investigators. As a result, mouse aging–genetics studies use genotypes selected for other characteristics (behavior, cancer, body weight) or mutants. The major recent exception to this rule is the ongoing series of Japanese studies reported below using genotypes selected for longevity.

Other genetic studies, that is, studies with the objective of understanding gene control of aging processes, include studies of sister chromatid exchange (SCE) and gene regulation.

REVIEW OF RECENT RESEARCH, MARCH, 1986–MARCH, 1989

The frequency of SCE formation in replicating cells has been suggested as a measure of DNA repair capacity. A decreased rate of SCE formation with advancing age would provide specific evidence in favor of a DNA repair theory of aging. In a paper described in volume 2 of this series, Schneider [1982] reported that bromoxyuridine-induced SCE formation rates did not decrease in aging C57BL/6J (B6) mice and Wistar rats. In the same review [Sprott, 1985] we also described a paper by Reiner and Singh [1983] that showed decreased rates of cyclophosphamide-induced SCE formation in C3H/s, BALB/c, and B6 mice. More recently, Bond and Singh [1988] report a genotype-dependent decrease in methylnitrosourea (MNU)-induced SCE formation in aged BALB/c, C3H/HeSnJ, B6, Cs[b], 129/ReJ, and BALB $\times$ B6 F$_1$ hybrid mice. This observation of strain differences in age-dependent SCE formation rate

National Institute on Aging, National Institutes of Health, Bethesda, Maryland 20892

decreases explains the earlier discrepancies between labs. However, it also suggests that the value of this change as supporting evidence for the DNA repair theory is not as strong as these authors suggest. Positive evidence for the DNA repair theory was reported by DeSousa et al. [1986] in a simple strain comparison study of repair of UV-induced damage in neurons of DBA/15, B6, and SJL/J mice.

Pursuing developmental changes in gene regulation of catalase activity in various tissues of young and older mice as a model for a developmental approach to aging processes, Singh [Schisler and Singh, 1987] has shown genotype-specific, age-related changes in this system as well. Again, these results will probably contribute more to our understanding of individual differences in aging than they will to our understanding of basic aging processes.

A rapidly expanding interest in diet restriction (DR) as a tool to increase longevity and perhaps alter aging rate and as a focus of research in its own right was beginning to be apparent during the period of this review. As in other areas of research, this rapid expansion in interest is tied to increased availability of appropriate models and consensus among investigators that the questions being asked are amenable to analysis and worth the effort. This has clearly occurred in the DR area, and we can undoubtedly expect an even greater increase in research in this area in the next review of this series.

In an example of DR studies being pursued to understand the mechanism of DR's effect on longevity, Ingram and Reynolds [1987] reanalyzed a large number of experiments in a post hoc test of the hypothesis that departures from optimal body weight in either direction (over or underweight) reduce life span. They found that an every-other-day diet regimen and weight loss increased longevity in Wistar rats. Results in mouse genotypes (A/J and B6 along with their F_1 hybrid, B6AF$_1$, and the obese mutants of B6, *ob/ob*, and *AY/a*) varied in magnitude across genotypes and with age of diet initiation. Increase in longevity was not necessarily due to decrease in body weight across mouse genotypes. Further intragroup and interstrain comparisons showed inconsistent correlations that could neither support nor refute their hypothesis.

In another study of the effects of DR on mouse longevity and organ function, Harrison and Archer [1987] compared genotypes B6CBAF1, B6, and B6-*ob/ob* to test the effects of DR on biological systems, overall activity, neuromuscular performance, and longevity. They reported increased longevity in all but B6 and mixed biological function results. It was noted that a wider variety of genotypes should be studied, taking into account individual characteristics of biological changes. The failure to find increased longevity in B6 mice in this experiment is an important departure from the results of other laboratories and needs to be replicated.

Koizumi et al. [1987] reported interesting enzyme changes resulting from DR. Using long-lived C3B1ORF$_1$ females, they found a significant increase in liver catalase activity comparing 12- (42%) and 24-month-old (64%) mice along with a respective decrease (30% and 13%) in lipid peroxidation. The results show a possible cause and effect relationship that is compatible with the free radical theory of aging. It is interesting that liver catalase activity in this genotype did not decrease with age in the way that the Singh group reported in studies described above. In another study using the same mouse strain, this group [Weindruch et al., 1986] tested the effects of six different diets on longevity, cancer, immunity, and lifetime energy intake. The results showed a positive correlation between DR and increased longevity, especially in heavier mice, a beneficial influence on tumor patterns, and a decrease in T-lymphocyte proliferation. They suggest that increased longevity in DR mice may be due to increased metabolic efficiency.

In a related study by Canada et al. [1986] that looked at the metabolism of pentobarbital after exposure to ozone in DBA-2 and Sprague–Dawley rats, metabolism was inhibited in older animals. The older animals weighed more, had increased fat stores, and increased volume of distribution, therefore prolonging inhibition of metabolism. It would have been interesting to have introduced DR as an added variable in this study.

Since the genetically defined mouse remains the primary model for studies of mammalian immunologic function, it is not surprising that this area continues to be the most active and best understood area of research on functional age-related change. Since a full section of this volume is devoted to this topic, we will simply describe the range of studies and provide a few examples here.

Studies of age-dependent decline in T-cell function included Ershler et al. [1986], using B6 mice and bone marrow transplantation, Zharhary [1986], using BALB/c nudes and bone marrow "chimeric" mice to study T cells' down regulation of B-cell responsiveness, Vissinga et al. [1987], using inbred strains (C57BL/Ka and CBA/Rij) and measures of spleen cell function to study quantitative and qualitative T-cell decline, and Francus et al. [1986] using congenic strains to study B-cell, T-cell, and spleen function. Autoimmune disease abnormalities and T-cell function changes with age in MRL/*lpr* mice of a variety of subtypes are being actively studied in several laboratories. The use of a mutation-like lymphoproliferation (*lpr*) as a modifier of age-related change continues to be one of the most productive techniques for characterizing age-related changes. Studies of this model included those by Cohen et al. [1988], T. Koizumi et al. [1986], and Okuyama et al. [1986].

The NZB mouse model of autoimmunity continues to be a common tool for investigation of immune function. Whether one agrees that this model represents premature aging (progeroid syndrome) or not, it is apparent that the model is useful [Seoane et al., 1986; Seldin et al., 1987; Sekigawa et al., 1987; and Jung et al., 1986]. Romball and Weigle [1987] used inbred mice (CBA/CaJ and A/J) to demonstrate autoimmune phenomena (e.g., thyroiditis and delayed-type hypersensitivity) in mice without obvious autoimmune mutations.

Harrison [Harrison et al., 1988], is continuing his series of studies of the interaction of age and transplantation in order to characterize changes in proliferative capacity in vivo and determine whether loss of immune competence is intrinsic to cells or derives from other sources. Similar studies using bone marrow "chimeras" are being conducted at the Tokyo Metropolitan Institute of Gerontology (TMIG) by Hirokawa et al. [1986].

Like Harrison, Cinader and his group are comparing age changes in a variety of tissues. During this period, they report that age-related changes in beta-andrenoceptor density and antagonist dissociation constant occur at different rates in different mouse strains (BALB/cJ, C3H/HeJ, and B6) and in different tissues within a single strain [Kohno et al., 1986].

The group led by Takeda at the Chest Disease Research Institute at Kyoto University has grown and is rapidly expanding its characterization of their collection of long- and short-lived mice. There are six senescence-accelerated mouse prone lines, designated SAM-P/1, SAM-P/2, SAM-P/3, SAM-P/4, SAM-P/6, and SAM-P/8, and three senescence-accelerated mouse resistant lines, designated SAM-R/1 through SAM-R/3 [Kunisada et al., 1986]. These lines appear to be essentially a set of recombinant inbred lines derived from AKR breeding pairs and perhaps from some other genotype as well. While the genetic origins of these lines are problematic, the results, which are now appearing with rapidly increasing frequency, are indicative of the tremendous value of this sort of resource. Takeda's group is characterizing the behavior and morphology of these lines and providing both a stimulus and a model for a wide variety of aging research at the same time. The range of this research during the period of this report includes defective immune responses in short-lived SAM mice [Hosokawa et al., 1987], studies of amyloid in all of the lines [Kunisada et al., 1986], studies of learning and memory in two long-lived and two short-lived lines [Miyamoto et al., 1986], studies of accelerated osteoporosis in senescence-prone lines [Matsushita et al., 1986], and studies of pathology in all of the lines [Akiyama et al., 1986; Shino et al., 1987]. Dr. Takeda has begun to make these lines available to investigators outside of Japan, which will increase the pace of characterization and the utility of these lines. The research emerging from this group is interesting in its own right but also suggests measures other investigators could use in efforts to develop biomarkers of aging and to assess the heuristic value of new animal models of aging.

In their behavioral investigation of SAM mice, Miyamoto et al. [1986] observed SAM-P/8/Ta (senescence-prone) and SAM-R/1/Ta (senescence-resistant) substrains of senescence-accelerated mice to study age-related changes in memory and learning ability. Compared with the senescence-resistant substrain, SAM-P/8/Ta exhibited impaired passive and active avoidance responses with aging with no change in response to shock sensitivity and prolonged water maze performance time with increased latency to escape. They suggest that SAM-P/8/Ta could serve as a useful model for further studies on memory and learning impairment with aging.

Akiyama et al. [1986] necropsied brains from SAM-P/8, SAM-R/1, and DDD to determine if there was a morphological basis for the above results. They discovered abnormal PAS-positive granular structures clustered in restricted brain regions of SAM-P/8, particularly in the hippocampus. Very few granular structures appeared in SAM-R/1 and DDD, although their presence increased with age. Histochemical evaluation suggested that the structures contained glucose polymers and were associated with astrocytes.

Another morphological parameter, autoimmune disorder, was used by Forster et al. [1988] to study acquisition and 48 hour retention of a step-up active avoidance learning response in aging mouse strains. Strains with an early onset of autoimmune disorder (NZB/B1NJ, MRL/MpJ-1pr, and BXSB/MpJ) showed a decrease in acquisition between 1.5 and 3.5 months of age. Strains with a mild, late onset of autoimmune disorder showed stable or improved acquisition. B6 showed a decrease in acquisition after age 3.5 months and a decrease in 48 hour retention (also exhibited by NZB/B1NJ) by 12 months of age. Results suggested that retention deficits were related to age, and avoidance acquisition deficits were related to onset of autoimmunity.

Fuhrmann et al. [1986] investigated genotypic differences between B6 and BALB/cJ due to effects of chronic alcohol exposure. Striatal and hippocampal somatostatin levels remained unchanged in B6 until 27 months of age, and there were no significant changes with chronic alcohol exposure. BALB/cJ exhibited an age-dependent decrease in striatal and hippocampal somatostatin but a significant increase in somatostatin levels with chronic alcohol exposure. The different reactions were probably related to the basic pre-existing levels of somatostatin in each genotype. In a related study, Ebel et al. [1987] compared aging B6 and BALB/c and also found that neurotransmitter responses in striatum and hippocampus were genotype-specific. In B6 there was an increase in serotonin in the hippocampus with no change in the striatum and no change in somatostatin levels. BALB/c showed defective dopaminergic mechanisms in the striatum and defective cholinergic and somatostatin levels in the striatum and hippocampus, a condition that may negatively affect memory retention.

Age changes at limb neuromuscular junctions were investigated in B6 and BALB/C by Anis and Robbins [1987]. The only reported significant strain difference was "diverse age changes in spontaneous transmitter release." Observed age changes that were consistent across strains were not due to disuse or denervation. In this case, two inbred strains and previous results with the appropriate F_1 hybrid were compared, not to isolate genetic components, but rather to assess generality.

In a strain comparison study Talan and Ingram [1986] measured age-related changes in body temperature and cold tolerance in B6, AJ, and B6AF$_1$/J and compared these with a pen-bred "wild" strain of *Mus musculus*; they found an age-related decline in thermoregulation among all strains. The degree of decline in aged mice of each strain compared with the others was inversely proportional to longevity. Although the aged pen-bred strain showed a significant decrease in the ability to thermoregulate, its overall thermoregulatory responses were superior to the domesticated strains. The whole issue of using "wild" mice in aging genetic research needs to be systematically developed. This study suggests that there might be real value to developing strategies to include "wild" genes in the gene pools currently available in laboratory mice. The value of finding other genes is also suggested by the surprising increase in rate of wound repair with advancing age in *Peromyscus* reported by Cohen et al. [1987].

Willott [1986] used B6, a strain with progressive sensorineural hearing loss, and CBA, a strain with good hearing sensitivity into the second year of life, to test aging effects on hearing loss and thresholds of inferior colliculus neurons. Aging in B6 resulted in a disruption of the dorsoventral arrangement of frequency sensitivity in the inferior colliculus central nucleus, a lowering of best frequencies and elevation of thresholds, and a decrease in multiple-unit threshold curves at 14 months of age, when hearing loss is severe at all frequencies. In contrast, CBA showed minimal or no changes until 22 months old, when a moderate loss of sensitivity occurred at all frequencies. The study suggested that age-related changes in multiple-unit threshold curves depend upon the pattern of hearing loss and the location of neurons within the inferior colliculus central nucleus.

CONCLUSIONS

The tremendous productivity already shown by the Takeda group in Kyoto and the promise of similar levels of productivity around new animal resources such as the National Institute on Aging/National Center for Toxicology Research (NIA/NCTR) biomarker colony in Arkansas clearly demonstrates the degree to which much aging research is dependent upon the quality and availability of relevant model systems. The potential development of one or more colonies of lines of mice selected for longevity from defined heterogeneous stocks has become a topic of corridor conversation at recent meetings. This topic is quite important to the future of mammalian genetic research on aging. We hope it will become a high priority issue for discussion and development. The field could also benefit from implementation of other ideas such as practical ways to broaden gene pools to include genes that may be lost during normal adaptation to laboratory environments and inbreeding.

REFERENCES

Akiyama H, Kameyama M, Akiguchi I, Sugiyama H, Kawamata T, Fukuyama H, Kimura H, Matsushita M, Takeda T (1986): Periodic acid-Schiff (PAS)-positive, granular structures increase in the brain of senescence accelerated mouse (SAM). Acta Neuropathol 72:124–129.

Anis NA, Robbins N (1987): General and strain-specific age changes at mouse limb neuromuscular junctions. Neurobiol Aging 8:309–318.

Bond SL, Singh SM (1988): Genotype-specific reduction in methyl nitrosourea (MNU) induced sister chromatid exchanges (SCE) in vivo during aging. Experientia 44:782–785.

Canada AT, Calabrese EJ, Leonard D (1986): Age-dependent inhibition of pentobarbitol sleeping time by ozone in mice and rats. J Gerontol 41:587–589.

Cohen BJ, Cutler RG, Roth GS (1987): Accelerated wound repair in old deer mice (*Peromyscus maniculatus*) and white-footed mice (*Peromyscus leucopus*). J Gerontol 42:302–307.

Cohen MG, Pollard KM, Schrieber L (1988): Relationship of age and sex to autoantibody expression in MRL-+/+ and MRL-*lpr/lpr* mice: Demonstration of an association between the expression of antibodies to histones, denatured DNA and Sm in MRL-+/+ mice. Clin Exp Immunol 72:50–54.

De Sousa J, De Boni U, Cinader B (1986): Age-related decrease in ultraviolet induced DNA repair in neurons but not in lymph node cells of inbred mice. Mech Ageing Dev 36:1–12.

Ebel A, Strosser MT, Kempf E (1987): Genotypic differences in central neurotransmitter responses to aging in mice. Neurobiol Aging 8:417–427.

Ershler WB, Robbins DL, Moore AL, Hebert JC (1986): The age-related decline in antibody response is transferred by old to young bone marrow transplantation. Exp Gerontol 21:45–53.

Forster MJ, Popper MD, Retz KC, Lal H (1988): Age differences in acquisition and retention of one-way avoidance learning in C57BL/6NNia and autoimmune mice. Behav Neural Biol 49:139–151.

Francus T, Chen YW, Staiano-Coico L, Hefton JM (1986): Effect of age on the capacity of the bone marrow and the spleen cells to generate B lymphocytes. J Immunol 137:2411–2417.

Fuhrmann G, Strosser MT, Besnard F, Kempf E, Kempf J, Ebel A (1986): Genotypic differences in age and chronic alcohol exposure effects on somatostatin levels in hippocampus and striatum in mice. Neurochem Res 11:625–636.

Harrison DE, Archer JR (1987): Genetic differences in effects of food restriction on aging in mice. J Nutr 117:376–382.

Harrison DE, Astle CM, DeLaittre J (1988): Effects of transplantation and age on immunohemopoietic cell growth in the splenic microenvironment. Exp Hematol 16:213–216.

Hirokawa K, Kubo S, Utsuyama M, Kurashima C, Sado T (1986): Age-related change in the potential of bone marrow cells to repopulate the thymus and splenic T cells in mice. Cell Immunol 100:443–451.

Hosokawa T, Hosono M, Hanada K, Aoike A, Kawai K, Takeda T (1987): Immune responses in newly developed short-lived SAM mice II. Selectively impaired T-helper cell activity in *in vitro* antibody response. Immunology 62:425–429.

Ingram DK, Reynolds MA (1987): The relationship of body weight to longevity within laboratory rodent species. Basic Life Sci 42:247–282.

Jung LKL, Good RA, Fernandes GA (1986): Studies on lymphocyte homing in autoimmune prone NZB mice. Immunol Invest 15:11–23.

Kohno A, Seeman P, Cinader B (1986): Age-related changes of beta-adrenoceptors in aging inbred mice. J Gerontol 41:439–444.

Koizumi T, Nakao Y, Matsui T, Katakami Y, Nakagawa T, Fujita T (1986): Synergistic induction by calcium ionophore and phorbol ester of interleukin-2 (IL-2) receptor expression, IL-2 production, and proliferation in autoimmune MRL/MP-lpr mice. Immunology 59:43–49.

Koizumi A, Weindruch R, Walford RL (1987): Influences of dietary restriction and age on liver enzyme activities and lipid peroxidation in mice. J Nutr 117:361–367.

Kunisada T, Higuchi K, Aota S, Takeda T, Yamagishi H (1986): Molecular cloning and nucleotide sequence of cDNA for murine senile amyloid protein: Nucleotide substitutions found in apolipoprotein A-II cDNA of senescence accelerated mouse (SAM). Nucleic Acids Res 14:5729–5740.

Matsushita M, Tsuboyama T, Kasai R, Okumura H, Yamamuro T, Higuchi K, Higuchi K, Kohno A, Yonezu T, Utani A, Umezawa M, Takeda T (1986): Age-related changes in bone mass in the senescence-accelerated mouse (SAM). Am J Pathol 125:276–283.

Miyamoto M, Kiyota Y, Yamazaki N, Nagaoka A, Matsuo T, Nagawa Y, Takeda T (1986): Age-related changes in learning and memory in the senescence-accelerated mouse (SAM). Physiol Behav 38:399–406.

Okuyama H, Yamamoto K, Matsunaga T, Kobayashi S, Tashiro A (1986): Analysis of defective delayed-type hypersensitivity in autoimmune mice bearing *lpr* gene. Clin Exp Immunol 63:87–94.

Reiner DL, Singh SM (1983): Cyclophosphamide-induced *in vivo* sister chromatid exchange in *Mus musculus*. II: Effect of age and genotype on sister chromatid exchange, micronuclei and metaphase index. Mech Ageing Dev 21:59–68.

Romball CG, Weigle WO (1987): The effect of aging on the induction of experimental autoimmune thyroiditis. J Immunol 139:1490–1495.

Schisler NJ, Singh SM (1987): Inheritance and expression of tissue-specific catalase activity during development and aging in mice. Genome 29:748–760.

Schneider EL, Bickings CK, Sternberg H (1982): Aging and sister chromatid exchange. Cytogenet Cell Genet 33:249–253.

Sekigawa I, Okada T, Noguchi K, Ueda G, Hirose S, Sato H, Shirai T (1987): Class-specific regulation of anti-DNA antibody synthesis and the age-associated changes in (NZB × NZW)F$_1$ hybrid mice. J Immunol 138:2890–2895.

Seldin MF, Conroy J, Steinberg AD, D'Hoosteleare LA, Raveche ES (1987): Clonal expansion of abnormal B cells in old NZB mice. J Exp Med 166:1585–1590.

Seoane R, Faro J, Eiras A, Lareo I, Couceiro J, Regueiro BJ (1986): Effects of antigen and internal environment on anti-phosphorylcholine immune responses of autoimmune aged NZB/W FI mice. Immunology 58:329–334.

Shino A, Tsukuda R, Omori Y, Matsuo T (1987): Histopathologic observations on the senescence-accelerated mice (SAM) reared under specific pathogen free conditions. Acta Pathol Jpn 37:1465–1475.

Sprott RL (1985): Genetic aspects of aging in *Mus musculus*. In Rothstein M (ed): "Review of Biological Research in Aging," Vol 2. New York: Alan R. Liss, Inc., pp 99–104.

Talan MI, Ingram DK (1986): Age comparisons of body temperature and cold tolerance among different strains of *Mus musculus*. Mech Ageing Dev 33:247–256.

Vissinga CS, Dirven CJAM, Steinmeyer FA, Benner R, Boersma WJA (1987): Deterioration of cellular immunity during aging. Cell Immunol 108:323–334.

Weindruch R, Walford RL, Fligiel S, Guthrie D (1986): The retardation of aging in mice by dietary restriction: Longevity, cancer, immunity and lifetime energy intake. J Nutr 116:641–654.

Willott JT (1986): Effects of aging, hearing loss, and anatomical location on thresholds of inferior colliculus neurons in C57BL/6 and CBA mice. J Neurophysiol 56:391–408.

Zharhary D (1986): T cell involvement in the decrease of antigen-responsive B cells in aged mice. Eur J Immunol 16:1175–1178.

SECTION II
IMMUNOLOGY OF AGING

Mechanisms of Impaired T-Cell Function in the Elderly

Rajesh K. Chopra

INTRODUCTION

During the past few years considerable progress has been made in understanding the mechanisms by which T cells respond to antigens and mitogens. Following recognition of antigen by T cells, signals transduced by second messenger molecules cause the T cell to be activated and to secrete several lymphokines. Interactions between specific receptors and their lymphokines further evoke a variety of intracellular biochemical processes that ultimately result in cellular proliferation. It is well established that T lymphocytes from aged animals as well as elderly humans, relative to those of younger adults, have a dminished in vitro proliferative capacity. The mechanisms underlying this impaired proliferative response will be the focus of this review. These mechanisms include early transmembrane signaling, interleukin 2 (IL2) synthesis, and interleukin 2 receptor (IL2R) expression, along with their interactions.

The T-cell receptor (TCR) molecule is a highly polymorphic heterodimer made up of α and β chains covalently linked by disulfide bonds [Oettegen and Terhorst, 1987] (Fig. 1). The TCR is also noncovalently associated with another group of nonpolymorphic proteins designated the T3 complex [Brenner et al., 1985; Oettegen and Terhorst, 1987; Marrack and Kappler, 1987]. The T3 complex is composed of at least three polypeptide chains collectively designated as T3 or CD3 [Marrack and Kappler, 1987]. The TCR recognizes antigen in conjunction with major histocompatability complex (MHC) products expressed on appropriate antigen-presenting cells [Davis and Bjorkman, 1988]. Following recognition of antigen, the Ti–T3 complex undergoes modulation and disappears from the cell membrane [Tse et al., 1986]. T cells may be activated by several different mechanisms. For example, phytohemagglutinin (PHA) binds carbohydrate moieties of the TCR, while concanavalin A (Con

---

Clinical Immunology Section, Gerontology Research Center, National Institute on Aging, National Institutes of Health, Baltimore, Maryland 21224

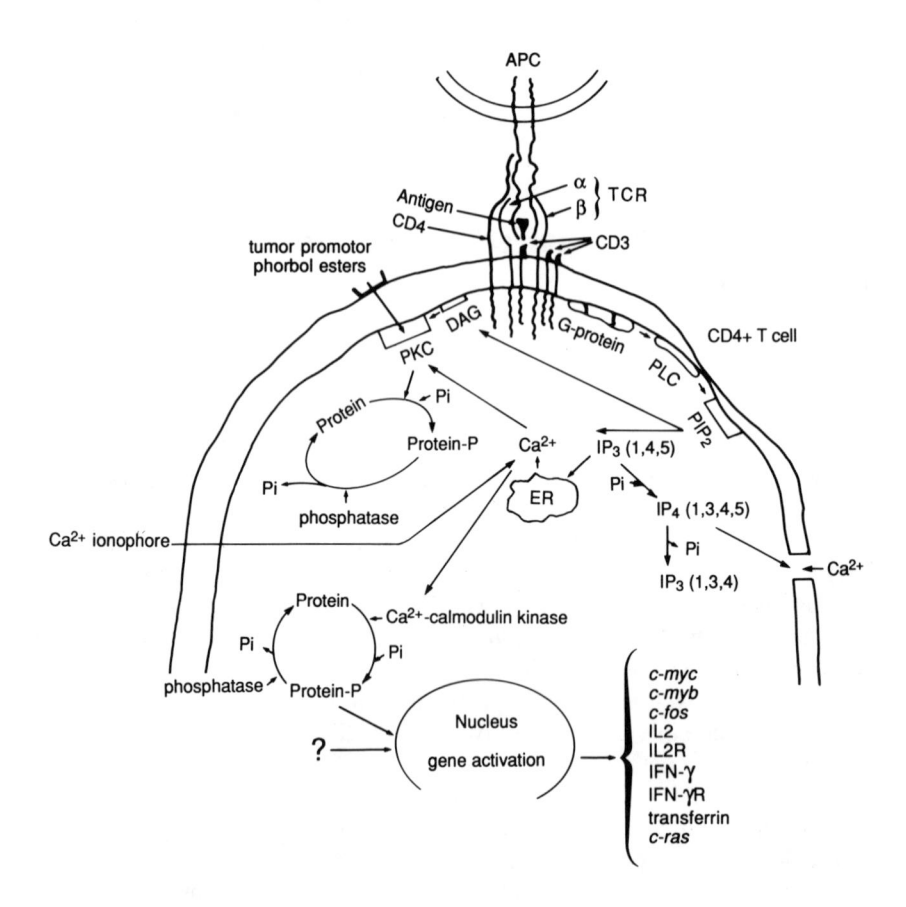

Fig. 1. The T-cell receptor (TCR)–CD3 complex on the T cell recognizes processed antigen along with MHC class II gene products on APC. Antigen under appropriate conditions activates PLC, presumably through GTP bound to the α subunit of G protein to hydrolyze PIP_2 to produce inositol 1,4,5-triphosphate (IP_3) and 1,2 diacylglycerol (DAG). IP_3 acts to increase cytosolic free Ca^{2+} ion concentration by mobilizing it from intracellular stores. Ca ionophore A23187 can also increase Ca^{2+} ion concentration by mobilizing it from an extracellular source. Elevated Ca^{2+} ion can activate Ca^{2+} calmodulin kinase. DAG activates the Ca^{2+} ion and phospholipid-dependent protein kinase C (PKC) which can also be activated by tumor promoter phorbol esters. PKC phosphorylates numerous proteins at serine and threonine residues. PKC-phosphorylated substrates, although not precisely characterized, appear to induce messages for nuclear activation. A complex series of reactions appears to lead to the appearance of specific transcripts for various genes in the cytoplasm, which are translated into their specific products. ER, endoplasmic reticulum; IL2R, interleukin 2 receptor; IFN-γR, interferon-gamma receptor; IP_4, 1,3,4,5-tetra-phosphate; Pi, PIP_2, phosphoinositol bis-phosphate.

A), along with several other glycoproteins, binds to the carbohydrate moieties on the CD3 molecule [Kanellopoulos et al., 1985]. Similarly, monoclonal antibodies to Ti–T3 cause modulation of the receptor complex and (under appropriate conditions) signal transduction [Alcover et al., 1987]. Ti–T3 recognition/modulation results in phospholipase C-induced hydrolysis of phosphoinositol bis-phosphate (PIP_2). Although there is at present only indirect evidence, it is thought that GTP binding proteins (G proteins) act as intermediaries between antigen, Ti–T3 recognition/modulation, and the phospholipase C-induced hydrolysis of PIP_2. This hypothesis is based on observations that the occupation of specific receptors on the cell membrane by agonists induces a rapid exchange of GDP for GTP on the α subunit of the G protein. This change in energy state activates the phospholipase C enzymes [Allende, 1988; Harnett and Klaus, 1988], which cause PIP_2 hydrolysis to produce two important secondary messenger molecules, inositol 1,4,5-triphosphate (IP_3) and 1,2 diacylglycerol (DAG). The IP_3 increases cytosolic free Ca^{2+} ion concentration by mobilizing Ca^{2+} from intracellular stores in the endoplasmic reticulum and mitochondria [Berridge, 1987]. Inositol 1,3,4,5-tetraphosphate (IP_4), a further product of inositol phosphate metabolism, is also thought to open Ca^{2+} channels [Hirasawa et al., 1982; Batty et al., 1985]. The membrane-embedded DAG binds, activates, and translocates the Ca^{2+} ion and phospholipid-dependent protein kinase C (PKC) enzymes from the cytosol to the cell membrane, where they phosphorylate numerous membrane proteins at their serine and threonine residues [Isakov et al., 1986; Cambier et al., 1985; Imboden and Stobo, 1985]. Although they are under intense investigation in a number of laboratories, the characterization of these phosphorylated proteins, as well as other second messenger molecules, remains largely incomplete.

Following the T cell's recognition of antigen and transduction of the initial cascade of biochemical signals across the cell membrane, progression from G1 through the S-G2 phases of the cell cycle is regulated by the biosynthesis of IL2 and expression of the IL2R [Green, 1987; Smith, 1988]. The binding of IL2 to its high-affinity IL2R on the cell membrane provides an essential growth-promoting signal [Greene, 1987; Smith, 1988]. The absence of IL2 or the high-affinity IL2R or the decreased expression of either results in a significant impairment of cellular proliferation. After transduction of the IL2 signal, further control of proliferation is regulated by the binding of transferrin to its specific membrane receptor. Such binding is necessary to provide iron to DNA-replicating enzymes [Lesley and Schulte, 1985]. Thus it can be readily appreciated that cells require a variety of heterogenous biochemical signals in order to undergo proliferation and differentiation. While a number of investigators have demonstrated minor alterations in the T-cell subpopulations of elderly individuals, such alterations do not appear to be of sufficient magnitude to explain the defects in T-cell function.

AGE DIFFERENCES IN INTERLEUKIN BIOSYNTHESIS

IL2 is a product of activated helper/inducer (CD4$^+$) T cells. The IL2 gene, located on chromosome 4 [Seigel et al., 1984], is divided into four exons [Fujita et al., 1983; Holbrook et al., 1984]. The 5' untranslated region and the first 49 amino acids of IL2, 20 of which constitute a signal peptide (not present in mature IL2) are encoded by exon 1. Exons 2 and 3, separated by a long intervening sequence, contain coding information for the next 20 and 48 amino acids respectively. The coding information for the remaining 36 amino acids and the 3' untranslated region is contained in the fourth exon [Fujita et al., 1983; Holbrook et al., 1984]. Mature IL2 is therefore composed of 133 amino acids. It has a single disulfide bond between amino acids 58 and 105 that is essential for bioactivity [Yamada et al., 1987]. The 20 amino acids at the NH$_2$ terminus and the 12 amino acids at the C terminus are also critical for bioactivity [Ju et al., 1987]. Furthermore, substitution of Asp^{20} or the deletion of Phe^{124} create subactive IL2 analogues that are able to bind only the low-affinity IL2R. Substitution of Trp^{121} or Leu^{17} creates IL2 analogues that do not bind any form of the IL2R [Collins et al., 1988] and hence lack bioactivity. IL2 transcripts are not able to be detected in unstimulated T cells. After cell activation by a variety of stimuli, IL2 mRNA can be readily detected within 6–24 hours [Kern et al., 1986].

The first report that decreased IL2 secretion might be related to impaired proliferation of PHA-stimulated peripheral blood lymphocytes (PBL) from elderly individuals appeared about 8 years ago [Gillis et al., 1981]. Several laboratories subsequently confirmed this observation in mouse splenocytes as well as in humans using many different cell-activating agents [Thoman and Weigle, 1981; Chang et al., 1982; Gilman et al., 1982, Vissinga et al., 1987; Canonica et al., 1988].

It would appear that the decreased IL2 secretion is due primarily to a defect within the T-cell population from old animals and humans (Table I). There are a number of ancillary observations that support this idea. T cells from young and elderly individuals have equal numbers of PHA receptors [Antel et al., 1980]. Less IL2 secretion is also found in allogenically stimulated T cells in mixed lymphocyte cultures [Gilman et al., 1982]. There is no evidence that suppressor T cells limit IL2 biosynthesis or decrease the bioavailability of IL2 in cell cultures [Thoman and Weigle, 1987]. Similarly, there is no evidence of decreased interleukin 1 (IL1) secretion by monocytes from elderly humans or old rats [Schwab et al., 1985a]. Impaired IL1 secretion by lipopolysaccharide (LPS)-stimulated peritoneal macrophages of aged mice has been reported by Bruley-Rosset and Vergnon [1984] and Inamizu et al. [1985]. However, impaired production of IL1 by old mice does not explain the decrease in IL2 synthesis, since in coculture macrophages from old mice efficiently support IL2 secre-

TABLE I. Effects of Age on Cellular IL2 Secretion

Species	Type of cell	Stimulus	Comments	Reference
Human	PBL	PHA	Decreased	Hara et al., 1987
				Froelich et al., 1988
				Nagel et al., 1988
				Canonica et al., 1988
Human	T	PMA	Decreased	Hara et al., 1987
		PMA/Ca ionophore	No change	Chopra et al., 1989a
Mouse	Spleen	Con A	Decreased	Vissinga et al., 1987
				Thoman and Weigle, 1988
				Fong and Makinodan, 1989
	T	PMA/Ca ionophore	No change	Thoman and Weigle, 1988
Rat	Spleen	Con A	Increased	Gilman et al., 1982
				Holbrook et al., 1989
		Allogenic cells	Decreased	Gilman et al., 1982
Rat	T	PMA/Ca ionophore	Decreased	Wu et al., 1986
			Increased	Holbrook et al., 1989

tion by the T cells of young mice [Bruley-Rosset and Vergnon, 1984; Chang et al., 1982; Thoman and Weigle, 1985]. Furthermore, macrophages from young adult mice do not correct deficient IL2 secretion by the T cells from old mice. The age-related inability to produce sufficient IL2 could be attributable to a defect in the expression of the IL2 gene following cell activation or the abnormal removal of bioactive IL2 from the cell (Table II). Recently, we examined the expression of IL2-specific mRNA in PHA-stimulated PBL from young and old humans [Nagel et al., 1988]. There was significantly less IL2-specific mRNA in cells from the elderly individuals as well as significantly less bioactive IL2 synthesized. Similar observations have also been made using Con A-stimulated spleen cells from old rats [Wu et al., 1986]. In contrast, Gilman et al. [1982] reported that Con A-stimulated splenocytes from aged rats produce more IL2 than their young counterparts. However, allogenic lymphocyte-stimulated spleen cells from old rats synthesized less IL2, effectively disproving the notion that the utilization of less IL2 by the cells of the old rats was the reason for the higher levels of IL2 detected in their cell cul-

TABLE II. Effects of Age on the Expression of IL2 mRNA

Species	Type of cell	Stimulus	Comments	Reference
Human	PBL	PHA	Decreased	Nagel et al., 1988
Human	T	PMA/Ca ionophore	No change	Chopra et al., 1989a
Mouse	Spleen	Con A	Decreased	Fong and Makinodan, 1989
Rat	Spleen	Con A	Decreased	Wu et al., 1986
			Increased	Holbrook et al., 1989
		PMA/Ca ionophore	Increased	Holbrook et al., 1989

ture supernatants. In a more recent study, Holbrook et al. [1989] examined expression of IL2 in splenocytes from two different strains of rats stimulated with Con A, as well as PMA/A23187. Spleen cells from both strains of old rats stimulated with Con A or PMA/A23187 showed higher, although not statistically significant in all cases, expression of IL2-specific mRNA and secretion of bioactive IL2. This situation may be different from that found in mice, since Fong and Makinodan [1989] have demonstrated decreased expression of IL2 mRNA in Con A-stimulated spleen cells of old mice. Furthermore, by in situ hybridization, their study also demonstrated that approximately 40% fewer cells in old mice express intermediate to high amounts of IL2 mRNA compared with the young mice. Earlier, Miller [1984] used the limiting dilution assay to demonstrate a lower frequency of IL2-producing cells in Con A-stimulated cells from old mice. However, the responder cells were found to be competent to produce amounts of IL2 equivalent to those produced by cells of young mice. Although the results of these two studies differ slightly, they suggest that there is an accumulation of nonresponding T cells in the spleens from old mice. Canonica et al. [1988], recently used the limiting dilution assay to study human T cells and did not find evidence of a decreased IL2-producing cell frequency but found that all cell populations from the elderly donors produced less bioactive IL2 compared with young controls. Thus, we are left with the question of whether there exists a "nonresponding" cell population that increases with age, or whether all cells gradually lose their ability to synthesize IL2. This remains an important issue for further studies.

Recently, we [Chopra et al., 1989a] studied IL2 mRNA expression in purified T cells stimulated with PMA/A23187. Purified T cells in this study were utilized to rule out a possible negative signal generated by monocytes releasing prostaglandin E-2 (PGE-2), which can effect cells from the elderly [Schwab et al., 1985b]. Furthermore, PMA/A23187 can optimally stimulate purified T cells by duplicating the effects of early secondary messenger molecules generated by the binding of mitogen to cells [Truneh et al., 1985; Chopra et al., 1987, 1989b]. Using PMA/A23187, this study did not find any difference in IL2 mRNA expression by the cells from young and elderly individuals. Similarly, at 24 or 72 hours after stimulation, there was a slight difference between the young and elderly for the secretion of bioactive IL2, but, unlike PHA stimulation, this difference was statistically not significant. The normal expression of the IL2 gene by PMA/A23187-stimulated T cells from the elderly clearly demonstrates that the level of expression of the IL2 gene per se is not age-associated and suggests that the decreased expression of IL2 mRNA in mitogen-stimulated cells might be due to insufficient or ineffective transmembrane signaling.

IL2 RECEPTOR EXPRESSION IN AGING

Binding of IL2 to membrane-associated high-affinity IL2 receptor is necessary for the exit of cells from the G_1 phase of the cell cycle [Cantrell and Smith, 1984]. Recent studies using human as well as animal models have demonstrated that at least three classes of IL2R are expressed on activated T cells: 1) a low-affinity IL2R represented by the 55 kd protein (also termed the α subunit or Tac protein); 2) an intermediate-affinity IL2R represented by a 70–75 kd protein(s) termed the β subunit; and 3) a high-affinity IL2R that is formed by the noncovalent association of α and β subunits [Sharon et al., 1986; Dukovich et al., 1987]. The β subunit is to some extent constitutively expressed on resting T cells, B cells, and large granular lymphocytes [natural killer (NK) cells] [Dukovich et al., 1987], where it mediates the induction of lymphokine-activated killer (LAK) cells and activated NK cells [Tsudo et al., 1987; Siegel et al., 1987]. The β subunit can independently transduce a growth-promoting signal and induce the expression of the α subunit [Bich-Thuy et al., 1987]. Simultaneous coexpression of both α and β subunits is essential for the expression of high-affinity IL2R. The high-affinity IL2R is the physiological form of the receptor that is internalized following binding to IL2 [Weissman et al., 1986; Robb and Kutny, 1987; Kumar et al., 1987]. After appropriate stimulation, T cells express several thousand molecules of the α subunit but only a few molecules of the β subunit [Smith, 1988]. Therefore, it is the β subunit that determines the number of high-affinity IL2R in a T-cell population.

Several studies have demonstrated an inability of even large quantities of exogenous IL2 to correct fully the lower proliferative response of the T cells from elderly humans [Gillis et al., 1981; Rabinowich et al., 1985]. Kennes et al. [1983] were able to increase DNA synthesis by PBL from elderly individuals to levels found with cells from young individuals when they stimulated cells with PHA and IL2 but not when IL2 was used alone. In mice, Thoman and Weigle [1985] were able to fully reconstitute in vitro as well as in vivo the primary and secondary cytolytic T-cell responses and humoral immune responses to T-independent antigens by exogenous IL2, but found only partial restoration of lymphocyte proliferative responses. In rats, Gilman et al. [1982] did not find any beneficial effect of exogenous IL2 in restoring the diminished proliferative response of allogenically stimulated lymphocytes. Together, these studies raise the possibility that T cells from the aged may have defective expression of the IL2R (Table III).

Vie and Miller [1986] were the first to demonstrate fewer IL2R[+] cells in Con A-stimulated spleen cell cultures from old mice. This finding was recently confirmed by Thoman and Weigle [1988]. Negoro et al. [1986], studying elderly humans, demonstrated fewer IL2R[+] cells in PHA-stimulated PBL, while

TABLE III. Effects of Age on Cellular IL2R Expression

Species	Type of cell	Stimulus	Comments	Reference
Human	PBL	PHA	Decreased	Negoro et al., 1986
				Nagel et al., 1988
Human	T	PMA/Ca ionophore	No change	Chopra et al., 1989a
Mouse	Spleen	Con A	Decreased	Vie and Miller, 1986
				Thoman and Weigle, 1988
	G_0 T	Con A	No change at 48 hr after stimulation	Proust et al., 1988
	Spleen	PMA/Ca ionophore	No change	Thoman and Weigle, 1988
Rat	T	Con A	No change	Holbrook et al., 1989
		PMA/Ca ionophore	No change	Holbrook et al., 1989

Chopra et al. [1987a] demonstrated fewer IL2R$^+$ cells in T cells stimulated with either PMA alone or PMA in combination with A23187-stimulated T cells from elderly humans. Nagel et al. [1988] examined IL2R mRNA expression in PHA-stimulated PBL from elderly donors and suggested the possibility that defective IL2R gene expression might be responsible for impairment of IL2R expression by stimulated cells from the elderly. Furthermore, this study also demonstrated less fluorescence when using an anti-IL2R antibody labeled with fluorescein to measure IL2R$^+$ cells from the elderly. This finding indicated that there was either lower IL2R density on the cells or decreased cell size. The findings of lower IL2R density supports the earlier findings of Negoro et al. [1986], which demonstrated significantly fewer low-affinity IL2R using a radiolabeled IL2 to study PHA-stimulated PBL from the elderly. The studies described above mainly detail the effect of aging on the expression of low-affinity IL2R. However, it should be recorded that the β subunit determines the number of high-affinity IL2R on activated T cells. What effect a decrease in α subunit expression may have on T-cell proliferation remain uncertain.

Most laboratories therefore focused their attention on the expression of high-affinity IL2R (Table IV). Because at present the antibody against the β subunit is not available, and neither is the c-DNA probe, the only way to effectively study the p70/75 subunit is to use a radiolabeled IL2. Using [^{125}I]IL2, Negoro et al. [1986] demonstrated significantly fewer high-affinity IL2R expressed on PHA-stimulated PBL from elderly donors. In a subsequent study this group of workers also demonstrated significantly less high-affinity IL2R bound-IL2 as well as IL2 internalization by PHA-stimulated T cells from elderly humans [Hara et al., 1988]. Furthermore, using crosslinking studies they found lower expression of α as well as β IL2R subunits in the cells of the elderly. However, in the first study these authors also reported a very low percentage of IL2R$^+$ cells identified with a fluorescent-labeled anti-IL2R (α subunit) antibody, making it possible that the low number of high-affinity IL2R they found might

TABLE IV. Effects of Age on Cellular High-Affinity IL2 Receptor Expression

Species	Type of Cell	Stimulus	Comments	Reference
Human	PBL	PHA	Decreased	Negoro et al., 1986
				Froelich et al., 1988
			No change	Nagel et al., 1989
Human	T	PMA/Ca ionophore	No change	Chopra et al., 1989a
Mouse	G_0 T	Con A	Decreased at 8 and 16 hr after stimulation	Proust et al., 1988
			No change at 48 hr after stimulation	Proust et al., 1988
Rat	Spleen	Con A	No change	Holbrook et al., 1989
		PMA/Ca ionophore	No change	Holbrook et al., 1989

reflect the presence of many nonresponding T cells (i.e., IL2R⁻) in the assay mixture. Proust et al. [1988] applied a unique approach to resolve this problem. They stimulated a purified G_0 murine T-cell population with Con A and exogeneous IL2 and did not find any difference between young and old mice in the expression of high-affinity IL2R. However, they also did not find any difference between cells from young and old mice in the number of IL2R⁺ cells using a fluorescent-labeled anti-IL2R antibody. These results are at variance with a previous report in mice by Vie and Miller [1986]. Furthermore, in the study of Proust et al., at 8 and 16 hours after Con A and IL2 stimulation, the T cells from the old mice did not express high-affinity IL2R while the cells from the young did. However, cells from both young and old mice showed the presence of low-affinity IL2R as revealed by their ability to bind [^{125}I]-IL2 under low-affinity conditions. This study demonstrated defects in the early expression as well as quality of the high-affinity IL2R on T cells from old mice. However, it should be noted that the T cells used in the study did not contain accessory cells, which are necessary for optimum Con A-induced stimulation of T cells [Larsson et al., 1980; Roosnek et al., 1985a]. Accessory cells as well as their products, possibly IL1, are necessary for optimal expression of high-affinity IL2R [Wakasugi et al., 1985; Malek et al., 1985; Lowenthal et al., 1986]. In a recent study, Holbrook et al. [1989] did not find any impairment in the expression of high-affinity IL2R on Con A- or PMA/A23187-stimulated splenic T cells from old rats even though there were fewer IL2R⁺ cells using a fluorescent anti-IL2 receptor antibody. When the numbers of high-affinity IL2R were calculated in terms of receptor-positive cells, there were twice the number of high-affinity IL2R on the splenocytes from the old rats.

Recently two additional reports have appeared on the expression of high-affinity IL2R on PHA-stimulated cells from elderly humans [Froelich et al.,

1988; Nagel et al., 1989]. While Froelich et al. [1988] demonstrated significantly fewer high-affinity IL2R on PHA-stimulated, percoll-fractionated lymphoblasts from the elderly, Nagel et al. [1989] did not find any differences. Interestingly when the numbers of high-affinity IL2R were expressed in terms of IL2R$^+$ cells determined with a monoclonal anti-IL2R antibody, any difference in the number of high-affinity IL2R was further decreased. Froelich et al. [1988] did not describe the number of receptor-positive cells for the young and elderly; therefore an exact comparison between these two studies cannot be made. However, it should be noted that the participants in the Froelich et al. [1988] study were ambulatory elderly who suffered from osteoarthritis and hypertension, whereas the study group used by Nagel et al. [1989] were healthy individuals chosen by exacting criteria [Nagel et al., 1982]. Using PMA/A23187-stimulated T cells from young and elderly donors, Chopra et al., [1989a] did not find any difference in the number of high-affinity IL2R. In the studies of Proust et al. [1988], Nagel et al. [1989], Holbrook et al. [1989], and Chopra et al. [1989a], the representation of high-affinity IL2R was not significantly different between the age groups. There were generally fewer viable cells and fewer receptor-positive cells at the end of the culture period, suggesting that decreased cell viability and an accumulation of a nonresponding cell population substantially contribute to the observed decrease in cellular proliferation.

A major problem in the assessment of high-affinity IL2R is that stimulated cells both express high-affinity IL2R and also secrete IL2. The secreted IL2 rapidly binds to the IL2R and results in internalization of the IL2–IL2R complex [Weissman et al., 1986; Kumar et al., 1987]. Extensive washing of cells to remove cell membrane-bound IL2 can induce the re-expression of high-affinity IL2R [Robb et al., 1985]. However, it is not clear if these receptors are simply recycled or if there is de novo synthesis of new high-affinity IL2R. Antibody against the CD3 complex has been shown to induce the expression of IL2R in the absence of IL2 secretion [Laing and Weiss, 1988]. This method of cell stimulation may be helpful in interpretation of the presence and amount of the high-affinity IL2R on cells from different age groups. Another method would be to utilize long-term culture of cells after mitogen stimulation with the reinduction of high-affinity IL2R on a synchronized cell population using a phorbol dibutyrate stimulation that also induces IL2R expression without IL2 synthesis. Smith and Cantrell [1985] have previously used this method to study high-affinity IL2R expression by human T cells.

In addition to the expression of IL2R on the cell plasma membrane, there is also a soluble form of the receptor (sIL2R) identified in the supernatant from cultures of activated T cells [Rubin et al., 1985, 1986; Diamantstein et al., 1986; Chopra et al., 1987a]. The soluble receptor is generally considered to be the P55 IL2R protein minus its intracytoplasmic and transmembrane amino acids [Treiger et al., 1987]. Because the sIL2R has the intact N terminus of the p55 IL2R, it retains the ability to bind IL2 [Rubin et al., 1986; Robb et al., 1988]. However, since it has a low affinity for IL2, extremely high concentrations of sIL2R are required to bind and limit the bioavailability of IL2 significantly [Robb and Kutny, 1987; Jacques et al., 1987]. While there is one report [Saadeh et al., 1986] that elderly humans have significantly higher concentrations of circulating sIL2R, other investigators have failed to confirm this. Nagel et al. [1988] measured sIL2R in culture supernatants from PHA-stimulated PBL from young and elderly people to determine if cells from the elderly release more sIL2R. However, they found that PHA-stimulated PBL from elderly people generally released less sIL2R compared with cells from the young. A similar observation was made by Froelich et al. [1988]. In both cases the concentrations of sIL2R in the supernatants were so low that they would not be expected to effect IL2-dependent cell responses.

T-CELL RECEPTOR IN AGING

The TCR plays a critical role in antigen recognition and the transduction of intracellular messengers. However, to date the effects of age on the TCR have not been extensively investigated. O'Leary and his colleagues have demonstrated small, but statistically significant decreases in $CD3^+$ T cells in elderly individuals [Hallgren et al. 1983, 1985; Jensen et al., 1986]. Recently, this group [O'Leary et al., 1988] used the antibody WT31, which recognizes the constant portion of the $Ti^{\alpha-\beta}$ heterodimer [Spits et al., 1985], to study the TCR of young and elderly humans. They found no age-related differences in the proportion of $WT31^+$ cells. This finding and the results of their previous work indicated that the decreased $CD3^+$ cells were confined to the subpopulation that does not express $Ti^{\alpha-\beta}$. $CD3^+$ cells were also noted to be especially deficient in elderly individuals with osteoarthritis and cardiopulmonary disorders.

Whisler and Newhouse [1986] used anti-CD3 antibody to analyze the integrity of the CD3 complex on the cells of elderly. By using low and high concentrations of anti-CD3 (OKT3) antibody to stimulate the cells, this study demonstrated significant impairment in the proliferative response of the elderly and agrees with similar findings of Schwab et al. [1985b]. Whisler and Newhouse [1986] also presented evidence that in OKT3 antibody-stimulated cells of the elderly there is a reduction in the capping of CD3 determinants and the elaboration of B-cell growth factors. However, the defect in CD3 capping did not appear directly responsible for the defective proliferation of the T cells of the elderly. This study suggested that cytoskeletal abnormalities or reduced levels of intracellular cAMP may be responsible for reduced CD3 capping. Canonica et al. [1988] failed to find any difference between young and elderly individuals in the OKT3 antibody-induced proliferative response. These results are at variance with those earlier reported by Schwab et al. [1985b] and Whisler and Newhouse [1986]. However, in the study of Canonica et al. [1988], some elderly subjects demonstrated impaired proliferation to autologous mixed lymphocyte reactions and PHA stimulation, suggesting the possibility that T cells from the elderly may have a defect in the Ti$^{\alpha\text{-}\beta}$ heterodimer reflected in a decreased ability to recognize antigen. The variability in different studies of the responses in elderly to anti-CD3 antibody may represent differences in the percentage of the monocytes in the culture system since monocytes are known to affect T-cell responses to anti CD3 antibody profoundly [Van Wauwe and Goossens, 1981; Ceuppens et al., 1985]. At present there is no evidence as to whether the impaired response to anti-CD3 antibody is due to defects in transmembrane signaling or to low CD3 density on the T cells of the elderly.

TRANSMEMBRANE SIGNALING IN T CELLS FROM ELDERLY INDIVIDUALS

One of the most important aspects of T-cell activation is the appropriate recognition of a stimulus (antigen) and the subsequent transduction of early biochemical signals into the cells. A defect in either of these can result in the lack of specific gene expression, the resultant release of growth factor, receptor expression, and the progression of the cells through the cell cycle. It is currently thought that protein kinase C (PKC) activation and an increase of the cytosolic free Ca^{2+} concentration are the two most important early events responsible for the initiation of cell activation [Dröge, 1986; Nishizuka, 1988]. It appears that a major effect of the C kinase protein is to phosphorylate numerous membrane proteins [Nishizuka, 1988]. However, the specific substrates that transduce the activation signals to the nucleus remain uncertain.

Kennes et al. [1981] demonstrated an increased sensitivity of T cells from the elderly to the Ca^{2+} chelating effects of EGTA and verapamil. This study indicated that the Ca^{2+} ion-dependent processes in lymphocyte activation are defective in the cells from the elderly. Recently Miller et al. [1987], using highly purified Thy-1$^+$ mouse cells, found that after Con A stimulation there is significantly less net increase in the cytosolic free Ca^{2+} ion concentration in cells from old mice. Before Con A stimulation, the cytosolic free Ca^{2+} ion concentration was similar in the T cells from both young and old mice. Furthermore, they demonstrated that a lower percentage of T cells from old mice responded to Con A by increasing their cytosolic free Ca^{2+} ion concentration. This impairment correlated with a decreased proliferative response by the cells. Subsequently Proust et al. [1987], using homogeneous resting G_0 T cells from young and old mice, found similar increases in the cytosolic free Ca^{2+} ion concentration and production of IP_3 after Con A stimulation. However, basal levels (before stimulation) of both these second messengers were consistently higher in the G_0 T cells from the old mice. Thus the net increase of the Ca^{2+} ion as well as IP_3 formation was significantly less in the T cells from the old mice. The finding of relatively higher basal concentrations of Ca^{2+} ion and IP_3 is at variance with that of Miller et al. [1987]. Although the basal levels (before Con A stimulation) of PKC activity were similar in T cells from young and old mice, Proust et al. [1987] also found reduced translocation of PKC in Con A-stimulated cells from old mice. These results indicated possible defects in the generation of secondary messengers by enzymes responsible for PIP_2 hydrolysis. To test this hypothesis, they examined levels of PKC activity before and after PMA stimulation of G_0 T cells from young and old mice and found similar levels [Proust et al., 1988]. Lerner et al. [1988] have recently also presented evidence that Con A induces comparable production of various derivatives of inositol phosphate (IP_1, IP_{2a}, IP_{2b}, IP_3, and IP_4) by splenic T cells of young and old mice. They did not find any differences between the cells from young and old mice in basal values for any of these inositol phosphate derivatives. They did find that the ability of cells from old mice to raise the cytosolic free Ca^{2+} ion concentration was significantly impaired [Miller et al., 1989]. One important consideration is that all these studies were conducted using purified G_0 T or T cells in the absence of accessory cells. Con A-stimulated T cells under such stringent assay conditions generally do not secrete detectable amounts of bioactive IL2 and do not proliferate optimally [Larsson et al., 1980; Hünig et al., 1983; Roosnek et al., 1985a,b]. Accessory cells also appear to be important for maximal PKC activation [Dröge, 1986].

To study possible defects in transmembrane signaling and to explore the responses of T cells in the absence of accessory cells, a number of laboratories have made use of the tumor-promoting phorbol esters and Ca^{2+} ionophores. The tumor-promoting phorbol esters directly bind and activate PKC [Castagna et al., 1982; Niedel et al., 1983; Kraft and Anderson, 1983]. The phorbol esters therefore mimic the effects of endogenously produced DAG in activating PKC [Kikkawa and Nishizuka, 1986] and thus provide a powerful tool for studying the role of PKC in cell function. Several studies have provided evidence that PKC functions as a receptor of the phorbol esters [Niedel et al., 1983; Kikkawa et al., 1983; Konig et al., 1985]. Calcium ionophores such as A23187 and ionomycin open Ca^{2+} channels on the plasma membrane and raises the cytosolic free Ca^{2+} ion concentration by mobilizing it from extracellular sources. The net effects of the Ca^{2+} ionophores appear to be like those of IP_3 and IP_4, which mobilize Ca^{2+} ion from intracellular or extracellular sources respectively [Tsien et al., 1982; Isakov and Altman, 1987]. Together PMA and Ca ionophores bypass early transmembrane signaling and directly activate PKC and increase cytosolic free Ca^{2+} ion concentrations, resulting in cellular proliferation [Truneh et al., 1985; Isakov et al., 1986; Chopra et al., 1987a, 1989b]. Miller [1986] used these agents to study the proliferative responses of purified T cells from young and old mice. He demonstrated an increasing ionomycin requirement by PMA-costimulated T cells from old mice in order to elicit a proliferative response identical to that seen with the cells from young mice. Interestingly, this study also suggested that approximately 70% of the T cells from old mice fail to leave the G_0–G_1 phase of the cell cycle after Con A stimulation. This compares with 50% for cells from young mice. On stimulation with PMA/ionomycin, a higher and comparable percentage of cells from young and old mice exit the G_0–G_1 phase of the cell cycle. This observation strengthens the importance of their subsequent findings describing defects in the ability of cells from old mice to increase cytosolic free Ca^{2+} ion concentration after Con A stimulation [Miller et al., 1987]. They showed that there is an increased representation of nonresponding cells in spleens from old mice, which causes a generalized lower level of proliferation. Subsequent studies by Thoman and Weigle [1988] have demonstrated that increasing concentrations of ionomycin only partially restore the lower proliferative responses by PMA-costimulated spleen cells from old mice. However, these studies further demonstrated that the addition of ionomycin to Con A-costimulated cells from old mice restored their IL2 levels and their percent $IL2R^+$ cells to levels seen with cells from young mice stimulated with Con A alone. A similar effect was not observed when PMA was added to Con A-costimulated cells from old mice. Similarly, PMA/ionomycin stimulation of cells restored IL2 secretion and IL2R expression but not the proliferative response. Taken together these findings showed a major defect in Ca^{2+}-related events that concern only IL2 secretion and IL2R expression and do not fully explain the defects in the proliferative response or genesis of the appearance of the nonresponsive population. Similar results were reported in a recent study in rats reported by Holbrook et al. [1989]. They found that PMA/A23187-stimulated splenocytes from old rats synthesize and secrete large amounts of bioactive IL2 as well as being able to express high numbers of high-affinity IL2R when compared with cells from young rats. This pattern of responses for cells from old rats is the same as that seen when the cells are stimulated with Con A. The interesting finding in this study is that the proliferative responses of cells from the old rats remained less if the cells from the old rats were stimulated with Con A or PMA/A23187. The lower proliferative response was demonstrated to be due to consistently fewer viable cells in the cultures 2–3 days after initiation of stimulation, with a higher representation of a nonresponding cell population in the spleen cell cultures from the old rats. It is possible that lower cell viability in the cultures may have contributed to the lower proliferative response found in a previous study by Thoman and Weigle [1988]. Chopra et al. [1987b] and Negoro et al. [1987] found that T cells from elderly humans display a significantly lower proliferative response after stimulation with PMA/A23187 or PMA/ionomycin, respectively. Both these studies demonstrated that increasing concentrations of PMA or Ca^{2+} ionophores do not correct the lower proliferative responses of the cells from the elderly. This finding would appear to be at variance with those described in cells from mice [Miller, 1986]. However, the human studies used a relatively large concentration of PMA (10 ng/ml) and A23187 (0.1–1 μM) for stimulating IL2 secretion and proliferation and membrane IL2R expression [Chopra et al., 1987b]. Subsequently it was demonstrated that the constant stimulation of purified human T cells with PMA/A23187 does not induce high-affinity IL2R and produces a significantly lower proliferative response [Chopra et al., 1989b]. Similarly, Kumagai et al. [1987] had demonstrated a lower proliferative response of cells subjected to a constant stimulation with PMA/ionomycin. Subsequent studies using lower but optimal concentrations of these stimuli showed that T cells from young and elderly humans displayed no difference in their levels of IL2 and IL2R mRNA, their secretion of bioactive IL2, or their expression of low- and high-affinity IL2R [Chopra et al., 1989a]. However, the difference in the proliferative responses for cells from young and elderly individuals, although decreased, was still present in most cell populations from elderly donors. The decreased proliferative response appeared to be due to a lower percentage of cells surviving in cultures of cells from the elderly. Responding cells harvested from cultures of cells from young and old donors after 72 hours of culture and plated at equal densities with variable concentrations of exogenous IL2 added elicited proliferative responses that were identical. However, using a similar strategy to study PHA-responding cells (percoll-fractionated lymphoblasts), it was possible to increase substantially but not correct fully the lower proliferative response of cells from the elderly [Nagel et al., 1989]. This finding suggests that a greater percentage of cells from the elderly dies in culture after stimulation with mitogens. Also a variable number of potential responder cells do not respond optimally to the mitogens PHA or Con A possibly because of defects in transmembrane signaling. A number of investigators have demonstrated increased accumulation of the non-responding cell population in T-lymphocyte cultures from the elderly. These cells do not respond to an activating stimulus, as evidenced by their inability to leave the G_0 phase and move through the G_1–S phase of the cell cycle [Staiano-Coico et al., 1984; Kubbies et al., 1985; Negoro et al., 1986; Nagel et al., 1988]. The responses of responding cell populations in subsequent mitotic cell cycles are unknown. There is conflicting data on this point, with some reports of further defects in the responding cell populations [Joncourt et al.,

1981; Kubbies et al., 1985] and other evidence showing no defect [Abraham et al., 1977; Staiano-Coico et al., 1984]. Identification of populations of cells that die in culture or that do not respond at all needs to be explored. It is possible that these populations are not confined to a specific subset of T cells and represent a nonspecific effect of aging on cell physiology.

CONCLUSIONS

On the basis of the data supplied by many studies, it appears that defects in a complex series of reactions contribute to decreased T-cell activation and differentiation. While a few clues to the mechanisms involved have been obtained, overall knowledge of the control processes is not yet available because of their complexity. With the recent advances in knowledge of the structure and biochemistry of the T-cell receptor, transmembrane signaling, and cytokines and their receptors, it should be possible to gain additional insight into the alterations responsible for the defective signal transduction by T cells from the elderly. Until now most reports have used heterogeneous T-cell populations and did not focus on defects in single cell populations. It is quite possible that only a minor T-cell population in the elderly is defective in its ability to react with stimuli. Identification of such a cell population as well as those with limited in vitro life spans will be important in understanding the age-related defects in T-cell development and function.

ACKNOWLEDGMENTS

The author acknowledges the excellent secretarial support of Eleanor Wielechowski, the assistance of Drs. Nikki J. Holbrook, James E. Nagel, and William H. Adler in the preparation of this manuscript, and Dr. Richard A. Miller for providing unpublished information.

REFERENCES

Abraham C, Tal Y, Gershon H (1977): Reduced in vitro response to concanavalin A and lipopolysaccharide in senescent mice: A function of reduced number of responding cells. Eur J Immunol 7:301–304.

Alcover A, Ramarli D, Richardson NE, Chang HS, Reinherz EL (1987): Functional and molecular aspects of human T lymphocyte activation via T3-Ti and T11 pathways. Immunol Rev 95:5–36.

Allende JE (1988): GTP-mediated macromolecular interactions: The common features of different systems. FASEB J 2:2356–2367.

Antel J, Oger JF, Dropcho E, Richman DP, Kuo HH, Arnason BG (1980): Reduced T lymphocyte cell reactivity as a function of human aging. Cell Immunol 54:184–192.

Batty IR, Nakorski SR, Irvine RH (1985): Rapid formation of inositol 1,3,4,5-tetrakisphosphate following muscarinic receptor stimulation of rat cerebral cortical slices. Biochem J 232:211–215.

Mechanisms of Impaired T Cell Function   99

Berridge MJ (1987): Inositol triphosphate and diacylglycerol: Two interacting second messengers. Annu Rev Biochem 56:159–193.
Bich-Thuy LT, Dukovich M, Peffer NJ, Fauci AS, Kehrl JH, Greene WC (1987): Direct activation of human resting T cells by IL2: The role of an IL2 receptor distinct from the TAC protein. J Immunol 139:1550–1556.
Brenner MB, Trowbridge IS, Strominger JL (1985): Cross-linking of human T cell receptor proteins: Association between the T cell idiotype beta subunit and the T3 glycoprotein heavy subunit. Cell 40:183–10.
Bruley-Rosset M, Vergnon I (1984): Interleukin-1 synthesis and activity in aged mice. Mech Ageing Dev 242:247–264.
Cambier JC, Monroe JG, Coggeshall KM, Ransom JT (1985): The biochemical basis of transmembrane signaling by B lymphocyte surface immunoglobulin. Immunol Today 6:218–222.
Canonica GW, Caria M, Venuti D, Cipro G, Ciprandi G, Bagnasco M (1988): T cell activation through different membrane structures (T3/Ti,T11,T44) and frequency analysis of proliferating and interleukin-2 producer T lymphocyte precursors in aged individuals. Mech Ageing Dev 42:27–35.
Cantrell DA, Smith KA (1984): The interleukin T-cell system: A new cell growth model. Science 224:1312–1316.
Castagna M, Takai Y, Kaibuchi K, Sano K, Kikkawa U, Nishikzuka Y (1982): Direct activation of calcium-activated, phospholipid-dependent protein kinase by tumor-promoting phorbol esters. J Biol Chem 247:7847–7851.
Ceuppens JL, Meurs L, Van Wauwe JP (1985): Failure of OKT3 monoclonal antibody to induce lymphocyte mitogenesis: A familial defect in monocyte helper function. J Immunol 134:1498–1502.
Chang MP, Makinoden T, Peterson WJ, Strehler BL (1982): Role of T cells and adherent cells in age-related decline in murine interleukin production. J Immunol 129:2426–2430.
Chopra RK, Nagel JE, Chrest FJ, Boto WM, Pyle RS, Dorsey B, McCoy M, Holbrook N, Adler WH (1987a): Regulation of interleukin 2 and interleukin 2 receptor gene expression in human T cells: 1. Effect of Ca^{++} ionophore on phorbol myristate co-stimulated cells. Clin Exp Immunol 69:433–438.
Chopra RK, Nagel JE, Chrest FJ, Adler WH (1987b): Impaired phorbol ester and calcium ionophore induced proliferation of T cells from old humans. Clin Exp Immunol 70:456–462.
Chopra RK, Holbrook NK, Powers DC, McCoy MT, Nagel JE (1989a): Interleukin 2, interleukin 2 receptor and interferon-γ synthesis and mRNA expression in phorbol myristate acetate and calcium ionophore A23187 stimulated T cells from elderly humans. (submitted)
Chopra RK, Powers DC, Adler WH, Nagel JE (1989b): Phorbol myristate acetate and calcium ionophore A23187-stimulated human T cells do not express high-affinity IL-2 receptors. Immunology 66:54–60.
Collins S, Tsien WH, Seals C, Hakimi J, Weber D, Bailon P, Hoskings J, Green WC, Toome V, Ju G (1988): Identification of specific residues of human interleukin 2 that affect binding to the 70-kD α subunit (p 70) of the interleukin 2 receptor. Proc Natl Acad Sci USA 85:7709–7713.
Davis MM, Bjorkman PJ (1988): T cell antigen receptor genes and T cell recognition. Nature 334:395–402.
Diamantstein TH, Osawa H, Mouzaki A, Josimovic-Alasevic (1986): Regulation of interleukin 2 receptor expression and release. Mol Immunol 23:1165–1172.
Dröge W (1986): Protein kinase C in T-cell regulation. Immunol Today 7:340–343.
Dukovich M, Wano Y, Bich-Thuy L, Katz P, Cullen BR, Kehrl JH, Greene WC (1987): Identification of a second human interleukin 2 binding protein and its possible role in the assembly of the affinity of the high affinity IL-2 receptor. Nature 327:518–522.

Fong TC, Mankinodan T (1989): In situ hybridization analysis of the age-associated decline in IL-2 mRNA expressing murine T cells. Cell Immunol 118:199–207.

Froelich CJ, Burkett JS, Guiffaut S, Kingsl R, Brauner D (1988): Phytohemagglutinin induced proliferation by aged lymphocytes: Expression of high affinity interleukin-2 receptors and interleukin-2 secretion. Life Sci 43:1583–1590.

Fujita T, Takaoka C, Matsui H, Taniguchi T (1983): Structure of the human interleukin 2 gene. Proc Natl Acad Sci USA 80:7437–7441.

Gillis S, Kozak R, Durante M, Weksler ME (1981): Immunologic studies of aging. Decreased production and response to T cell growth factor by lymphocytes from aged humans. J Clin Invest 67:937–942.

Gilman S, Rosenberg J, Feldman J (1982): T lymphocytes of young and aged rats. II. Functional defects and the role of interleukin 2. J Immunol 128:644–650.

Greene WC (1987): The human interleukin 2 receptor: A molecular and biochemical analysis of structure and function. Clin Res 35:439–450.

Hallgren HM, Jackola DR, O'Leary JJ (1983): Unusual patterns of surface marker expression on lymphocytes from aged humans suggestive of a population of less differentiated T cells. J Immunol 131:191–194.

Hallgren HM, Jackola DR, O'Leary JJ (1985): Evidence for expansion of a population of lymphocytes with reduced or absent T3 expression in aged donors. Mech Ageing Dev 30:139–150.

Hara H, Tanaka T, Negoro S, Deguchi Y, Nishio S, Saiki O, Kishimoto S: (1988): Age-related changes of expression of IL-2 receptor subunits and kinetics of IL-2 internalization in T cells after mitogenic stimulation. Mech. Ageing Dev 45:167–175.

Harnett MM, Klaus GG (1988): G protein regulation of receptor signaling. Immunol Today 9:315–319.

Hirasawa K, Irvine RF, Dawson RMC (1982): Proteolytic activation can produce a phosphatidylinositol phosphodiesterase highly sensitive to Ca^{2+}. Biochem J 206:675–678.

Holbrook NJ, Smith KA, Fornace Jr AF, Comeau CM, Wiskocil RL, Crabtree GR (1984): T-cell growth factor: Complete nucleotide sequence and organization of the gene in normal and malignant cells. Proc Natl Acad Sci USA 81:1634–1638.

Holbrook NJ, Chopra RK, McCoy MT, Nagel JE, Adler WH, Schneider EL (1989): Expression of interleukin 2 and interleukin 2 receptor in aging rats. Cell Immunol (in press).

Hünig T, Loos M, Schimpl A (1983): The role of accessory cells in polyclonal T cell activation. I. Both induction of interleukin 2 production and of interleukin 2 receptor responsiveness are accessory cell dependent. Eur J Immunol 13:1–6.

Imboden JB, Stobo JD (1985): Transmembrane signaling by the T cell antigen receptor. Perturbation of the T3–antigen complex generates inositol phosphate and releases calcium ions from intracellular stores. J Exp Med 161:446–456.

Inamizu T, Chang MP, Makinoden T (1985): Influence of age on the production and regulation of interleukin 1 in mice. Immunology 55:447–455.

Isakov N, Altman A (1987): Human T lymphocyte activation by tumor promoters: Role of protein kinase C. J Immunol 138:3100–3107.

Isakov N, Scholz W, Altman A (1986): Signal transduction and intracellular events in T-lymphocyte activation. Immunol Today 7:271–277.

Jacques Y, LeMauff B, Boeffard F, Godard A, Olive D, Soulillou JP (1987): A soluble interleukin 2 receptor produced by a normal alloreactive human T cell clone binds interleukin 2 with low affinity. J Immunol 139:2308–2316.

Jensen TL, Hallgren HM, Yasmineh WG, O'Leary JJ (1986): Do immature T cells accumulate in advancd age? Mech Ageing Dev 33:237–245.

Joncourt F, Bettens F, Kristensen F, DeWeck AL (1981): Age-related changes in mitogen responsiveness in different lymphoid organs from outbred NMRI mice. Immunobiology 158:439–449.

Ju G, Collins L, Kaffka KL, Tsien WH, Chizzonite R, Crow R, Bhatt R, Kilman PL (1987): Structure–function analysis of human interleukin-2. Identification of amino acid residues required for biological activity. J Biol Chem 262:5723–5731.

Kanellopoulos JM, Petris SD, Leca G, Crumpton MJ (1985): The mitogenic lectin from *Phaseolus vulgaris* does not recognise the T3 antigen of human T lymphocytes. Eur J Immunol 15:479–486.

Kennes B, Hubert CL, Brohee D, Neve P (1981): Early biochemical events associated with lymphocyte activation in ageing I. Evidence that Ca^{++} dependent processes induced by PHA are impaired. Immunology 42:119–125.

Kennes B, Brohee D, Neve P (1983): Lymphocyte activation in human aging. V. Acquisition of response to T cell growth factor and production of growth factors by mitogen stimulated lymphocytes. Mech Ageing Dev 23:103–125.

Kern JA, Reed JC, Daniele RP, Nowell PC (1986): The role of accessory cells in mitogen stimulated human T cell gene expression. J Immunol 137:764–769.

Kikkawa U, Nishizuka Y (1986): The role of protein kinase C in transmembrane signaling. Annu Rev Cell Biol 2:149–178.

Kikkawa U, Takai Y, Tamaca Y, Miyake R, Nishizuka Y (1983): Protein kinase C as a possible receptor protein of tumor promoting phorbol esters. J Biol Chem 258:11442–11445.

Konig B, DiNitto A, Blumberg P (1985): Phospholipid and Ca^{++} dependency of phorbol ester receptors. J Cell Biochem 27:255–265.

Kraft AS, Anderson WB (1983): Phorbol esters increase the amount of Ca^{2+} phospholipid-dependent protein kinase associated with plasma membrane. Nature 301:621–623.

Kubbies M, Schindler D, Hoehn H, Rabinovitch PS (1985): BrdU-Hoechst flow cytometry reveals regulation of human lymphocyte growth by donor-age-related growth fraction and transition rate. J Cell Physiol 125:229–234.

Kumagai NK, Benedict SH, Mills GB, Gelfand EW (1987): Requirements for the simultaneous presence of phorbol esters and calcium ionophores in the expression of human T lymphocyte proliferation-related genes. J Immunol 139:1393–1399.

Kumar A, Moreau JL, Baran D, Theze J (1987): Evidence for negative regulation of T cell growth by low affinity interleukin 2 receptors. J Immunol 138:1485–1491.

Laing TJ, Weiss A (1988): Evidence for IL-2 independent proliferation in human T cells. J Immunol 140:1056–1062.

Larsson EL, Iscove NN, Coutinho A (1980): Two distinct factors are required for induction of T-cell growth. Nature 283:664–666.

Lerner A, Philosophe B, Miller RA (1988): Defective calcium influx and preserved inositol phosphate generation in T cells from old mice. Aging: Immunol Infect Dis 1:149–157.

Lesley JF, Schulte RJ (1985): Inhibition of cell growth by monoclonal anti-transferrin receptor antibodies. Mol Cell Biol 5:1814–1821.

Lowenthal JW, Cerottini J, MacDonald HR (1986): Interleukin T-dependent induction of both interleukin 2 secretion and interleukin 2 receptor expression by thymoma cells. J Immunol 137:1226–1231.

Malek TR, Chan C, Glimcher LH, Germain RN, Shevach EM (1985): Influence of accessory cell and T cell surface antigens on mitogen-induced IL2 receptor expression. J Immunol 135:1826–1832.

Marrack P, Kappler J (1987): The T cell receptor. Science 238:1073–1079.

Miller RA (1984): Age related decline in precursor frequency for different T cell mediated reactions, with preservation of helper or cytotoxic effect per precursor cell. J Immunol 132:63–68.

Miller RA (1986): Immunodeficiency of aging: Restorative effects of phorbol ester combined with calcium ionophore. J Immunol 137:805–808.

Miller RA, Jacobson B, Weil G, Simons ER (1987): Diminished calcium influx in lectin-stimulated T cells from old mice. J Cell Physiol 132:337–342.

Miller RA, Philosophe B, Ginnis I, Weil G, Jacobson B (1989): Defective control of cytoplasmic calcium concentration in T lymphocytes from old mice. J Cell Physiol 138:175–182.

Nagel JE, Chrest FJ, Adler WH (1982): Mitogenic activity of 12-0-tetradecanoyl phorbol-13-acetate on peripheral blood lymphocytes from young and aged adults. Clin Exp Immunol 49:217–224.

Nagel JE, Chopra RK, Chrest FJ, McCoy MT, Schneider EL, Holbrook NJ, Adler WH (1988): Decreased proliferation, interleukin 2 synthesis, and interleukin 2 receptor expression are accompanied by decreased mRNA expression in phytohemagglutinin-stimulated cells from elderly donors. J Clin Invest 81:1096–1102.

Nagel JE, Chopra RK, Powers DC, Adler WH (1989): Effect of age on the high affinity interleukin 2 receptor of phytohemagglutinin stimulated peripheral blood lymphocytes. Clin Exp Immunol (in press).

Negoro S, Hara H, Miyata S, Saiki O, Tanaka T, Yoshizaki K, Igarashi T, Kishomoto S (1986): Mechanism of age-related decline in antigen specific T cell proliferative response: IL2 receptor expression and recombinant IL2 induced proliferative purified TAC-positive T cells. Mech Aging Dev 36:223–241.

Negoro S, Hara H, Miyata S, Saiki O, Tanaka T, Yoshi K, Nishimoto N, Kishomoto S (1987): Age related changes of the function of T cell subsets: Predominant defects of the proliferative response in CD8 positive T cells subset in aged persons. Mech Ageing Dev 39:263–279.

Neidel JE, Kuhn LJ, Vandenbark GR (1983): Phorbol diester receptor copurifies with protein kinase C. Proc Natl Acad Sci USA 80:36–40.

Nishizuka Y (1988): The molecular heterogeneity of protein kinase C and its implications for cellular regulation. Nature 334:661–665.

Oettgen HC, Terhorst C (1987): A review of the structure and function of the T-cell receptor-T2 complex. CRC Crit Rev Immunol 7:131–166.

O'Leary J, Fox R, Bergh N, Rodysill KJ, Hallgren HM (1988): Expression of the human T cell antigen receptor complex in advanced age. Mech Aging and Dev 45:239–252.

Proust JJ, Filburn CR, Harrison SA, Buchholz MA, Nordin AA (1987): Age-related defect in signal transduction during lectin activation of murine T lymphocytes. J Immunol 139:1472–1478.

Proust JJ, Kittur DS, Buchholz MA, Nordin AA (1988): Restricted expression of mitogen-induced high affinity IL-2 receptors in aging mice. J Immunol 141:4209–4216.

Rabinowich H, Goses Y, Reshef T, Klajman A (1985): Interleukin-2 production and activity in aged humans. Mech Ageing Dev 32:213–226.

Robb RJ, Kutny RM (1987): Structure-function relationships for the IL2 receptor system. IV. Analysis of the sequence and ligand-binding properties of soluble Tac protein. J Immunol 139:855–862.

Robb RJ, Mayer PC, Garlick R (1985): Retention of biological activity following radioiodination of human interleukin 2: Comparison with biosynthetically labeled growth factor in receptor binding assays. J Immunol Methods 81:15–30.

Robb RJ, Rusk CM, Neeper MP (1988): Structure-function relationships for the interleukin 2 receptor: Location of ligand and antibody binding sites on the Tac receptor chain by mutational analysis. Proc Natl Acad Sci USA 85:5654–5658.

Roosnek EE, Brouwer MC, Aarden LA (1985a): T cell triggering by lectins. I. Requirements for interleukin 2 production; lectin concentration determines the accessory cell dependency. Eur J Immunol 15:652–656.

Roosnek EE, Brouwer MC, Aarden LA (1985b): T cell triggering by lectins. II. Stimulus for induction of interleukin 2 responsiveness and interleukin 2 production differ only in quantitative aspects. Eur J Immunol 15:657–661.

Rubin LA, Kuman CC, Fritz ME, Biddison WE, Boutin B, Yarchoan R, Nelson DL (1985): Soluble interleukin 2 receptors are released from activated human lymphoid cells in vitro. J Immunol 135:3172–3177.

Rubin LA, Jay G, Nelson DL (1986): The released interleukin 2 receptor binds interleukin 2 efficiently. J Immunol 137:3841–3844.

Saadeh C, Auzenne C, Nelson D, Orson F (1986): Sera from the aged contain higher levels of IL-2 receptor compared to young adults. Fed Proc 45:378.

Schwab R, Crow MK, Russo C, Weksler ME (1985a): Requirements for T cell activation by OKT3 monoclonal antibody: Role of modulation of T3 molecules and interleukin 1. J Immunol 135:1714–1718.

Schwab R, Hausman PB, Rinnooy-Kan E, Weksler ME (1985b): Immunological studies of aging. Impaired T lymphocytes and normal monocyte response from elderly humans to the mitogenic antibodies OKT3 and Leu4. Immunology 55:677–684.

Seigel LJ, Harper ME, Wong-Staal F, Gallo RC, Nash WB, O'Brien SJ (1984): Gene for T-cell growth factor: Location on human chromosome 4q and feline chromosome B1. Science 223:175–178.

Sharon M, Klausner RD, Cullen BR, Chizzonite R, Leonard WJ (1986): Novel interleukin-2 receptor subunit detected by cross-linking under high affinity conditions. Science 234:859–863.

Siegel JP, Sharon M, Smith PL, Leonard WJ (1987): The IL-2 receptor beta chain (p70): Role in mediating signals for LAK, NK, and proliferative activities. Science 238:75–78.

Smith KA (1988): Interleukin-2: Inception, impact, and implications. Science 240:1169–1176.

Smith KA, Cantrell DA (1985): Interleukin 2 regulates its own receptors. Proc Natl Acad Sci USA 82:864–868.

Spits H, Borst J, Tax W, Capel PJA, Terhorst C, deVries JE (1985): Characteristics of a monoclonal antibody (WT-31) that recognizes a common epitope on the human T cell receptor for antigen. J Immunol 135:1922–1928.

Staiano-Coico L, Darzynkiewicz Z, Melamed MR, Weksler ME (1984): Immunological studies of aging. IX. Impaired proliferation of T lymphocytes detected in elderly humans by flow cytometry. J Immunol 132:1788–1792.

Thoman ML, Weigle WO (1981): Lymphokines and aging: Interleukin 2 production and activity in aged animals. J Immunol 127:2102–2106.

Thoman ML, Weigle WO (1982): Cell-mediated immunity in aged mice: An underlying lesion in IL-2 synthesis. J Immunol 128:2358–2361.

Thoman ML, Weigle WO (1985): Reconstitution of in vivo cell mediated lympholysis responses in aged mice with interleukin 2. J Immunol 134:949–956.

Thoman ML, Weigle WO (1987): Age-associated changes in the synthesis and function of cytokines. In Goidl EA (ed): "Aging and the Immune Response." New York: Marcell Dekker, pp 199–223.

Thoman ML, Weigle WO (1988): Partial restoration of Con A-induced proliferation, IL-2 receptor expression, and IL-2 synthesis in aged murine lymphocytes by phorbol myristate acetate and ionomycin. Cell Immunol 114:1–11.

Treiger BF, Leonard WJ, Svetlik P, Rubin LA, Nelson DL, Greene WC (1987): A secreted form of human interleukin 2 receptor encoded by an "anchor minus" cDNA. J Immunol 136:4099–4105.

Truneh A, Albert F, Goldstein P, Schmitt-Verhulst AM (1985): Early steps of lymphocyte activation bypassed by synergy between calcium ionophores and phorbol ester. Nature 313:318–320.

Tse DB, Haideri MA, Pernis B, Cantor CR, Wang CY (1986): Intracellular accumulation of T-cell receptor complex molecules in a human T cell line. Science 234:748–751.

Tsien RY, Pozzan T, Rink TJ (1982): T-cell mitogens cause changes in cytoplasmic free Ca^{++} and membrane potential in lymphocytes. Nature 295:767–772.

Tsudo M, Kozak RW, Goldman CK, Waldmann TA (1987): Contribution of a p75 interleukin binding peptide to a high-affinity interleukin 2 receptor complex. Proc Natl Acad Sci USA 84:4215–4218.

Van Wauwe J, Goossens J (1981): Mitogenic actions of Orthoclone OKT3 on human peripheral blood lymphocytes: Effects of monocytes and serum components. Int J Immunopharmacol 3:203–208.
Vie H, Miller RA (1986): Decline with age in the proportion of mouse T cells that express IL-2 receptors after mitogen stimulation. Mech Ageing Dev 33:313–322.
Vissinga CS, Dirven CJ, Steinmeyer FA, Benner R, Boersma WJ (1987): Deterioration of cellular immunity during aging. The relationship between age-dependent impairment of delayed-type hypersensitivity reactivity, interleukin-2 production capacity and frequency of Thy-1$^+$, Lyt-2$^-$ cells in C57B1/Ka and CBA/Rij mice. Cell Immunol 108:323–334.
Wakasugi H, Bertoglio J, Tursz T, Fradelizi D (1985): IL2 receptor induction on human T lymphocytes: Role for IL2 and monocytes. J Immunol 135:321–327.
Weissman AJ, Harford J, Svetlik P, Leonard W, Depper J, Waldman TA, Greene WC, Klausner RD (1986): Only high affinity receptors for interleukin 2 mediate internalization of ligand. Proc Natl Acad Sci USA 83:1463–1466.
Whisler RL, Newhouse YG (1986): Function of T cells from elderly humans: Reductions of membrane events and proliferative responses mediated via T3 determinants and diminished elaboration of soluble T cell factors for B cell growth. Cell Immunol 99:422–433.
Wu W, Pahlavani M, Cheng HT, Richardson A (1986): The effect of aging on the expression of interleukin 2 messenger ribonucleic acid. Cell Immunol 100:224–231.
Yamada T, Fujishima A, Kawahara K, Kakto K, Nishimura O (1987): Importance of disulfide linkage for constructing the biologically active human interleukin-2. Arch Biochem Biophys 257:194–197.

Humoral Immunosenescence: An Update

David L. Ennist

INTRODUCTION

The previous reviews in this series [Nagel, 1983; Bender, 1985; Nagel and Proust, 1987] summarized data from descriptive studies aimed at documenting age-related changes in B lymphocytes. These studies have shown that changes in the B-cell compartment are primarily qualitative in nature. The total number of B cells and their total proportion appear to remain fairly constant with age [Weksler and Hutteroth, 1974; Callard et al., 1977; Adler and Nagel, 1981], although changes in serum immunoglobulins support the notion that changes in subpopulations of B cells take place [Makinodan and Kay, 1980; Kay and Makinodan, 1981]. Most groups have reported that serum IgG and IgA titers increase while serum IgM titers decline with age [Radl et al., 1975; Radl, 1981]. In humans, heightened IgG levels are due primarily to increases in IgG_1 and IgG_3 [Radl et al., 1975; Radl, 1981], while in mice IgG_1 and IgG_{2b} subclasses are preferentially increased [Haaijman et al., 1977]. Earlier studies have also catalogued age-associated increases in the prevalence of homogeneous antibody [Radl and Hollander, 1974; Radl et al., 1975; Radl, 1981; Weksler, 1981] and have illustrated the apparent paradox between decreased responsiveness to foreign antigens and the increased incidence of autoantibodies [Rowley et al., 1968; Hallgren et al., 1973; Barrett et al., 1980; Weksler, 1981].

The present review will update the data on humoral immunosenescence from 1986 through 1988 and will concentrate on studies delineating biochemical, cellular, and regulatory aberrations that are responsible for some of the phenomena seen in aging. Certainly, many earlier studies have outlined such mechanisms of immunosenescence, most notably in the area of auto-anti-idiotypic antibody regulation [Goidl et al., 1980; Szewczuk and Campbell, 1980;

Section on Molecular Genetics of Immunity, Laboratory of Developmental and Molecular Immunity, National Institute of Child Health and Human Development, NIH, Bethesda, Maryland 20892

Klinman, 1981]. Nevertheless, we are now in a position to delve deeper into the causes of the observed abnormalities and to use this knowledge in attempts to restore immune function in the elderly.

BIOCHEMICAL AND CELL BIOLOGICAL CHANGES IN B LYMPHOCYTES

Both translation and transcription decline with age [reviewed in Richardson and Semsei, 1987]. Tollefsbol and Cohen [1986], basing their hypothesis largely on studies involving phytohemagglutinin (PHA)-stimulated lymphocytes, have proposed that cellular aging is a genetically controlled process that results from decreased protein synthesis. In the immune system, this basic change would lead to declines in mitogenic potential, glycolysis [Tollefsbol and Cohen, 1985], and lymphokine and immunoglobulin production. The decline in protein synthesis is believed to be gradual, taking place over an extended period of time.

Cheung et al. [1987] examined the total actin content and its polymerization state in T and B lymphocytes from aging rats. They found that the total actin content did not change with age, but the ratio of polymeric to monomeric actin declined significantly in both cell types. This result provides a molecular explanation for the old observation of Woda and Feldman [1979] that there is a decline in capping of B-cell surface antigens in old rats. Diminished actin polymerization has broad implications for numerous other cellular functions that are dependent upon contractile elements, including motility and proliferation.

Parekh and colleagues [1988b] have reported that agalactosyl N-linked oligosaccharides of total serum IgG decreased from birth to a minimum at age 25 and then increased with age. They have also found an elevated incidence of agalactosyl IgG in both adult and juvenile rheumatoid arthritis [Parekh et al., 1988a]. Furthermore, galactosylation of IgG fluctuates with disease activity, and they have speculated that the altered immunoglobulin plays a seminal role in the inflammatory process. Recent studies by Casali et al. [1987] and Hardy et al. [1987] have shown in humans that rheumatoid factor-producing cells belong to the CD5 positive B-cell subpopulation. Finally, Stall et al. [1988] have found clonal CD5 bearing B cells in aging mice of the immunologically normal Balb/c, C56BL/6, and CBA strains. Although rheumatoid factor is predominantly of the IgM isotype and CD5 B cells are primarily IgM secretors, there may be a connection between these cells and agalactosyl IgG.

The production of homogeneous antibody, known as idiopathic paraproteinemia or monoclonal gammopathy, is a benign condition commonly seen in old mice and people. Radl [1981] has argued that the onset of this condition is due to a regulatory T-cell defect because its appearance generally accompanies a deficiency in the T-cell system. The maintenance of the condition, however, appears to be a function of the B cell itself because it persists when the paraprotein-producing clone is transferred to young, nonirradiated syngeneic recipients. Radl and colleagues have recently extended these observations by showing a genetic basis for the development of the condition in F1 hybrid mice derived from strains with high and low frequencies for the development of idiopathic paraproteinemia [Radl et al., 1985]. They found that the level of homogeneous immunoglobulin in the F1 mice was intermediate to the levels expressed by the parental strains and that the heavy chain allotype of the homogeneous antibody was predominantly of the high-frequency strain. It is possible that CD5 B cells, because they arise clonally [Stall et al., 1988], are responsible for producing the paraprotein.

Sidman and colleagues [1987] used the cytofluorograph to determine not only the proportions of splenocytes bearing specific antigens but also the density distributions of the antigens on the positive cells. They confirmed previous results showing no apparent changes in the proportions of B and T cells and also showed that the expression of class I and class II major histocompatibility complex (MHC) antigens increases throughout life. The data suggested that the elevated class I expression occurs on the majority of splenocytes. In contrast, the elevated class II expression was clearly attributable to B cells. The increased MHC antigen expression was found to be functionally significant as assessed by mixed lymphocyte culture and long term T-cell lines reactive to self MHC determinants. A number of studies have shown that the level of MHC antigen expression is important to the generation of specific immune responses [Heber-Katz et al., 1982; Conrad et al., 1982; Janeway et al., 1983]. Thus, antigen presentation may become more efficient with age, leading to the activation of T-cell clones that are silent in young animals.

The abilities of B cells from young and old people to form colonies in semisolid cultures was investigated by Whisler et al. [1985]. They found that about 75% of aged adults displayed decreased colony responses. Approximately half of the poor responses were improved by supplementing the cultures with indomethacin, indicating that the B-cell responses in these people were sensitive to prostaglandins. In the remaining individuals, indomethacin had no effect, but the capping of B-cell surface immunoglobulin was impaired. In related experiments, impaired B-cell differentiation from some, but not all, elderly subjects could be improved by allogeneic T cells from young subjects [Ennist et al., 1986]. These data illustrate the multicentric nature of immunosenesence. For any given immunologic assay, there are likely to be several mechanisms responsible for a deficient response.

Ennist et al. [1986] used *Staphylococcus aureus* Protein A (SpA)-stimulated allogeneic cocultures to assess B-cell growth and differentiation under conditions that minimized in vitro suppressor cell activity. B cells from old people demonstrated proliferative responses that were equivalent to those of young people. In contrast, IgM production by B cells from most elderly subjects was substantially impaired even when cocultured with T cells from young donors. Thus, B-cell activation and proliferation in the elderly are dissociated from the ability of the cells to differentiate into IgM-secreting cells.

Hara et al. [1987] studied the ability of highly purified B cells from young and aged subjects to repeat proliferation following stimulation with killed *S. aureus* Cowan I (SAC). They used colchicine block and tritiated thymidine uptake to show that the original clone size of SAC-responding B cells was unimpaired. In contrast, B cells from aged subjects showed a twofold lower proliferative response. They concluded that B cells from the elderly showed a decreased capacity for repeated replication. The discrepancies between the results of Ennist et al. [1986] and Hara et al. [1987] can be explained by the different culture systems employed. In the study by Ennist and colleagues, SpA was used to assess B-cell proliferation in the presence of T cells and monocytes. In contrast, Hara et al. used SAC to stimulate relatively pure populations of B cells.

Snow (1987) purified trinitrophenyl antigen binding cells (TNP-ABC) from three strains of young and old mice and compared their responses to T-independent and T-dependent antigens. In agreement with previous studies, B-cell proliferative responses showed little change with age [Kim et al., 1982; Ennist et al., 1986]. Surprisingly, TNP-ABC from young and old mice also failed to differ in their TNP-specific plaque-forming capabilities. Thus, at this level of analysis, B cells appear to maintain normal function with age, and deficient humoral responses may be the result of alterations in T-cell function and regulatory networks. One should be careful not to overinterpret this data, as there are a few technical difficulties with the TNP-ABC isolation procedure. It is possible that the procedure selects for fully functional cells with high affinity for TNP. Low-affinity TNP-specific B cells may have been discarded. In this regard, a number of investigators have demonstrated an age-associated decline in antibody avidity [Goidl et al., 1976; Kishimoto et al., 1976; Doria et al., 1978; Kay and Makinodan, 1981; Weksler, 1981]. In addition, Snow confined his studies of B-cell differentiation to plaque-forming cells. It remains unclear whether there are any age-associated changes in the amount of antibody secreted by individual TNP-specific plasma cells in this system.

The effects of aging on the ability of the bone marrow to generate B cells were examined by Francus et al. [1986] and Zharhary [1988]. Francus et al. [1986] used allotype congenic mouse strains to study the sIgD-positive B-cell population following bone marrow reconstitution of irradiated, thymectomized recipients. They found that bone marrow from aged donors could replenish the peripheral B-cell population as effectively as bone marrow from young donors. However, in competition experiments in which a single recipient was reconstituted with bone marrow from young and old donors, cells of young donor origin predominated in the peripheral B-cell population even though the bone marrow contained similar numbers of B cells bearing young or old donor allotypes. Zharhary [1988] used in vitro techniques to come to a similar result. By depleting mature B cells from bone marrow and then following their reappearance in culture, she found that bone marrow cells from aged mice were significantly inferior to young cells. Removal of T cells from the bone marrow of aged mice did not restore the frequency of newly generated B cells to young adult levels. Finally, culture-generated B cells from old mice demonstrated reduced lipopolysaccharide responses. These studies support the idea that the age-associated reduction in antigen-responsive B cells is caused by the generation of defective B cells in the aged.

This conclusion has recently been strengthened by two studies using 7-methyl-8-oxoguanosine (7m8oGuo), a potent B-cell activator. Weigle et al. [1987] found that 7m8oGuo improved the anti-human gamma globulin antibody response of aged mice to the levels obtained in young mice injected with the immunogen alone. In another report [Weigle et al., 1988] they examined the age-related resistance to tolerance induction. They showed that 7m8oGuo could retrieve a population of B cells that had become functionally deleted during aging and provided evidence that aged mice contain two populations of B cells. One population is resistant to tolerance induction and remains functional during aging. A second population of B cells that is easily tolerized becomes unresponsive during aging and can be restored by 7m8oGuo.

Zharhary and Klinman [1986a] found evidence for a selective increase in the frequency of phosphorylcholine-specific splenic B cells in aging mice. In contrast, most other studies have shown that splenic precursor frequencies for specific antigens decline with age [Zharhary and Klinman, 1983, 1986b]. This study raises the possibility that aging may be accompanied by changes in variable gene usage. It also shows that there are exceptions to the general rule of age-associated diminished immune responsiveness.

CHANGES IN THE REGULATION OF HUMORAL IMMUNITY

Goidl et al. [1980], Szewczuk and Campbell [1980], and Klinman [1981] documented increases in auto-anti-idiotypic antibody production in aged mice and suggested that this phenomenon results in the down-regulation of B-cell responses [reviewed in Wade and Szewczuk, 1984]. Current evidence indicates that the B-cell repertoire expressed by the bone marrow of mice does not change appreciably with age. Rather, changes in the expressed B-cell repertoire appear to take place in the spleens of old mice under the influence of peripheral T cells [Kim et al., 1985; Doria et al., 1987; reviewed in Wade and Szewczuk, 1984]. In fact, the frequencies of antigen-specific B cells in T-cell-deficient nude mice do not change with age [Zharhary, 1986]. Kim et al. [1985]

irradiated old mice with bone marrow shielding, allowed the peripheral lymphoid system to regenerate, and found that the resulting auto-anti-idiotypic antibody response resembled the pattern observed in young mice. If the mice were reconstituted with old T cells during recovery from the shielded irradiation, the auto-anti-idiotypic antibody response was similar to the pattern seen in aged mice.

The discovery of CD5 B-cell clones in aging mice [Stall et al., 1988] is a significant finding, as these cells exhibit an apparent preference for self antigens and have been associated with the development of autoimmunity [reviewed in Hardy and Hayakawa, 1986]. CD5 B cells are further distinguished by their early appearance during ontogeny. Whereas conventional B cells are continuously replenished from immunoglobulin-negative precursors, CD5 B cells represent a long-lived self-replenishing population that is maintained by division of immunoglobulin-positive cells. It has been suggested that the long-lived nature of the CD5 B cells predisposes them toward growth deregulation and neoplasia [Stall et al., 1988]. Okumura et al. [1982] have presented evidence in the murine system that a Thy-1 negative, Ly-1 (i.e., CD5)-positive immunoglobulin-bearing cell functions in idiotype regulation, and others have speculated that CD5 B cells could constitute an idiotype network [Hardy and Hayakawa, 1986]. Thus, disregulation of CD5 B cells may be responsible for the age-associated increase in auto-anti-idiotypic regulation.

Most studies have shown only minor age-associated changes in the relative proportions of total T cells, CD4 positive (helper/inducer) T cells, or CD8 positive (suppressor/cytotoxic) T cells, and it is apparent that alterations in T-cell numbers are not sufficient to account for the decline in T-cell regulatory functions commonly seen in the aged [reviewed in Nagel, 1983; Wade and Szewczuk, 1984; Jones and Ennist, 1985; Gottesman, 1987]. This has made investigators search for other causes of the age-associated aberrations in helper and suppressor T-cell function.

Suppressor cell function has variously been reported to be increased [Goidl et al., 1976; Segre and Segre, 1976; Antel et al., 1978; Callard et al., 1980; DeKruyff et al., 1980; Delfraissy et al., 1982; Kim et al., 1982], decreased [Hallgren and Yunis, 1977; Pahwa et al., 1981; Thoman and Weigle, 1983; Gottesman et al., 1984], or unaffected [Barrett et al., 1980; Hollingsworth and Otte, 1981] by age. Gottesman [1987] has suggested that these discrepancies have arisen from the diversity of assay systems used. It is likely that the conflicting results reflect the activities of different populations of suppressor cells. Gottesman and colleagues [1988] have uncovered three populations of suppressor cells that are active in the mixed lymphocyte response (MLR) and in the generation of cytotoxic cells. In these systems, the in vitro generation of specific suppressor T cells and nonspecific, non-T suppressor cells declines with age. In contrast, the activity of a pre-existing suppressor cell isolated from the spleen without in vitro culture was shown to be higher in older animals. At present, it is unknown whether these results can be extended to the suppression of antibody responses.

Some potentially useful information may be gained by considering recent studies that have examined regulatory changes in the autoimmune NZB and (NZB × NZW)F_1 (B/W) strains. McCoy et al. [1985] have found evidence for a decline in the ability of suppressor T cells to specifically induce B-cell tolerance as NZB mice age. The same group has also found that old NZB mice undergo a decline in helper T-cell activity and an increase in amplifier T-cell activity [Baker et al., 1986]. In their hands, helper T cells are required to initiate an immune response to some antigens. These cells are distinguished from amplifier T cells that drive antigen-stimulated B cells to proliferate and differentiate further. Interestingly, the decline in suppressor T-cell activity and the increase in amplifier T-cell activity coincided with the onset of autoimmunity [Baker et al., 1986].

Sekigawa et al. [1987] found evidence for immunoglobulin class-specific suppressor T cells in B/W hybrid mice. The in vitro production of anti-DNA antibody of the IgG class was inhibited by suppressor T cells from young but not aged B/W mice. Conversely, IgM anti-DNA antibody production was inhibited by aged but not young suppressor T cells. If this finding can be applied to immunologically normal mice, it may provide an explanation for the observation that serum IgG levels increase with age while serum IgM levels decline.

In a recent series of papers using young adult mice, Coico et al. [1985a, 1985b, 1987a] have shown that IgD, gamma interferon (IFN), and interleukin-2 (IL2) induce IgD receptors on a population of CD4-positive T cells that is responsible for the augmentation of antibody responses in vivo. In contrast, they found old mice to be deficient in their ability to generate IgD receptor-bearing T cells following exposure to IgD, gamma IFN, or IL2. In addition, IgD injections failed to augment the antibody response of old mice. Therefore, it seems that reduced antibody responses in old age may, at least in part, be due to deficient induction of IgD receptor-bearing helper T cells.

One way in which helper T cells provide a stimulus for the generation of humoral responses is by the elaboration of soluble helper factors or lymphokines. Recent studies [reviewed by O'Garra et al., 1988] have shown that virtually all of these factors exhibit significant effects on the growth and/or differentiation of B lymphocytes. It is now apparent that IL2, formerly known as T-cell growth factor, has direct effects on the growth and differentiation of B cells in both murine and human systems. Coico et al. [1987b] have shown that IL2 can improve helper T-cell function in old mice and thus indirectly aid in the restoration of humoral immunity. It is also likely that the age-associated decline in IL2 production [Miller and Stutman, 1981; Thoman and Weigle, 1981; Gillis et al., 1981; Gilman et al., 1982; Wu et al., 1986] is in part directly responsible for the deficient humoral responses of the aged. In addition to decreased IL2 production, T cells from old animals exhibit impaired expression of high-affinity IL2 receptors [Proust et al., 1988] and a defective signal transduction mechanism [Proust et al., 1987]. It remains to be seen whether aged B cells also have defective receptor expression and signal transducing mechanisms.

Reports of age-associated changes in lymphokines other than IL2 have been scarce, particularly for B-cell stimulatory factors. Whisler and Newhouse [1986] found that human SpA-stimulated T cells from aged subjects produced lower levels of soluble B-cell growth factors than similarly stimulated T cells from young donors. Winchurch et al. [1987] found that supplemental zinc in vitro increased the production of a B-cell growth factor (BSF-1) by splenocytes from 24-month-old mice. In contrast, Hara et al. [1987] found that phytohemagglutinin-stimulated T cells from elderly people produced threefold higher B-cell differentiation (BCDF) activity when compared with young controls. These data imply that aging is accompanied by decreased production of B-cell growth factors and increased production of B-cell differentiation factors. Interestingly, Hara et al. [1987] also reported that aged T cells produced tenfold less IL2 activity and showed an inverse correlation between IL2 and BCDF when both activities were measured in the same samples. This finding brings up the interesting possibility that aging is associated with differential effects on two lymphokine-producing helper T-cell subpopulations. Murine helper T-cell clones have recently been classified on the basis of the pattern of lymphokines produced [reviewed in Coffman et al., 1988]. Hara et al. [1987] also reported that the spontaneous secretion of BCDF activity increases with age. The increased secretion of BCDF may contribute to the monoclonal gammopathy and, more significantly, may also result in the increased incidence of autoantibody production seen in aging individuals. In this regard, the autoantibody secretion associated with autoimmune mice and certain human tumors has been attributed to heightened levels of BCDF secretion or increased sensitivity of aged B cells to the lymphokine [Dobashi et al., 1987; Herron et al., 1988; Hirano et al., 1986, 1987].

In addition to IL2 and the B-cell stimulatory factors described above, the production of IL1, IL3, and gamma IFN have also been studied in aging animals. All three of these lymphokines affect the development of humoral immunity. IL1 promotes the growth and differentiation of activated B cells, IL3 supports the growth of certain pre-B cell lines, and gamma IFN has numerous effects on the growth and differentiation of B cells [reviewed in O'Garra et al., 1988].

As was the case in a previous review in this series [Proust, 1987], the data on gamma IFN production remain controversial. In rodents, in vivo IFN production declined [DeMaeyer and DeMaeyer-Guignard, 1968], and lectin-induced gamma IFN production increased [Heine and Adler, 1977; Saxena et al., 1988] with age. Elderly humans have been reported to exhibit normal [Canonica et al., 1985; Weifeng et al., 1986] or decreased [Abb et al., 1984] gamma IFN production. The data on IL1 production in aging also remain unclear. Peritoneal macrophages from old mice appear to produce less IL1 than cells from young mice [Bruley-Rosset and Vergnon, 1984; Inamizu et al., 1985]. Winchurch et al. [1987] found that supplemental zinc increased IL1 production by 300% in cultures derived from aged mouse splenocytes. In contrast, the production of IL1 by human peripheral blood monocytes does not change appreciably with age [Jones et al., 1984; Whisler et al., 1985]. The reasons for these contradictory results are unclear, although differences in cell sources and assay conditions may be responsible. These findings may be resolved by examining messenger RNA levels rather than relying on biologic assays of lymphokine production. Chang et al. [1988] found that spleen cells from old mice produced significantly less IL3 than cells from young mice. Furthermore, by using young–old cell mixtures and antibody treatments, they were unable to attribute the decline in IL3 production to suppressor cells or adherent cells. Instead, the age-associated decline in IL3 production appeared to be due to changes in IL3-producing T cells. They reported a strong positive correlation between the production of IL2 and IL3, indicating that these lymphokines are affected in a similar manner by age.

CONCLUSIONS

The studies summarized in this review have shown numerous examples of biochemical and cell biological changes in B lymphocytes derived from aging animals. In particular, the ratio of polymeric to monomeric actin in B cells from old rats is lower than in young B cells [Cheung et al., 1987]. This finding has broad implications for cellular functions that are dependent upon contractile elements, including capping of surface antigens, motility, and division. Agalactosyl IgG has been reported to increase in humans over the age of 25 [Parekh et al., 1988b]. The significance of this finding for IgG function is presently unclear, although the level of agalactosyl IgG was also found to fluctuate with inflammatory episodes in rheumatoid arthritis [Parekh et al., 1988a]. The study by Sidman and colleagues [1987] found that the densities of splenocyte cell surface MHC class I and class II antigens increased with age. This may result in the activation of otherwise silent T-cell clones and may have an effect on the development of autoimmunity in the elderly. Certainly, the efficiency of antigen presentation by B cells derived from old animals needs to be evaluated.

Several recent studies have illustrated the multicentric nature of the decline in immunocompetence. Ennist et al. [1986] found that impaired IgM production in some, but not all elderly subjects could be improved by coculture with

T cells derived from young donors. Likewise, Whisler et al. [1985] found that indomethacin supplementation improved the colony-forming responses of B cells from about half of their elderly donors. In most of the remaining individuals, the capping of cell surface immunoglobulin was impaired. Radl et al. [1985] found a genetic basis for the expression of idiopathic paraproteinemia. Thus, immunosenescence is very much an individual phenomenon such that the expression of a given deficiency can result from any one of several possible biochemical or cellular causes.

Auto-anti-idiotypic antibody levels are higher in old than in young animals [Goidl et al., 1980; Szewczuk and Campbell, 1980; Klinman, 1981]. Kim et al. [1985] showed that the altered pattern of anti-idiotypic antibody levels in old mice is not due to altered repertoire expression among bone marrow precursor B cells. Rather, the increased expression of anti-idiotypic antibody seen in old animals results from the interaction of emerging B cells with peripheral T cells.

Most studies have shown little change in bone marrow precursor frequencies for specific antigens [Zharhary and Klinman, 1983, 1986b]. These studies also found that the frequency of mature antigen-specific B cells declines significantly in old age, and this effect has been attributed to increased auto-anti-idiotypic antibody regulation. In contrast, it appears that the frequencies of both precursor and mature phosphorylcholine-specific B cells increase with age [Zharhary and Klinman, 1986a]. It is thus possible that alterations in variable gene usage contribute to the age-associated decline in humoral immunity.

Except for IL2, reports of age-associated changes in the production and effects of lymphokines have been few and far between. This is something of a mystery, as most of the known factors have been cloned and are commercially available. Studies at the level of messenger RNA are notably lacking. Nevertheless, a number of tentative conclusions can be drawn from the few published reports. The data show that production of B-cell growth factors declines while the production of B-cell differentiation factors increases with age [Whisler and Newhouse, 1986; Hara et al., 1987]. Although the exact significance of these findings is difficult to evaluate, decreased levels of growth factors may result in deficient responses to new antigens. In contrast, heightened secretion of differentiation factors coupled with alterations in suppressor T-cell-mediated tolerance may be related to the induction of autoantibodies in the elderly.

Several investigators have suggested that zinc depletion plays an important role in the expression of immunosenescence [reviewed in Antonaci et al., 1987]. Indeed, even mild zinc deficiency in adolescent rhesus monkeys was found to depress immunoglobulin production [Golub et al., 1988]. This possibility is particularly significant for the elderly because many are taking diuretics that are known to induce zinc depletion [Rikans, 1986]. Thus, the finding that zinc supplementation can boost the production of IL1 and a B-cell growth factor may be quite significant [Winchurch et al., 1987].

A major finding of the past few years is the apparent differential effects of aging on subpopulations of B cells, suppressor cells, and helper T cells. The production of IL2 and IL3 declines with age (Chang et al., 1987), whereas the production of BCDF increases with age (Hara et al., 1987). Is it possible that IL2 and IL3 are produced by Baker's helper T cell, whose activity declines in aged NZB mice (Baker et al., 1986)? Conversely, the activity of Baker's amplifier T cell increases in aging NZB mice. Do these cells produce BCDF? Gottesman et al. (1988) have presented evidence that three types of suppressor cells, differentially affected by aging, regulate mixed lymphocyte and cytotoxic responses. Can these observations be extended to the regulation of humoral immunity in aging animals? Finally, several investigators have shown that B-cell populations change with age. The effects of CD5 B cells on idiotype regulation and autoantibody production need to be explored. The roles of these cell populations in humoral immunosenescence should prove to be a fertile field of study during the next few years.

REFERENCES

Abb J, Abb H, Deinhardt F (1984): Age-related decline of human interferon alpha and interferon gamma production. Blut 48:285–289.
Adler WH, Nagel JE (1981): Studies of immune function in a human population. In Segre D, Smith L (eds): "Immunological Aspects of Aging." New York; Marcel Dekker, pp 295–311.
Antel JP, Weinrich M, Arnason BGW (1978): Circulating suppressor cells in man as a function of age. Clin Immunol Immunopathol 9:134–141.
Antonaci S, Jirillo E, Bonomo L (1987): Immunoregulation in aging. Diagn Clin Immunol 5:55–61.
Baker PJ, Fauntleroy MB, Stashak PW, McCoy KL, Chused TM (1986): Increased amplifier T cell activity in autoimmune NZB mice and its possible significance in the expression of autoimmune disease. Immunobiology 171:400–411.
Barrett DJ, Stenmark S, Wara DW, Ammann AJ (1980): Immunoregulation in aged humans. Clin Immunol Immunopathol 17:203–211.
Bender BS (1985): B lymphocyte function in aging. In Rothstein M (ed): "Review of Biological Research in Aging," Vol. 2. New York: Alan R. Liss, Inc., pp 143–154.
Bruley-Rosset M, Vergnon I (1984): Interleukin-1 synthesis and activity in aged mice. Mech Ageing Dev 24:247–264.
Callard RE, Basten A, Waters LK (1977): Immune function in aged mice. II. B-cell function. Cell Immunol 31:26–36.
Callard RE, de St Groth BF, Basten A, McKenzie IF (1980): Immune function in aged mice. V. Role of suppressor cells. J Immunol 124:52–58.
Canonica GW, Ciprandi G, Caria M, Dirienzo W, Shums A, Norton-Koger B, Fudenberg HH (1985): Defect of autologous mixed lymphocyte reaction and interleukin-2 in aged individuals. Mech Ageing Dev 32:205–212.
Casali P, Burastero SE, Nakumura M, Inghirami G, Notkins AL (1987): Human lymphocytes making rheumatoid factor and antibody to ssDNA belong to Leu-I+ subset. Science 236:77–81.

Chang MP, Utsuyama M, Hirokawa K, Makinodan T (1988): Decline in production of IL3 with age in mice. Cell Immunol 115:1–12.
Cheung HT, Rehwaldt CA, Twu JS, Liao NS, Richardson A (1987): Aging and lymphocyte cytoskeleton: Age-related decline in the state of actin polymerization in T lymphocytes from Fischer F344 rats. J Immunol 138:32–36.
Coffman RL, Seymour BWP, Lebman DA, Hiraki DD, Christiansen JA, Shrader B, Cherwinski HM, Savelkoul HFJ, Finkelman FD, Bond MW, Mosmann TR (1988): The role of helper T cell products in mouse B cell differentiation and isotype regulation. Immunol Rev 102:5–28.
Coico RF, Xue B, Wallace D, Pernis B, Siskind GW, Thorbecke GJ (1985a): T cells with surface receptors for IgD. Nature 316:744–746.
Coico RF, Xue B, Wallace D, Siskind GW, Thorbecke GJ (1985b): Physiology of IgD. VI. Transfer of the immunoaugmenting effect of IgD with T-delta containing helper cell populations. J Exp Med 162:1852–1861.
Coico RF, Berzofsky JA, York-Jolley J, Ozaki S, Siskind GW, Thorbecke GJ (1987a): Physiology of IgD. VII. Induction of receptors for IgD on cloned T cells by IgD and interleukin-2. J Immunol 138:4–6.
Coico RF, Gottesman SRS, Siskind GW, Thorbecke GJ (1987b): Physiology of IgD. VIII. Age-related decline in the capacity to generate T cells with receptors for IGD and partial reversal of the defect with IL2. J Immunol 138:2776–2781.
Conrad PJ, Lerner EA, Murphy DB, Jones PP, Janeway CA Jr (1982): Differential expression of Ia glycoprotein complexes in F1 hybrid mice detected with alloreactive cloned T cell lines. J immunol 129:2616–2620.
DeKruyff R, Kim YT, Siskind GW, Weksler ME (1980): Age-related changes in the *in vitro* immune response: Increased suppressor activity in immature and aged mice. J Immunol 125:142–147.
Delfraissy JF, Galanaud P, Wallon C, Balavoine JF, Dormont J (1982): Abolished *in vitro* antibody response in the elderly: Exclusive involvement of prostaglandin-induced T suppressor cells. Clin Immunol Immunopathol 24:377–385.
DeMaeyer E, DeMaeyer-Guignard J (1968): Influence of animal genotype and age on the amount of circulating IFN induced by Newcastle disease virus. J Gen Virol 2:445–447.
Dobashi K, Ono S, Murakami S, Takahama Y, Katoh Y, Hamaoka T (1987): Polyclonal B cell activation by a B cell differentiation factor, B151-TRF2. III. B151-TRF2 as a B cell differentiation factor closely associated with autoimmune disease. J Immunol 138:780–787.
Doria G, D'Agostaro G, Poretti A (1978): Age-dependent variations of antibody avidity. Immunology 35:601–611.
Doria G, Mancini C, Frasca D, Adorini L (1987): Age restriction in antigen-specific immunosuppression. J Immunol 139:1419–1425.
Ennist DL, Jones KH, St Pierre RL, Whisler RL (1986): Functional analysis of the immunosenescence of the human B cell system: Dissociation of normal activation and proliferation from impaired terminal differentiation into IgM immunoglobulin secreting cells. J Immunol 136:99–105.
Francus T, Chen YW, Staina-Coico L, Hefton JM (1986): Effect of age on the capacity of bone marrow and spleen cells to generate B cells. J Immunol 137:2411–2417.
Gillis S, Kozak R, Durante M, Weksler ME (1981): Immunological studies of aging. Decreased production of and response to T cell growth factor by lymphocytes from aged humans. J Clin Invest 67:937–942.
Gilman SC, Rosenberg JS, Feldman JD (1982): T lymphocytes of young and aged rats. II. Functional defects and the role of interleukin 2. J Immunol 128:644–650.
Goidl E, Innes J, Weksler M (1976): Immunologic studies of aging. II Loss of IgG and high avidity plaque-forming cells and increased suppressor cell activity in aging mice. J Exp Med 144:1037–1048.

Goidl E, Thorbecke GJ, Weksler M, Siskind GW (1980): Production of auto-anti-idiotypic antibody during the normal immune response: Changes in the auto-anti-idiotypic antibody response and idiotype repertoire associated with aging. Proc Natl Acad Sci USA 77:6788–6792.

Golub MS, Gershwin ME, Hurley LS, Hendrickx AG (1988): Studies of marginal zinc deprivation in rhesus monkeys. VIII. Effects in early adolescence. Am J Clin Nutr 47:1046–1051.

Gottesman SRS (1987): Changes in T cell mediated immunity with age: An update. In Rothstein M (ed): "Review of Biological Research in Aging," Vol. 3. New York: Alan R. Liss, Inc., pp 95–127.

Gottesman SRS, Walford RL, Thorbecke GJ (1984): Proliferative and cytotoxic immune functions in aging mice. II. Decreased generation of specific suppressor cells in alloreactive cultures. J Immunol 133:1782–1787.

Haaijman J, Berg P, Brinkhof J (1977): Immunoglobulin class and subclass levels in the serum of CBA mice throughout life. Immunology 32:923–927.

Hallgren HM, Yunis EJ (1977): Suppressor lymphocytes in young and aged humans. J Immunol 118:2004–2008.

Hallgren HM, Buckley CE III, Gilbertsen VA, Yunis EJ (1973): Lymphocyte phytohemagglutinin responsiveness, immunoglobulins and autoantibodies in aging humans. J Immunol 111:1101–1107.

Hara H, Negoro S, Miyata S, Saiki O, Yoshizaki K, Tanaka T, Igarashi T, Kishimoto S (1987): Age associated changes in proliferative and differentiative response of human B cells and production of T cell-derived factors regulating B cell functions. Mech Ageing Dev 38:245–258.

Hardy RR, Hayakawa K (1986): Development and physiology of Ly-1 B and its human homolog Leu-1 B. Immunol Rev 93:53–79.

Hardy RR, Hayakawa K, Shimizu M, Yamasaki K, Kishimoto T (1987): Rheumatoid factor secretion from human Leu-1+ B cells. Science 236:81–83.

Heber-Katz E, Schwartz RH, Matis LA, Hannum C, Fairwell T, Appella E, Hansburg D (1982): Contribution of antigen-presenting cell major histocompatibility complex gene products to the specificity of antigen-induced T cell activation. J Exp Med 155:1086–1099.

Heine JW, Adler WH (1977): The quantitative production of interferon by mitogen stimulated mouse lymphocytes as a function of age and its effect on the lymphocyte proliferative response. J Immunol 118:1366–1369.

Herron LR, Coffman RL, Bond MW, Kotzin B (1988): Increased autoantibody production by NZB/NZW B cells in response to IL5. J Immunol 141:842–848.

Hirano T, Yasukawa K, Harada H, Taga T, Watanabe Y, Matsuda T, Kashiwamura S, Nakajima K, Koyama K, Iwamatsu A, Tsunasawa S, Sakiyama F, Hatsui H, Takahara Y, Taniguchi T, Kishimoto T (1986): Complementary DNA for a novel human interleukin (BSF-2) that induces B lymphocytes to produce immunoglobulin. Nature 324:73–76.

Hirano T, Taga T, Yasukawa K, Nakjima K, Nakano N, Takatsuki F, Shimizu M, Murashima A, Tsunasawa S, Sakiyama F, Kishimoto T (1987): Human B cell differentiation factor defined by an anti-peptide antibody and its possible role in autoantibody production. Proc Natl Acad Sci USA 84:228–231.

Hollingsworth JW, Otte RG (1981): B lymphocyte maturation in cultures from blood of elderly men: A comparison of plaque-forming cells, cells containing intracytoplasmic immunoglobulin and cell proliferation. Mech Ageing Dev 15:9–18.

Inamizu T, Chang M-P, Makinodan T (1985): Influence of age on the production and regulation of interleukin-1 in mice. Immunology 55:447–455.

Janeway CA, Conrad PJ, Tite J, Jones B, Murphy D (1983): Efficiency of antigen presentation differs in mice differing at the Mls locus. Nature 306:80–82.

Jones KH, Ennist DL (1985): Mechanisms of age-related changes in cell-mediated immunity. In Rothstein M (ed): "Review of Biological Research in Aging," Vol. 2. New York: Alan R. Liss, Inc., pp 155–157.

Jones PG, Kauffman CA, Bergman AG, Hayes CM, Kluger MJ, Cannon JG (1984): Fever in the elderly. Production of leukocytic pyrogen by monocytes from elderly persons. Gerontology 30:182–187.

Kay MMB, Makinodan T (1981): Relationship between aging and the immune system. Prog Allergy 29:134–181.

Kim YT, Siskind GW, Weksler ME (1982): Cellular basis of the impaired immune response in elderly people. In Fauci AS, Ballieux RE (eds): "Human B Lymphocyte Function: Activation and Immunoregulation." New York: Raven Press, pp 129–139.

Kim YT, Goidl EA, Samarut C, Weksler ME, Thorbecke GW (1985): Bone marrow function. I. Peripheral T cells are responsible for the increased auto-anti-idiotypic antibody response of older mice. J Exp Med 161:1237–1242.

Kishimoto S, Takahama T, Mizumachi H (1976): *In vitro* immune response to the 2,4,6-trinitrophenyl determinant in aged C57B1/6J mice: Changes in the humoral immune response to, avidity for the TNP determinant and responsiveness to LPS effect with aging. J Immunol 116:294–300.

Klinman NR (1981): Antibody specific immunoregulation and the immunodeficiency of aging. J Exp Med 154:547–551.

Makinodan T, Kay MMB (1980): Age influence on the immune system. Adv Immunol 29:287–330.

McCoy KL, Baker PJ, Stashak PW, Chused TM (1985): Suppressor T cells in old New Zealand Black mice are involved in the loss of low dose paralysis to type III pneumococcal polysaccharide. J Immunol 135:2438–2442.

Miller RA, Stutman O (1981): Decline, in aging mice, of the anti-2,4,6-trinitrophenyl cytotoxic T cell response attributable to loss of Lyt 2-, interleukin-2 producing helper cell function. Eur J Immunol 11:751–756.

Nagel JE (1983): Immunology. In Rothstein M (ed): "Review of biological Research in Aging," Vol. 1. New York: Alan R. Liss, Inc., pp 103–160.

Nagel JE, Proust JJ (1987): Age-related changes in humoral immunity, complement, and polymorphonuclear leukocyte function. In Rothstein M (ed): "Review of Biological Research in Aging," Vol. 3. New York: Alan R. Liss, Inc., pp 147–159.

O'Garra A, Umland S, DeFrance T, Christiansen J (1988): "B cell factors" are pleiotropic. Immunol Today 9:45–54.

Okumura K, Hayakawa K, Tada T (1982): Cell-to-cell interaction controlled by immunoglobulin genes. Role of Thy-1-, Lyt-1 + , Ig + (B') cell in allotype-restricted antibody production. J Exp Med 156:443–453.

Pahwa SG, Pahwa RN, Good RA (1981): Decreased *in vitro* humoral response in aged humans. J Clin Invest 67:1094–1102.

Parekh RB, Roitt IM, Isenberg DA, Dwek RA, Ansell BM, Rademacher TW (1988a): Galactosylation of IgG associated oligosaccharides: Reduction in patients with adult and juvenile onset rheumatoid arthritis and relation to disease activity. Lancet 1:966–969.

Parekh RB, Roitt IM, Isenberg DA, Dwek RA, Rademacher TW (1988b): Age-related galactosylation of the N-linked oligosaccharides of human serum IgG. J Exp Med 167:1731–1736.

Proust JJ (1987): Signal transduction, lymphokine production, receptor expression and the immunology of aging. In Rothstein M (ed): "Review of Biological Research in Aging," Vol. 3. New York: Alan R. Liss, Inc., pp 139–146.

Proust JJ, Filburn CR, Harrison SA, Buchholz MA, Nordin AA (1987): Age-related defect in signal transduction during lectin activation of murine T lymphocytes. J Immunol 139:1472–1478.

Proust JJ, Kittur DS, Buchholz MA, Nordin AA (1988): Restricted expression of mitogen-induced high affinity IL2 receptors in aging mice. J Immunol 141:4209–4216.

Radl J (1981): Immunoglobulin levels and abnormalities in aging humans and mice. In Adler WH, Nordin AA (eds): "Immunological Techniques Applied to Aging Research." Boca Raton, Florida: CRC Press, pp 121–139.

Radl J, Hollander CF (1974): Homogeneous immunoglobulins in sera of mice during aging. J Immunol 112:2271–2273.

Radl J, Sepers JM, Skvaril F, Morell A (1975): immunoglobulin patterns in humans over 95 years of age. Clin Exp Immunol 22:84–90.

Radl J, Hollander CF, van den Berg P, de Glopper E (1978): Idiopathic paraproteinemia. I. Studies in an animal model—the aging C57/KaLwRij mouse. Clin Exp Immunol 33:395–402.

Radl J, Vieveen MHM, van den Akker TW, Benner R, Haaijman JJ, Zurcher C (1985): Idiopathic paraproteinaemia. V. Expression of Igh1 and Igh5 allotypes within the homogeneous immunoglobulins of ageing (C57B1/LiARij × CBA/BrARij)F1 mouse. Clin Exp Immunol 62:405–411.

Richardson A, Semsei I (1987): Effect of aging on translation and transcription. In Rothstein M (ed): "Review of Biological Research in Aging," Vol. 3. New York: Alan R. Liss, Inc., pp 467–483.

Rikans LE (1986): Drugs and nutrition in old age. Life Sci 39:1027–1036.

Rowley MJ, Buchanan H, Mackay IR (1968): Reciprocal change with age in antibody to extrinsic and intrinsic antigens. Lancet Jul 2:24–26.

Saxena RK, Saxena QB, Adler WH (1988): Lectin-induced cytotoxic activity in spleen cells from young and old mice. Age-related changes in types of effector cells, lymphokine production and response. Immunology 64:457–461.

Segre D, Segre M (1976): Humoral immunity in aged mice. II. Increased suppressor T cell activity in immunologically deficient old mice. J Immunol 116:735–738.

Sekigawa I, Okada T, Noguchi K, Ueda G, Horose S, Sato H, Shirai T (1987): Class-specific regulation of anti-DNA antibody synthesis and the age-associated changes in (NZB × NZW)F1 hybrid mice. J Immunol 138:2890–2895.

Sidman CL, Luther EA, Marshall JD, Nguyen K-A, Roopenian DC, Worthen SM (1987): Increased expression of major histocompatibility complex antigens on lymphocytes from aged mice. Proc Natl Acad Sci USA 84:7624–7628.

Snow EC (1987): An evaluation of antigen-driven expansion and differentiation of hapten-specific B lymphocytes purified from aged mice. J Immunol 139:1758–1762.

Stall AM, Farinas MC, Tarlington DM, Lalor PA, Herzenberg LA, Strober S, Herzenberg LA (1988): Ly-1 B-cell clones similar to human chronic lymphocytic leukemias routinely develop in older mice and young autoimmune (NZB-related) animals. Proc Natl Acad Sci USA 85:7312–7316.

Szewczuk MR, Campbell RJ (1980): Loss of immune competence with age may be due to anti-idiotypic antibody regulation. Nature 286:164–166.

Thoman ML, Weigle WO (1981): Lymphokines and aging: Interleukin-2 production and activity in aged animals. J Immunol 127:2102–2106.

Thoman ML, Weigle WO (1983): Deficiency in suppressor T cell activity in aged animals. J Exp Med 157:2184–2189.

Tollefsbol TO, Cohen HJ (1985): Carbohydrate metabolism in transforming lymphocytes from the aged. J Cell Physiol 123:417–424.

Tollefsbol TO, Cohen HJ (1986): Expression of intracellular biochemical defects of lymphocytes in aging: Proposal of a general aging mechanism which is not cell specific. Exp Gerontol 21:129–148.

Wade AW, Szewczuk MR (1984): Aging, idiotype repertoire shifts, and compartmentalization of the mucosal-associated lymphoid system. Adv Immunol 36:143–188.

Weifeng C, Shulin L, Xiamei G, Zuewen P (1986): The capacity of lymphokine production by peripheral blood lymphocytes from aged humans. Immunol Invest 15:575–583.

Weigle WO, Thoman ML, Goodman MG (1987): Augmentation of the antibody response in aged mice with an 8-derivatized guanosine nucleoside and its effect on immunological memory. Cell Immunol 109:332–337.

Weigle WO, Thoman ML, Goodman MG (1988): The effect of aging on the induction of tolerance in a subpopulation of B lymphocytes. Cell Immunol 111:253–257.

Weksler ME (1981): The senesence of the immune system. Hosp Pract 16:53–64.

Weksler ME. Hutteroth TH (1974): Impaired lymphocyte function in aged humans. J Clin Invest 53:99–104.

Whisler RL, Newhouse YG (1985): Immunosenescence of the human B cell system: Impaired activation/proliferation in response to autologous monocytes pulsed with Staph protein A and the effects of interleukins 1 and 2 compared to interferon. Lymphokine Res 4:331–337.

Whisler RL, Newhouse YG (1986): Function of T cells from elderly humans: Reductions of membrane events mediated by T3 determinants and diminished elaboration of soluble T cell factors for B cell growth. Cell Immunol 99:422–423.

Whisler RL, Newhouse YG, Ennist DL, Lachman LB (1985): Human B cell colony responses: Suboptimal colony responsiveness in aged humans associated with defective function of B cells and monocytes. Cell Immunol 94:133–146.

Winchurch RA, Togo J, Adler WH (1987): Supplemental zinc (Zn^{2+}) restores antibody formation in cultures of aged spleen cells. II. Effects on mediator production. Eur J Immunol 17:127–132.

Woda BA, Feldman JD (1979): Density of surface immunoglobulin and capping on rat B lymphocytes. I. Changes with age. J Exp Med 149:416–423.

Wu W, Pahlavani M, Cheung HT, Richardson A (1986): The effect of aging on the expression of interleukin 2 messenger ribonucleic acid. Cell Immunol 100:224–231.

Zharhary D (1986): T cell involvement in the decrease of antigen-responsive B cells in aged mice. Eur J Immunol 16:1175–1178.

Zharhary D (1988): Age related changes in the capability of the bone marrow to generate B cells. J Immunol 141:1863–1869.

Zharhary D, Klinman NR (1983): Antigen responsiveness of the mature and generative B cell populations of aged mice. J Exp Med 147:1300–1308.

Zharhary D, Klinman NR (1986a): Selective increase in the generation of phosphorylcholine-specific B cells associated with aging. J Immunol 136:368–370.

Zharhary D, Klinman NR (1986b): The frequency and fine specificity of B cells responsive to (4-hydroxy-3-nitrophenyl)acetyl in aged mice. Cell Immunol 100:452–461.

SECTION III
NEUROLOGY OF AGING

The Genetics of Alzheimer's Disease: A Review of Recent Work

Jeremy M. Silverman, Richard C. Mohs, Linda M. Bierer, Richard S.E. Keefe, and Kenneth L. Davis

INTRODUCTION

Although the etiology of Alzheimer's disease (AD) is unknown, it has long been suspected that genetic factors play a role. Family history studies [Sjogren et al., 1952; Larsson et al., 1963; Akesson, 1969; Heston et al., 1981; Heyman et al., 1983; Amaducci et al., 1986] showing an increased risk to family members of AD probands compared with those of controls, as well as the identification of large pedigrees with numerous cases of AD [Feldman et al., 1963; Cook et al., 1979; Nee et al., 1983] have strongly suggested the involvement of genetic factors in at least certain subtypes of AD.

Recent work has attempted to characterize more fully the genetic mechanisms that appear to be involved in AD. Studies to identify genetic linkage markers in high density families, epidemiologic studies designed to determine the role of genetic factors in typical late onset AD, and studies designed to identify characteristics that might be related to genetic versus nongenetic subtypes of AD are all under way. Most of this work has been conducted through the use of three major methodologies: genetic linkage, twin, and family history studies. A review of work conducted in the last 3 years using these methodologies will be the subject of this chapter.

GENETIC LINKAGE STUDIES

The strongest evidence for a genetic etiology in any disease, short of identifying the gene itself, is the identification of a genetic marker for the disease. A genetic marker is a location on a specific chromosome in close proximity to the disease gene. Linkage studies examine the cosegregation of a genetic marker

Psychiatry Service, Mt. Sinai School of Medicine, New York, New York 10029 and Bronx VA Medical Center, Bronx, New York 10468

with the disease in question. A recent comprehensive discussion of the major methodological issues of linkage studies appears elsewhere [Payne and Roses, 1988]. Certain issues will be briefly addressed as they relate to studies of AD.

Methodological Issues of AD Linkage Studies

Families with multiply affected as well as unaffected members (i.e., multiplex families) are needed, in order to provide the statistical power required to evaluate the likelihood that a particular marker apparently associated with the disease is not simply the result of chance (independent assortment) [Ott, 1985]. The greater the number of affected relatives who show the genetic marker, as well as the number of unaffected relatives without the marker, the greater the likelihood (i.e., the lod score) that the marker tends to segregate with the disease and therefore is located near the disease gene.

In late onset diseases identifying multiplex families becomes problematic even if the disease is fully determined by a single gene because there will be few living affected cases for study. In addition, increased confidence that a disease-free individual truly does not carry the gene can be present only in very long-lived individuals (though misclassifying individuals in this way is not as detrimental to a linkage analysis as misclassifying an individual who does not carry the disease gene as "affected"). Etiologic heterogeneity, when it exists, is also likely to hamper progress in identifying any particular gene that may be involved.

Linkage to Chromosome 21

Prior to 1987 no linkage marker had been identified for AD, although some areas of the genome had been excluded. The HLA region on chromosome 6 was excluded in one large kindred [Goldsmit et al., 1981], as were small regions on chromosomes 1 and 4 [Spence et al., 1986] in a group of 18 families. Another study [Weitkamp et al., 1983], however, raised the possibility that the HLA region of chromosome 6 and a region (Gm) of chromosome 14 might contain two or more genes that interact to increase susceptibility for AD.

St. George-Hyslop et al. [1987a] studied four large pedigrees with high rates of AD among the biological members of these kindreds. Alzheimer's disease in these families was linked to two chromosome 21q21 loci, D21S1/D21S11 and D21S16. The identification of these markers on chromosome 21 was the first real proof of genetic factors in some patients with AD and as such was a major advance in the investigation of the etiology of AD. It nevertheless remains unclear whether these results can be generalized to include the more typical variants of AD.

The four kindreds studied by St. George-Hyslop et al. [1987a] had several unusual characteristics. Although families with multiple affected AD cases are not rare, very few show patterns of illness with the high density of those studied by St. George-Hyslop et al. The segregation ratio in these families resembled an autosomal dominant mode of genetic transmission even without using age correction techniques. Also atypical was the very early age at onset of disease that characterized most of the identified cases. The mean ages at onset of AD in each of the four families where such information could be obtained was 52 ± 6.23 years, 48.7 ± 5.73 years, 49.8 ± 4.84 years, and 39.9 ± 7.18 years, far lower than estimates of the mean age at onset for AD cases in the general population [Jorm et al., 1987]. However, to the degree that age of onset may be related within a kindred [Huff et al., 1988; Breitner et al., 1988a,b], the high density of affected cases in such families may simply be the result of early onset.

One study found evidence of duplication of the gene for beta-amyloid in a series of "sporadic" cases of AD [Delabar et al., 1987]. This protein is prominently found in the neuritic plaques that characterize the histopathology of AD and of trisomy 21 (Down's syndrome) individuals who live beyond young adulthood. The possibility that duplication of this gene might be involved in the etiology of AD was suggested when the location of the beta-amyloid gene also mapped to chromosome 21 in the general vicinity of the marker identified by St. George-Hyslop et al. [Tanzi et al., 1987a]. Several recent studies, however, have failed to demonstrate duplication of the beta-amyloid gene in patients with AD [Podlisny et al., 1987; St. George-Hyslop et al., 1987b; Tanzi et al., 1987b].

Linkage to Chromosome 21q21 Excluded

To date no other linkage studies in AD have similarly been able to identify the two 21q21 loci in an independent study. Two studies using pooled data from relatively small families have excluded linkage at these loci [Roses et al., 1988; Schellenberg et al., 1988].

Roses et al. [1988] studied eight AD families with a mean age at onset for AD greater than 60 years and excluded the possibility of linkage for the chromosome 21 markers identified by St. George-Hyslop et al. Two additional families with mean ages at onset for AD at less than 60 years showed nonsignificant though positive lod scores. These findings, combined with those of St. George-Hyslop et al. [1987a], have prompted some investigators to interpret the results as providing at least weak evidence for the possibility that early onset ($<$ 60 years) and late onset ($>$ 60 years) AD may identify two distinct subtypes of AD.

Schellenberg et al. [1988] studied 15 families with up to 17 identified cases of AD. Seven of these families were Americans of Volga German ethnicity originating from the same village on the Volga River in the Soviet Union [Bird et al., 1988]. The 82 affected individuals in the Volga German families are thought to have a common genetic founder and therefore to have a homogeneous disease. Schellenberg et al. [1988], like Roses et al. [1988], also excluded linkage between AD and the genetic markers identified by St. George-Hyslop et al. using their entire sample of 15 families and also using the 7 Volga German families alone. Mean age at onset by family ranged from 41.1 ± 5.1 years to 77.7 ± 2.3 years; among the Volga German families alone the range was 41.1 ± 5.1 years to 63.1 ± 5.9 years. Since with one exception, all the Volga German families had mean onsets at less than 60 years, these findings do not support the suggestion that families with a mean onset less than 60 years may belong to a single genetically homogeneous subtype of AD. However, Schellenberg notes that even his early onset families have, in general, developed the disease later than the four studied by St. George-Hyslop et al. Thus the possibility of a homogeneous subtype of AD with *very* early onset remains, though even this more restricted hypothesis can be questioned because of the failure to observe any neuropathologic difference among these differing groups of families [Schellenberg et al., 1988].

Taken together, the three recent studies of linkage provide the first strong evidence for genetic heterogeneity in AD. Although additional linkage markers at different loci may ultimately be found for many of the large kindreds with high-density AD in the families, such families are not typical. An AD gene or genes found for them may or may not also be associated with the more commonly observed types of AD. This does mean that the later onset forms of AD are not genetic, although of course that possibility must remain. Nevertheless, obtaining the information for linkage from more typical but at the same time potentially informative families is likely to be a far more arduous task. Compounding the uncertainties relating to diagnosis of AD in demented family members and defining true unaffected individuals, which have always made linkage studies in AD difficult, there will be the added problem that a great many more AD families will be needed, since most commonly observed multiplex families for AD do not provide very much information for linkage. Possible heterogeneity in these more typical cases may also obscure genetic factors in some subgroups of them.

TWIN STUDIES

Studies of identical twins have reported both concordant [Kallman, 1956; Sharman et al., 1979; Cook et al., 1981; Embry and Lippman, 1985] and discordant [Davidson and Robertson, 1955; Hunter et al., 1972] twin pairs. In some cases, however, reported discordant pairs were followed up sometimes over a decade later and found to be concordant [Cook et al., 1979]. Nee et al. [1987] studied 22 twin pairs in which at least one member met NIH consensus criteria for probable AD. Seven of 17 monozygotic twin pairs were concordant for AD, a ratio similar to the 2 pairs concordant for AD of 5 dizygotic twin pairs. Among the discordant pairs were several unaffected individuals that lived to 80 and beyond. Nee et al. discuss several important limitations of twin studies including selection biases, censored data, small samples, and environmental differences in twins compared with singletons. Nevertheless, the observation of discordance in long-lived MZ twin pairs proves a nongenetic factor in the onset of AD and offers significant evidence that genetic factors alone will not fully explain the causes of AD.

FAMILY HISTORY STUDIES

Despite the recent progress made using molecular biologic techniques to examine genetic factors in AD, the nature and extent of possible genetic causes in more commonly observed later onset AD remain unclear, and, with few exceptions [Cook et al., 1979; Nee et al., 1983], an obvious pattern of inheritance cannot be directly discerned from studies of these AD proband families.

For many diseases, the absence of such patterns argues against the likelihood of a simple genetic cause. For example, in a disease caused by a single autosomal dominant gene, the expectation is that approximately 50% of all first degree relatives of an identified proband would eventually manifest the disease; the observation of lower rates of disease tends to suggest either reduced penetrance of the gene or a causal mechanism other than autosomal dominance.

These expectations change, however, when the disorder under consideration is a late onset disease. In such diseases many or most relatives may die from other causes before reaching the age at which the risks are realized. Thus Davies [1986] has noted that the genetics of AD would be far better understood if humans all lived to 150—well beyond, that is, the presumed period of risk. Study of the patterns of inheritance in late onset diseases is possible, however, provided that at least some of the population under investigation, independent of their genetic susceptibility for the disease, live through the risk period. The appropriate conduct of such studies, typically employing family history interviews of knowledgable informants to ascertain AD proband relatives, requires the use of statistical methods that take into account the decreasing numbers of individuals at risk for the disease with increasing age [Chase et al., 1983]. Such procedures can then provide an assessment of the cumulative risk throughout the human life span. This is the approach that has been taken in a number of studies investigating the genetic factors of AD as it is found in the clinic and other medical and institutional settings.

Methodological Issues Related to Family History Studies

Numerous methodological issues must be taken into consideration to ensure the utility and generalizability of family history studies of AD. The diagnosis of the proband is a major potential source of error. In the absence of autopsy-confirmed AD diagnoses, other dementing processes may be misdiagnosed as AD. The inclusion of non-AD demented probands dilutes the homogeneity of the sample of relatives under investigation and thus decreases the accuracy of the estimation of risk to relatives of true AD cases. This concern has been mitigated to a large extent with the advent of rigorous clinical research diagnostic criteria for AD [McKhann et al., 1984; Tierney et al., 1988]. Nevertheless, the combined use of clinical evaluation and neuropathology remains the most accurate method of identifying AD cases. Although these studies are now under way at some centers, they await the accumulation of a sufficiently large pool of autopsy-confirmed AD probands for analysis.

Another source of possible error in family history studies concerns the reliability and accuracy of the presumed AD diagnosis and age at onset information in secondary cases. In most studies, data gathered from direct family study and health records are generally considered more accurate than those obtained by the family history method [Andreasen et al., 1977]. Practically, however, the former methods often impose unrealistic requirements of manpower, resources, and cooperation from families and institutions. Furthermore, in the case of investigations of dementing illnesses there are additional difficulties with these methods. Because AD is typically a late onset disease, most secondary cases, virtually all parents for example, will be dead. Health records are often of limited use in the evaluation of dementia, since until recently clinicians often failed to recognize or diagnose dementia as a clinical condition, or they ascribed it to undocumented causes. For these reasons, the family history method is often the only practical investigational technique to examine genetic factors in the most common forms of AD, and a number of recent investigations have employed it [Mohs et al., 1987; Huff et al., 1988; Martin et al., 1988; Farrer et al., in press]. It has therefore been of concern that the methods used for these studies show good reliability.

Silverman et al. [1986] examined inter-informant reliability by independently interviewing two family informants of 27 AD probands and 22 nondemented controls. These investigators used the Breitner and Folstein Alzheimer's Dementia Risk Questionnaire [Breitner and Folstein, 1984] to identify all first degree relatives and obtain basic demographic information on them and to screen for dementia of any type. Informants were further questioned about the probands and those relatives identified as having some form of memory loss or dementia using the Dementia Questionnaire [Silverman et al., 1986], which attempts to characterize the nature and progression of a dementia and thus identify those with a primary progressive AD-like dementia and those with other dementing conditions. Age at onset information was also collected. Independent informants were found to show very good agreement in their estimates of the age at onset in the proband [intraclass correlation coefficient (ICC) = 0.91]. In addition, informants agreed on whether their relatives were demented or not [kappa

(k) = 1.0 AD proband informants; kappa = 0.92 control informants], and in distinguishing among primary dementia, stroke, other CNS disease and the absence of cognitive symptoms (k = 0.82 AD proband informants; k = 0.92 control informants).

In another study by the same group [Silverman et al., in press], inter-rater reliability of the family history method was assessed by examining the agreement between two raters who interviewed the identical family informants of 30 AD probands. These interviews were conducted at least 1, and in most cases more than 1 1/2 years apart. The second rater was kept blind to the evaluations of the first rater. Raters showed good agreement on the presence and type of dementia in the relatives (k = 0.94) and on the age at onset in the secondary cases (ICC = 0.91). In addition, there was excellent agreement on the age of all nonaffected cases (ICC = 0.99).

Huff et al. [1988] conducted test-retest reliability on seven family history informants interviewed in person and nine informants interviewed on the telephone by the same interviewer on two occasions 1 year apart. In both conditions, the proportion of agreement for all data items was 99%. Reliability was also excellent for agreement on who in the family was "forgetful and confused in their later years" (k = 0.90 for in-person interviews; k = 1.0 for telephone interviews). This group assessed inter-rater reliability from eight family informants interviewed 1 year apart by different raters and found a 96% rate of agreement for all data items and good agreement regarding dementia status in relatives (k = 0.84).

The results of Silverman et al. [1986] and Huff et al. [1988] provide evidence that reliable information about dementing illness can be obtained using the family history method when standardized techniques are employed. However, these data do not guarantee that the secondary cases even reliably identified through this method truly have AD; thus the validity of this method remains unproven.

Patterns of Inheritance in AD From Family History Studies

Mohs et al. [1987] investigated the cumulative risk for a primary progressive (i.e., AD-like) dementia (PPD) to 244 first degree relatives (45+ years of age) of 50 patients meeting NIH consensus criteria [McKhann et al., 1984] for "probable" AD entered into studies without regard for family history status. In addition, they assessed the risk for PPD to the 211 relatives of 45 nondemented elderly controls (mostly spouses or other nonbiologically related relatives of the AD probands). Using the Kaplan–Meier life-table method [Kaplan and Meier, 1958] to estimate the age-specific cumulative incidence of PPD in the two groups, they found a cumulative morbid risk of 45.9% (± SE 9.8%) among relatives of AD probands by 86 years of age (the age of the oldest incident case). In contrast, the cumulative morbid risk for PPD to the relatives of the controls rose to 12.1% ($\pm$ SE 7.5%) by 85 years of age, far less than the relatives of AD probands. These results were similar to an earlier study by Breitner and Folstein [1984], who collected data on the first degree relatives of a sample of nursing home residents with an AD-like illness whose clinical features included agraphia (cumulative risk = 55.3%). Unlike linkage studies, family history studies can offer no sure proof for underlying genetic mechanisms.

Breitner et al. [1988b], using an AD proband sample overlapping with the Mohs et al. [1987] sample, but enlarged by approximately 60% (n = 79), found that the cumulative morbid risk for PPD to their 379 relatives (45+ years of age) rose to 49.3% ($\pm$ SE 8.4%) by age 87, compared with 9.8% ($\pm$ SE 6.1%) among the 271 comparable relatives of controls. Thus the cumulative risk to the first degree relatives in this expanded sample rose slightly higher and closer to 50%.

Huff et al. [1988] collected family history data on 50 AD probands meeting NIH consensus criteria and on 47 nondemented age-matched controls. The probands had been consecutively entered into a longitudinal study of AD without regard to family history status. Using life-table methods identical to those described in the Mohs et al. [1987] and Breitner et al. [1988b] studies, these investigators found that the cumulative risk to the first degree relatives of AD probands rose to 45% ($\pm$ 11%) by age 99. The risk to relatives of controls rose to 11% ($\pm$ SE 4%), significantly lower than relatives of AD probands and similar to previously reported estimates.

In another recent family history study, Martin et al. [1988] evaluated the 130 first degree relatives of a group of 22 patients meeting rigorous clinical criteria for AD; 3 of these probands were ultimately neuropathologically verified. The 144 first degree relatives of 24 nondemented spouses of demented patients were also investigated by identical family history methods. Relatives were categorized by two sets of criteria for dementia. The first, called dementia of Alzheimer type (DAT), included mild to severe cases of reported dementia excluding those with an onset or course inconsistent with AD or with a dementia attributable to another identifiable cause. The second included relatives reported to have "consistently evident decline but not clearly exhibiting impairment in performance" in five major areas of functioning. Members of this second group were called cases of senescent forgetfulness (SF). The risk for DAT to first degree relatives of AD probands was 40.8% ($\pm$ 9.42%) by age 83 and was significantly higher than the cumulative risk found among controls, 23.3% ($\pm$ 10.8%) by age 85. Including the cases of SF in addition to DAT, the differences in cumulative risk between the groups became even more pronounced. The cumulative risk to relatives of AD probands rose to 66.8% (± 10.8%), and it remained the same among the relatives of controls since no cases of SF were observed in this group.

Farrer et al. [1989] studied the family histories of 128 clinically diagnosed AD probands and used the family histories of 84 subjects with Parkinson's disease, a late onset neurodegenerative disease with no apparent genetic cause [Ward et al., 1983], as controls. The risk to relatives was primarily assessed by a procedure that weighted the confidence of the presence of AD in each of the secondary cases. Using this procedure the risk to AD relatives rose to 24% compared with 16% among relatives of controls. Although the age-specific risks between the two groups of relatives differed significantly, a lower risk to relatives of AD probands was observed in this study than in those studies reviewed above. The authors suggest that this discrepancy may be the result of an ascertainment bias in the earlier studies, which might have favored the selection of AD probands with a positive family history. For example, they report that in their center a group of probands were referred for study because of a positive family history. Had these nonrandomly selected families been included among their random sample the risk for an AD-like dementia to these relatives would have been artificially increased. Although this possibility cannot be ruled out for the other independent series studied, the AD samples in those studies appear to have been obtained in a manner without bias for or against a positive family history for dementia. Another explanation for the lower risk observed in this study may be decreased sensitivity in the family history method due to the reliance on a single informant.

Fitch et al. [1988] divided the relatives of 91 AD probands according to the presence or absence of at least one secondary case in the family. The authors then examined the hypothesis that the pattern of inheritance in the relatives of the familial cases fit an autosomal dominant genetic mode of transmission by examining the ratio of affected secondary cases to unaffected relatives in this group, which for this mode should be 1:1. Since in AD, one's status as unaffected may change with increased age, only unaffected individuals age 80 or over were counted. Using this method, the ratio 24 affected:31 unaffected did not depart significantly from 1:1. As the authors note, this method suffers from the inherent uncertainties regarding the confidence that unaffected relatives will not go on to develop the disease, or would not have had they lived longer. Their choice therefore to treat family history negative cases as nongenetic may not be wholly justified. The onset of AD is not uncommon in the ninth decade, and in fact fully one-third of the secondary cases (8/24) identified in this study developed AD after they turned 80. These investigators did not assess the cumulative morbid risk for the entire group of relatives.

CHARACTERISTICS POSSIBLY RELATED TO GENETIC LOADING FOR AD

Age at Onset

Some early family history studies [Heston, 1981; Heston et al., 1981] found the risk for dementia to relatives of AD probands was increased for early onset (< 65 years, by convention) probands, but not for those with a late onset (> 65 years). This finding has not been a consistent one, however, and recent studies using current research criteria for AD have generally seen no significant difference between the risk to relatives of early and late onset AD probands [Breitner et al., 1988b; Huff et al., 1988; Farrer et al., in press]. One exception is Chandra et al. [1987], who conducted a case-control study with late onset AD probands (n = 64) only and found no increased risk for dementia compared with the relatives of age-matched controls.

A number of investigators have suggested that onset may itself be in part a heritable trait [Powell and Folstein, 1984; Breitner et al., 1986; Breitner et al., 1988a; Huff et al., 1988]. If this is so it may help to explain the conflicting results that have occasionally been obtained in assessing the heritability of dementia in relatives of early and late onset AD. As Farrer et al. [1989] have suggested, there may be a significant censoring bias in those studies having a high preponderance of *very* late onset probands, such as in Chandra et al. [1987] (mean onset = 83.2 years); if age at onset is inherited, then relatives from late onset families who carry the genetic liability for AD may nevertheless die unaffected at a disproportionately high rate—before the disease is expressed.

Risk for Dementia in Males Versus Females

There does not appear to be a difference in the age-specific cumulative risk for dementia to male versus female relatives of AD probands [Breitner et al., 1988b, Farrer et al., in press]. Women have been found to have a higher prevalence for AD than men in most epidemiological surveys [Jorm et al., 1987]. Although this is likely to be due to the greater longevity in females [Farrer et al., 1989], Breitner et al. [1988b] has suggested the possibility of differential age-specific expression of AD in females versus males.

Risk for Dementia in Differing Generations

Significant generational differences in the risk for dementia among relatives of AD probands (in most cases parents versus the siblings; children are generally too young to be at risk) have not been observed in recent studies [Heyman et al., 1983; Breitner et al., 1988b, Farrer et al., 1989], although some have found that siblings show a slightly higher risk compared with parents [Breitner and Folstein, 1984; Breitner et al., 1988b, Farrer, 1989]. It is possible that these occasionally observed differences may be the result of the inevitably poorer quality of data concerning relatives deceased sometimes for many decades.

Platelet Membrane Fluidity

The familial risk for AD in a substantial subgroup of AD patients showing increased platelet membrane fluidity, have been investigated by Zubenko et al. [1988]. This trait has been found to be increased in AD patients compared with patients with depression and in early onset versus late onset patients [Zubenko et al., 1987a]. In addition, increased fluidity appears to be present in about half of the first degree relatives of AD patients who have the trait themselves [Zubenko et al., 1987b], a rate 3.5 to 11.5 times higher than the prevalence in the general population. In their family study, 22 or 43 (51%) AD probands showed increased platelet membrane fluidity. Relatives of both groups showed cumulative risks near 50% by the early 90s, but relatives of probands with increased fluidity showed comparable cumulative incidence rates about 5 years earlier, and the mean age at onset in these secondary cases was also significantly lower in this group. Thus, it is possible that increased membrane fluidity may be a marker for a genetic trait that reduces the age at onset of AD, rather than a marker for AD itself.

CONCLUSIONS AND PROSPECTS FOR THE FUTURE

Linkage studies have clearly demonstrated that some forms of AD are genetic and linked to chromosome 21, but they also appear to indicate genetic heterogeneity. Most family studies have indicated a cumulative morbid risk to first degree relatives that reaches 45–50% by the ninth decade. Subtypes with greater or lesser genetic contribution have not been clearly identified although some evidence suggests age of onset may distinguish genetically homogeneous groups.

Conclusive proof of genetic factors in late as well as early onset AD will ultimately derive from linkage analyses. These methods, however, rely on the ascertainment of kindreds with multiple affected as well as unaffected relatives, which are particularly rare in kindreds with illness of later onset. The identification of a pre- or subclinical marker associated with a gene for AD would aid linkage studies by permitting the classification of more family members as "affected" (or putative gene carriers) or "unaffective," thus increasing the informativeness of a given family. Twin studies will offer the possibility of identifying cases that appear less likely to be genetic, as well as of exploring the potential contributions of environmental and personal historical factors as they may influence the expression of AD. Because of the extraordinary difficulty in ascertaining large kindreds suitable for linkage studies, or of identifying monozygotic twins with genetic vulnerability for AD, family history techniques still offer the most accessible methods for pursuing the question of genetic risk. The combination of prospective clinical as well as autopsy-confirmed diagnoses of AD probands will increase the validity of these methods. Moreover, the assimilation of National Institute of Neurological and Communicative Disorders and Stroke (NINCDS)–Alzheimer's Disease and Related Disorders Association (ADRDA) criteria for the diagnosis of AD into standard medical practice will increase the confidence with which secondary cases can be correctly identified as presumptive AD through family history interviews. In addition, the replication of these studies with increasing sample sizes will result in smaller standard errors, permitting more precise estimates of cumulative risk to relatives, as well as allowing for investigations of possible contributors to the variance in these estimates, such as age of onset, sex of the proband, and environmental exposure. Finally, population-based studies of AD utilizing the family history method will mitigate against possible nonrandom ascertainment due to proband recruitment through tertiary referral centers.

REFERENCES

Akesson HO (1969): A population study of senile and arteriosclerotic psychoses. Hum Hered 19:546–566.

Amaducci LA, Fratiglioni L, Rocca WA, Fieschi C, Livrea P, Pedone D, Bracco L, Lippi A, Gandolfo C, Bino G, Prencipe M, Bonatti ML, Girotti F, Carella F, Tavolato B, Ferla S, Lenzi GL, Carolei A, Gambi A, Grigoletto F, Schoenberg BS (1986): Risk factors for clinically diagnosed Alzheimer's disease: a case-control study of an Italian population. Neurology 36:922.

Andreasen NC, Endicott J, Spitzer RL, Winokur G (1977): The family history method using diagnostic criteria: Reliability and validity. Arch Gen Psychiatry 34:1229–1235.

Bird TD, Lampe TH, Nemens EJ, Miner GW, Sumi SM, Schellenberg GD (1988): Familial Alzheimer's disease in American descendants of the Volga Germans: Probable genetic founder effect. Ann Neurol 23:25–31.

Breitner JCS, Folstein MF (1984): Familial Alzheimer's disease: A prevalent disorder with specific clinical features. Psychol Med 14:63–80.

Breitner JCS, Murphy EA, Folstein MF (1986): Familial Alzheimer's dementia—II: Clinical genetic implications of age-dependent onset. J Psychiatr Res 20:45–55.

Breitner JCS, Murphy EA, Silverman JM, Mohs RC, Davis KL (1988a): The age-dependent expression of familial risk in Alzheimer's disease. Am J Epidemiol 128:536–548.

Breitner JCS, Silverman JM, Mohs RC, Davis KL (1988b): Familial aggregation in Alzheimer's disease: Comparison of risk among relatives of early- and late-onset cases, and among male and female relatives in successive generations. Neurology 38:207–212.

Chandra V, Philipose V, Bell PA, Lazaroff A, Schoenberg BS (1987): Case-control study of late onset "probable Alzheimer's disease." Neurology 37:1295–1300.

Chase GA, Folstein MF, Breitner JCS, Beaty TH, Self SG (1983): The use of life tables and survival analysis in testing genetic hypotheses, with an application to Alzheimer's disease. Am J Epidemiol 117:590–597.

Cook RH, Ward BE, Austin JH (1979): Studies in aging of the brain IV. Familial Alzheimer's disease: Relation to transmissible dementia, aneuploidy, and microtubular defects. Neurology 29:1402–1412.

Cook RH, Schneck SA, Clark DB (1981): Twins and Alzheimer's disease. Arch Neurol 38:300–301.

Dalebar J-M, Goldgabor M, Lamour Y, et al. (1987): Beta amyloid gene in Alzheimer's disease and karyotypically normal Down syndrome. Science 235:1390–1392.

Davidson EA, Robertson EE (1955): Alzheimer's disease with acne rosacea in one of identical twins. J Neurol Psychiatr 18:72–77.

Davies P (1986): The genetics of Alzheimer's disease: A review and a discussion of the implications. Neurobiol Aging 7:459–466.

Embry C, Lippman S (1985): Presumed Alzheimer's disease beginning at different ages in two twins. J Am Geriatr Soc 33:61–62.

Farrer LA, O'Sullivan DM, Cupples LA, Growden JH, Myers R (1989): Assessment of genetic risk for Alzheimer's disease among first degree relatives. Neurology 25:485–493.

Feldman RG, Chandler KA, Levy LL, Glaser GH (1963): Familial Alzheimer's disease. Neurology 13:811–824.

Fitch N, Becker R, Heller A (1988): The inheritance of Alzheimer's disease: A new interpretationn. Ann Neurol 23:14–19.

Goldsmit J, White BJ, Weitkamp LR, Keats BJB, Morrow CH, Gajdusek DC (1981): Familial Alzheimer's disease in two kindreds of the same geographic and ethnic origin: A clinical and genetic study. J Neurol Sci 49:79–89.

Heston LL (1981): Genetic studies of dementia with emphasis on Parkinson's disease and Alzheimer's neuropathology. In Mortimer JA, Schuman LM (eds): "The Epidemiology of Dementia." New York: Oxford University Press.

Heston LL, Mastri AR, Anderson VE, White J (1981): Dementia of the Alzheimer type: Clinical genetics, natural history, and associated conditions. Arch Gen Psychiatry 38:976–981.

Heyman A, Wilkenson WE, Hurwitz BJ, Schmechel D, Sigmon AH, Weinberg T, Helms MJ, Swoft M (1983): Alzheimer's disease: Genetic aspects and associated clinical disorders. Ann Neurol 14:507–515.

Huff FJ, Auerbach J, Chakravarti A, Boller F (1988): Risk of dementia in relatives of patients with Alzheimer's disease. Neurology 38:786–790.

Hunter R, Dayan AD, Wilson J (1972): Alzheimer's disease in one monozygotic twin. J Neurol Neurosurg Psychiatry 35:707–710.

Jorm AF, Korten AE, Henderson AS (1987): The prevalence of dementia: A quantitative integration of the literature. Acta Psychiatr Scand 76:465–479.

Kallman FJ (1956): Genetic aspects of mental disorders in later life. In Kaplan OJ (ed): "Mental Disorders in Later Life," 2nd ed. Stanford, CA: Stanford University Press.

Kaplan EL, Meier P (1958): Non-parametric estimation from incomplete observations. J Am Stat Assoc 53:457–481.

Larsson T, Sjorgren T, Jacobson G (1963): Senile dementia: A clinical, sociomedical and genetic study. Acta Psychiatr Scand 39(Suppl 67):1–259.

Martin RL, Gerteis G, Gabrielli Jr WF (1988): A family-genetic study of dementia of Alzheimer type. Arch Gen Psychiatry 45:894–900.

McKhann G, Drachman D, Folstein MF, Katzman R, Price D, Stadlan E (1984): Clinical diagnosis of Alzheimer's disease: Report of the NINCDS–ADRDA Work Group under the aus-

pices of Department of Health and Human Services Task Force on Alzheimer's disease. Neurology 34:939–944.
Mohs RC, Breitner JCS, Silverman JM, Davis KL (1987): Alzheimer's disease: Morbid risk among first-degree relatives approximates 50% by 90 years of age. Arch Gen Psychiatry 44:405–408.
Nee LE, Polinsky RJ, Eldridge R, Weingartner H, Smallberg S, Ebert M (1983): A family with histologically confirmed Alzheimer's disease. Arch Neurol 40:203–208.
Nee LE, Eldrigge R, Sunderland T, Thomas CB, Katz D, Thompson BS, Weingartner H, Weiss H, Julian C, Cohen R (1986): Dementia of the Alzheimer type: Clinical and family study of 22 twin pairs. Neurology 37:359–363.
Ott, J (1985): "Analysis of Human Genetic Linkage." Baltimore: Johns Hopkins University Press.
Payne CS, Roses AD (1988): The molecular genetic revolution: Its impact on clinical neurology. Arch Neurol 45:1366–1376.
Podlisny MB, Lee G, Slekoe DJ (1987): Gene dosage of the amyloid beta protein in Alzheimer's disease. Science 238:669–671.
Powell D, Folstein MF (1984): Pedigree study of familial Alzheimer's disease. J Neurogenet 1:189–197.
Roses AD, Pericak-Vance MA, Haynes CS, Gaskell LH, Yamaoka LH, Hung WY, Clark CM, Alberts MJ, Lee JE, Siddique T, Heyman A (1988): Genetic linkage studies in Alzheimer's disease. Neurology 38(Suppl 1):173.
Schellenberg GD, Bird TD, Wijsman EM, Moore DK, Boehnke M, Bryant EM, Lampe TH, Nochlin D, Sumi SM, Deeb SS, Beyreuther K, Martin GM (1988): Absence of linkage of chromosome 21q21 markers to familial Alzheimer's disease. Science 241:1507–1510.
Sharman MG, Watt DC, Janota I, Carrasco L (1979): Alzheimer's disease in a mother and identical twin sons. Psychol Med 9:771–774.
Silverman JM, Breitner JCS, Mohs RC, Davis KL (1986): Reliability of the family history method in genetic studies of Alzheimer's disease and related dementias. Am J Psychiatry 143:1279–1282.
Silverman JM, Keefe RSE, Mohs RC, Davis KL: A study of the reliability of the family history method in genetic studies of Alzheimer's disease. Alzheimer's Disease and Associated Disorders, in press.
Sjogren T, Sjogren H, Lundgren GH (1952): Morbus Alzheimer and Morbus Pick: A genetic, clinical and patho-anatomical study. Acta Neurol Psychiatr Scand 82:1–152.
Spence AS, Heyman A, Marazita ML, Sparkes RS, Weinberg T (1986): Genetic linkage studies in Alzheimer's disease. Neurology 36:581–584.
St. George-Hyslop Ph, Tanzi RE, Polinsky RJ, Haines JL, Nee L, Watkins PC, Myers RH, Feldman RG, Pollen D, Drachman D, Growden J, Bruni A, Fancin JF, Samon D, Frommelt P, Amaducci L, Sorbi S, Piacentini S, Stewart GD, Hobbs WJ, Conneally PM, Gusella JF (1987a): The genetic defect causing familial Alzheimer's disease maps on chromosome 21. Science 235:885–890.
St. George-Hyslop PH, Tanzi RE, Polinsky RJ, Neve RL, Pollen D, Drachman S, Growden J, Cupples LA, Nee L, Myers R, O'Sullivan D, Watkins PC, Amos JA, Deutsch CK, Bodfish JW, Kinsbourne M, Feldman RG, Bruni A, Amaducci L, Foncin JF, Gusella JF (1987b): Familial and sporadic Alzheimer's disease. Science 238:664–666.
Tanzi RE, Gusella JF, Watkins PC, Bruns GAP, St George-Hyslop PH, Van Keuren ML, Patterson D, Pagan S, Kurnit DM, Neve RL (1987a): Amyloid beta protein gene: cDNA, mRNA distribution and genetic linkage near the Alzheimer locus. Science 235:880–884.
Tanzi RE, Bird ED, Latt SA, Neve RL (1987b): The amyloid beta protein gene is not duplicated in brains from patients with Alzheimer's disease. Science 238:666–669.

Tierney MC, Fisher RH, Lewis AJ, Zorzitto ML, Snow WG, Reid DW, Nieuwstraten P (1988): The NINCDS–ADRDA Work Group criteria for the clinical diagnosis of probable Alzheimer's disease: A clinicopathologic study of 57 cases. Neurology 38:359–364.

Ward CD, Duvoisin RC, Ince SE, Nutt JD, Eldridge R, Calane DB (1983): Parkinson's disease in 65 pairs of twins and in a set of quadruplets. Neurology 33:815–824.

Weitkamp LR, Nee L, Keats B, Polinsky RJ, Guttormsen S (1983): Alzheimer's disease: Evidence for susceptibility loci on chromosome 6 and 14. Am J Hum Genet 35:443–453.

Zubenko GS, Cohen BM, Boller F, Reynolds CF, Malinakova I, Keefe N, Chojnacki B (1987a): Platelet membrane fluidity in Alzheimer's disease and major depression. Am J Psychiatry 144:860–868.

Zubenko GS, Wusylko M, Boller F, et al. (1987b): Family study of platelet membrane fluidity in Alzheimer's disease. Science 238:539–542.

Zubenko GS, Huff FJ, Beyer J, Auerbach J, Teply I (1988): Familial risk of dementia associated with a biologic subtype of Alzheimer's disease. Arch Gen Psychiatry 45:889–893.

Transgenic Mouse Models of Amyloidosis in Alzheimer's Disease

Axel Unterbeck, Richard M. Bayney, George Scangos, and Dana O. Wirak

INTRODUCTION

Alzheimer's disease (AD) is the most common single cause of dementia in late life and is characterized by progressive memory impairments, loss of language and visuospatial skills, and behavior deficits [McKhann et al., 1986]. The cognitive impairment of AD patients is the result of degeneration of neuronal cells located in the cerebral cortex, hippocampus, basal forebrain, and other brain regions [for reviews, see Kemper, 1984; Price, 1986]. Histologic analyses of AD brains obtained at autopsy demonstrated the presence of neurofibrillary tangles (NFT) in perikarya and axons of degenerating neurons, extracellular neuritic (senile) plaques, and amyloid plaques inside and around some blood vessels of affected brain regions [Alzheimer, 1907]. Neurofibrillary tangles are abnormal filamentous structures containing fibers (about 10 nm in diameter) that are paired in a helical fashion; thus they are also called paired helical filaments [Kidd, 1963; Wisniewski et al., 1976; Selkoe et al., 1982; Brion et al., 1985; Grundke-Iqbal et al., 1986; Wood et al., 1986; Kosik et al., 1986, Goedert et al., 1988; Wischik et al., 1988a,b]. Neuritic plaques are located at degenerating nerve terminals (both axonal and dendritic) and contain a core composed of amyloid protein fibers [Masters et al., 1985a, b]. Cerebrovascular amyloid protein material is found in blood vessels in the meninges and the cerebral cortex [Glenner and Wong, 1984a,b; Wong et al., 1985].

During the past several years, significant progress has been made in the characterization of the primary pathological markers associated with AD. Biochemical analyses of three forms of Alzheimer brain lesions [for reviews, see Kemper, 1984; Wurtman, 1985; Katzman 1986; Price, 1986; Selkoe, 1989; Muller-Hill and Beyreuther, 1989], tangles, neuritic plaques, and cerebrovascular plaques, have revealed protein sequence information and have facilitated

Molecular Therapeutics Inc., Miles Research Center, West Haven, Connecticut 06516

subsequent cDNA cloning and chromosomal mapping of some of the corresponding genes. Immunological studies have identified several candidates for protein constituents of the paired helical filaments (PHF), including microtubule-associated protein 2 (MAP-2), tau, ubiquitin, and the amyloid protein (A4). Degenerating nerve cells express specific antigens such as A68, a 68 kd protein. This abnormal antigen is detectable with the monoclonal antibody ALZ-50 [Wolozin et al., 1986,1988; Wolozin and Davies, 1987].

A central feature of the pathology of AD is the deposition of amyloid protein within plaques. The 4kd amyloid protein (also referred to as A4, β-amyloid, or BAP) is a truncated form of a larger amyloid precursor protein (APP), which is encoded by a gene localized on chromosome 21 [Goldgaber et al., 1987; Kang et al., 1987; Robakis et al., 1987; Tanzi et al., 1987a]. Genetic linkage analysis, using DNA probes that detect restriction fragment length polymorphisms (RFLPs), [Botstein et al., 1980], has resulted in the localization of a candidate gene (FAD, familial AD) on human chromosome 21 in families with high frequencies of Alzheimer's disease [St. George-Hyslop et al., 1987a]. Patients with Down's syndrome, caused by trisomy of chromosome 21, develop Alzheimer-like pathology after the second decade of life [Masters, personal communication; Cork and Price, personal communication]. However, analysis of multiple Alzheimer pedigrees demonstrated that the A4-amyloid gene did not segregate with familial Alzheimer's disease [Von Broeckhoven et al., 1987; Tanzi et al., 1987b]. Furthermore, two recent studies with new families demonstrated absence of linkage of chromosome 21 markers to familial Alzheimer's disease [Schellenberg et al., 1988; Roses et al., 1988].

Age, genetic elements, and (possibly) environmental factors appear to contribute to the cellular pathology of AD (Fig. 1). A fundamental question in the pathogenesis of AD is the relationship between abnormalities of neurons and the deposition of amyloid. The 40–42 residue β-amyloid protein (A4-peptide) derives from a larger precursor, the amyloid protein precursor (APP), which has the typical features of a cell-surface receptor [Kang et al., 1987]. The amyloid peptide is part of the putative transmembrane domain (Fig. 2), thus implicating abnormal processing of this membrane-spanning protein in the subsequent extracellular degradation and amyloid formation. The structure of the APP gene has been recently characterized [Lemaire et al., 1989]; the amyloid peptide region of the precursor is encoded by two different exons (Fig. 3). Since the primary defect of AD is not yet known, mutant genes are currently not available. The familial AD locus (FAD) has not been localized precisely, and very little is known about its function. In this review, we discuss the application of transgenic mouse technology to the study of biochemical mechanisms involved in the pathogenesis of AD. As an alternative to classical mutants, the generation of transgenic mice carrying APP genes altered in

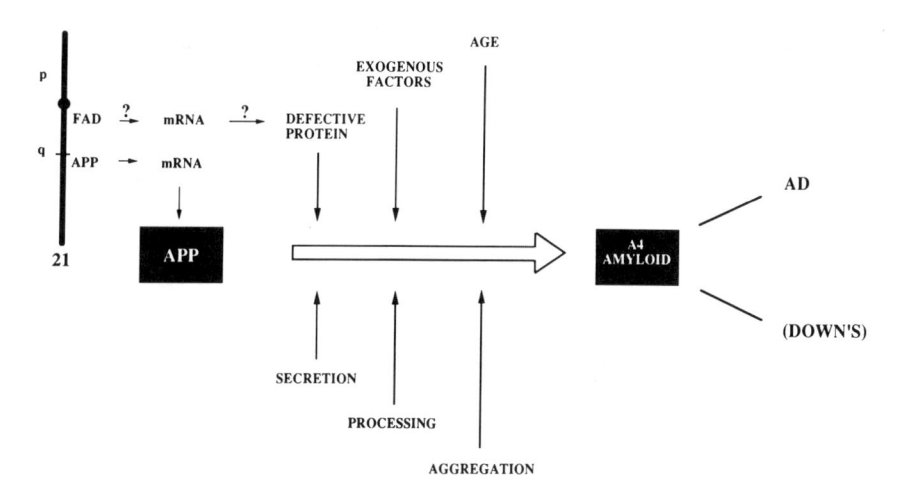

Fig. 1. Factors involved in the amyloidosis of AD and Down's syndrome [for review, see Price, et al., 1989]. The A4 amyloid peptide is derived from a larger amyloid precursor protein (APP). The APP gene is located on human chromosome 21. The process by which A4 derives from APP is not understood. Theoretically, a cascade of events may lead to the formation of amyloid, including the release of precursor protein(s) from the cell, abnormal processing or degradation of the precursor, and subsequent aggregation of A4 in plaques. Several factors are involved in the etiology of AD: age, environmental factors, and genetic elements. One gene (FAD), located on chromosome 21, is genetically linked to the familial form of AD and acts in an autosomal dominant fashion. The role of the FAD locus in the amyloidosis of AD is unknown.

their protein coding or regulatory regions may lead to dominant mutant phenotypes mimicking aspects of the AD pathology.

MOLECULAR BIOLOGY OF THE AMYLOID PRECURSOR PROTEIN GENE

One of the most important questions in AD is the cellular origin of pathological events leading to the deposition of amyloid fibrils adjacent to some areas of the blood–brain barrier (cerebrovascular amyloid) and in the proximity of nerve terminals (neuritic plaques) in specific brain regions as well as the deposition of extracellular amyloid in the plaque cores. Glenner and Wong have addressed this question by biochemical purification and analysis of meningeal amyloid from brains of Alzheimer [Glenner and Wong, 1984a] and Down's patients [Glenner and Wong, 1984b]; they have determined N-terminal peptide sequences. The two amyloid peptides showed only one difference in amino acid position 11 (Gln in AD amyloid versus Glu in Down's amyloid) among 24 residues analyzed. Subsequent studies by Masters and Beyreuther's group

AMYLOID PRECURSOR PROTEIN

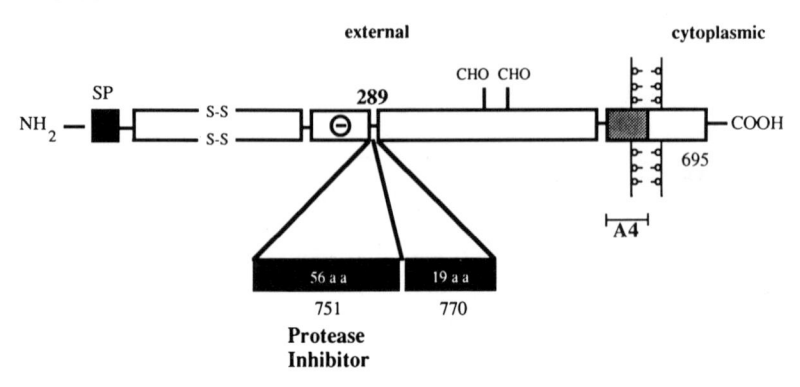

Fig. 2. Schematic drawing of APP-695,-751,-770. The putative cell-surface proteins span the membrane once. The amyloid A4 peptide partially resides in the transmembrane domain and in the extracellular domain, as indicated by the shaded region. There are two possible N-linked glycosylation sites (CHO) in the extracellular domain. The 56 amino acid putative protease inhibitor domain is inserted at residue 289 of the precursor protein, resulting in APP-751, and the additional insertion of the 19 amino acid domain results in APP-770. A highly negative-charged domain and a cysteine-rich domain are symbolized by a minus sign and S-S bridges respectively. The signal peptide (SP) is located at the N terminus.

of amyloid from Alzheimer brain plaque cores revealed amino acid sequences identical to the reported Down's cerebrovascular amyloid data [Masters et al., 1985b]. Copy-DNA analysis of APP transcripts from both normal tissue and Alzheimer brain material determined the codon for glutamic acid at this position [Kang et al., 1987; Goldgaber et al., 1987; Robakis et al., 1987; Tanzi et al., 1987a; Zain et al., 1988; Vitek et al., 1988].

The availability of protein sequence information from the amyloid protein in Alzheimer brains allowed the design of synthetic oligonucleotides complementary to the putative messenger RNA transcripts. Four groups independently reported successful cloning of cDNAs including the region of the amyloid protein sequence [Goldgaber et al., 1987; Kang et al., 1987; Robakis et al., 1987; Tanzi et al., 1987a], and one group [Kang et al., 1987] cloned the apparent full-length transcript (approximately 3.4 kb) for APP from a human fetal brain cDNA library. The 695-residue APP shows typical features of a glycosylated cell-surface transmembrane protein (Fig. 2). The C-terminal 12–14 residues of the A4-protein reside in the putative transmembrane domain of the precursor, and the 28 N-terminal residues are in the "extracellular domain" [Dyrks et al., 1988]. Genomic mapping localized the APP gene on human chromosome 21 using human–rodent somatic cell hybrids [Goldgaber et al.,

APP-GENE STRUCTURE

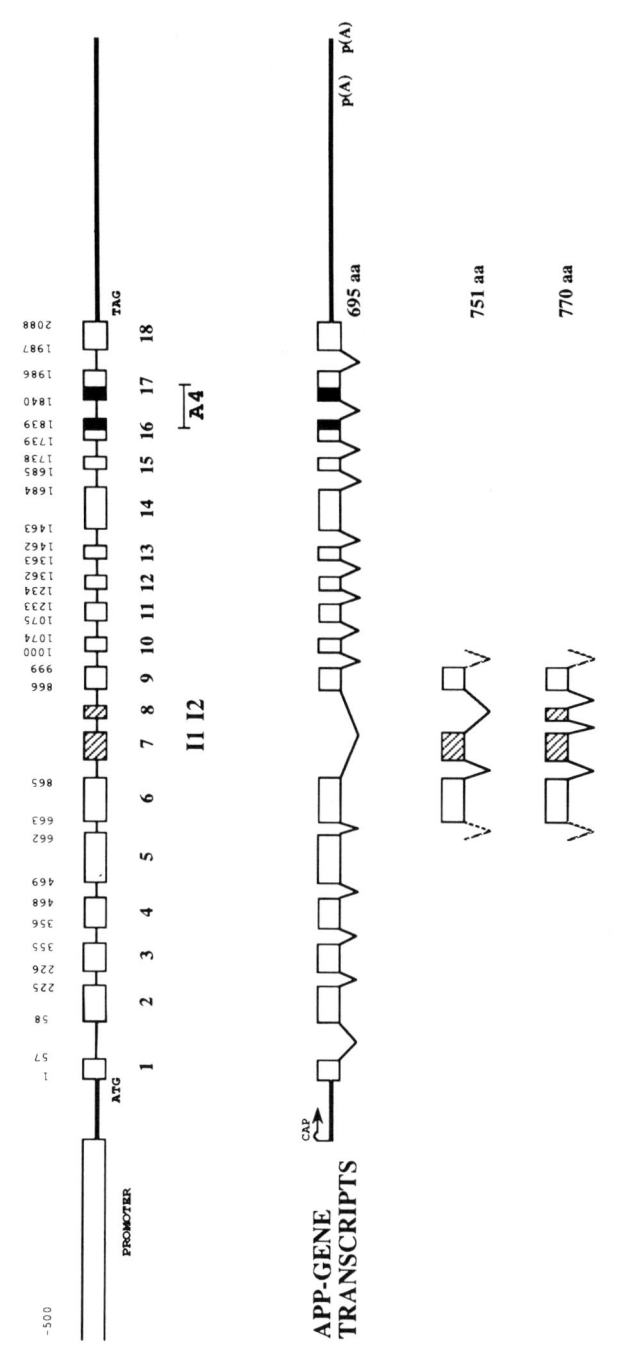

Fig. 3. Schematic outline of the APP gene structure and three alternatively spliced gene transcripts (APP-695,-751,-770). The numbers above the exon boxes 1–18 refer to coordinates determined by Lemaire et al. [1989] and indicate positions of exon/intron borders in the original APP-695 cDNA [Kang et al., 1987]. The exons encoding the alternatively spliced "aprotinin-like" domain (I1) and the 19 aa domain (I2) are hatched. The A4 amyloid region is indicated in black in boxes 16 and 17.

1987; Kang et al., 1987; Tanzi et al., 1987a]; by applying in situ hybridization techniques, this gene was sublocalized at chromosome 21q21 [Robakis et al., 1987] and more recently at the border of 21q21-22 [Blanquet et al., 1987; Patterson et al., 1988].

Chromosome 21 has been the subject of intensive studies because of its involvement in Down's syndrome (trisomy 21). Ninety-five percent of Down's cases are trisomic for the entire chromosome 21, 2–3% are mosaics (i.e., trisomic in only some cells), and the remainder (3–4%) are caused by triplication (translocation) of the distal part of the long arm (21q22) of chromosome 21 [Crome and Stern, 1972]. The occurrence of such translocations has led to the conclusion that this syndrome can be attributed to trisomy of the distal part (the "pathological region") of chromosome 21 [Summitt, 1981]. To date it is not known precisely where the breakpoint on the q arm of chromosome 21 is located, and it is not known whether Down's cases, which have partial trisomy, are demented and develop Alzheimer pathology. In this context it will be of particular interest to determine if the APP gene maps within the obligate Down's region. The localization of the APP gene on the long arm of chromosome 21, together with the development of AD pathology in Down's patients, provides a potential mechanism for the formation of amyloid on the basis of overexpression of a number of genes on 21, including the APP gene and the FAD gene locus (Fig. 1). Studies of genomic DNA from sporadic (nonfamilial) AD cases and "karyotypically normal" Down's cases implied the presence of a microduplication of a segment of chromosome 21 including the APP gene [Delabar et al., 1987; Schweber et al., 1987]. Subsequent analyses of large numbers of AD cases by several laboratories did not confirm these findings [Tanzi et al., 1987c; St George-Hyslop et al., 1987b; Podlisny et al., 1987; Warren et al., 1987].

Chromosomal mapping experiments, using human APP probes in human–rodent cell hybrids, have shown cross-hybridization with mouse and hamster genomic DNA [Kang et al., 1987; Tanzi et al., 1987a; Goldgaber et al., 1987]. Southern blot analysis of DNA from various species indicated that the APP gene is highly conserved during evolution. Comparison of the mouse APP sequence [Yamada et al., 1987] with the sequence from rat [Shivers et al., 1988] indicated 99% homology on the protein level; furthermore, the human sequence is 96.8% homologous to the mouse sequence and 97.3% homologous to the rat sequence. Based on the striking conservation of APP proteins, Yamada et al. [1987] have calculated the evolutionary rate of changes at the amino acid level to be 0.1×10^{-9}/site/year (which is comparable to that of cytochrome C), thus suggesting an essential biological function for APP proteins. K. White and colleagues have cloned a *Drosophila* gene (vnd locus) that is highly homologous to large regions of the APP sequence [K. White, personal communication]. Northern blot experiments have confirmed these data at the level of mRNA and have demonstrated in various mammalian species the ubiquitous expression of APP transcripts in a number of different tissues [Manning et al., 1988].

Kang et al. [1987] reported the presence of two distinct bands (3.2 and 3.4 kb) by Northern analysis of human fetal brain mRNA using APP cDNA as probe. This finding suggests either differential splicing of mRNA or alternative usage of polyadenylation sites. Both post-transcriptional events were found to be operative following detailed investigation by several groups. First, Kang et al. indicated a potential polyadenylation signal (AATAAA tandem repeat) 259 bp upstream of the 3' end of the reported APP full-length cDNA. The analysis of eight other full-length APP cDNA clones obtained from a human fetal brain cDNA library [Unterbeck, 1986] resulted in a 1:1 ratio of shorter cDNAs (3.2 kb) using the first polyadenylation signal versus the original cDNA forms (3.4 kb) using the second polyadenylation signal [Unterbeck, unpublished observations]. Interestingly, all eight clones encoded for 695 residues of APP. The alternative use of different polyadenylation signals in APP transcripts was confirmed by other laboratories [Goldgaber, 1988; Johnson et al., 1988]. Second, a number of groups have screened several tumor cell line-derived cDNA libraries for the presence of APP transcripts and have identified clones coding for new APP precursor molecules containing an additional domain that shows striking homology to the Kunitz family of serine protease inhibitors [Tanzi et al., 1988; Ponte et al., 1988; Kitaguchi et al., 1988]. Two independent laboratories reported cDNA sequences containing an additional 167 bp insert at residue 289 of the APP-695 precursor, which encodes 56 amino acids showing a high degree of homology to bovine aprotinin [Laskowski and Kato, 1980], a well-characterized trypsin inhibitor (Fig. 2). The sequences flanking the region of insertion of this domain are identical to the original APP-695 clone, resulting in an open reading frame encoding 751 residues (APP-751). Kitaguchi et al. [1988] isolated a third APP form with another addition of a 19 amino acid domain at the C-terminal end of the 56 amino acid "aprotinin-like" region of APP-751, thus resulting in a larger protein of 770 residues (APP-770). Transient expression of APP-770 in COS-1 cells conferred a marked inhibition of trypsin activity in cell lystates. Both additional domains have been found to be encoded by discrete exons [Kitaguchi et al., 1988], and all three transcripts are generated by differential splicing of a single gene on chromosome 21 (Fig. 3). These protease inhibitor domains have recently been found to be present also in mouse [Yamada et al., 1989] and rat [Kang and Muller-Hill, 1989] species.

The existence of at least three different amyloid precursor forms provoked the immediate question of the relationship between these precursors and the formation of amyloid in AD. It will be necessary to determine whether a specific form of APP contributes to A4 deposition. It is possible that either an imbalance in the relative expression levels of the three APP forms or their overexpression might be involved in AD pathology. Initial in situ hybridization analyses using APP cDNA probes in human CNS sections indicated that many neuronal cell types express these mRNAs [Bahmanyar et al., 1987; Goedert, 1987; Cohen et al., 1988; Higgins et al., 1988; Lewis et al., 1988; Schmechel et al., 1988], but because of the nature of the probes used, these studies did not allow a differential analysis of the various APP transcripts. Furthermore, there is little documented correlation between APP mRNA levels, amyloid deposition, and neuronal degeneration in AD. What seems certain, however, is that high levels of APP mRNAs alone do not form a sufficient prerequisite for cellular pathology in either the aging or AD brain [Higgins et al., 1988]. Northern analyses using specific probes that discriminate between the APP transcripts suggest a developmental and tissue-specific pattern of expression of these mRNAs [Tanzi et al., 1988; Kitaguchi et al., 1988; Neve et al., 1988]. Using polymerase chain reaction (PCR) technology with specific primers, the relative levels of all three mRNAs can be assessed within a given mRNA population [Bayney, Lee, and Unterbeck, unpublished observations]. In the tissues examined, all APP transcripts show a ubiquitous pattern of expression, with highest levels of APP-695 detected in the brain while APP-751 and APP-770 display significant expression in peripheral tissues. From a developmental viewpoint, APP-695 appears to be the most abundant species in fetal brain, while APP-751 shows marked expression in aged brain specimens. It is not clear if there is an aberration of a finely tuned ratio of APP transcripts in the progression from a healthy state to that of AD, ultimately leading to the disease phenotype. A number of investigators have addressed the question of potential gene dosage effects of APP expression in AD, reflecting, in part, the phenotype of Down's syndrome. While results of initial studies are to some extent controversial, semiquantitative in situ hybridization studies using specific probes for the APP transcripts seem to be the method of choice. Such studies allow a comparison of affected neurons (which may show abnormal immunoreactivities to tau or A68 in perikarya) with nonaffected neurons, in the same field [Palmert et al., 1988].

The APP Gene Promoter

The availability of full-length APP cDNA clones allowed the isolation, using 5'-end cDNA probes [Kang et al., 1987], of genomic clones containing the 5' end of the APP (PAD) gene [Salbaum et al., 1988; La Fauci et al., 1989; Bayney, Lee, and Unterbeck, unpublished observations]. Approximately 3.7 kb of sequences upstream of the strongest RNA start site have been extensively analyzed [Salbaum et al., 1988]. By a combination of primer extension and S1 protection analyses, five putative transcription initiation sites have been determined within a ten bp region. This 3.7 kb region lacks a typical

TATA box and displays a 72% GC-rich content in a region (−1 to −400) that confers promoter activity to a reporter gene in an in vivo assay system [Salbaum et al., 1988]. The absence of a typical TATA and CAAT box, the presence of multiple RNA start sites, and the ubiquitous expression of transcripts derived from the APP gene [Bayney, Lee, and Unterbeck, unpublished observations] is suggestive of its function as a housekeeping gene but does not imply constitutive gene expression [Salbaum et al., 1988]. The regulatory region contained within 400 bp upstream of the strongest RNA start site shows a variety of typical promoter-binding elements (Fig. 4). These include two AP-1 consensus sites [Lee et al., 1987], a single heat shock recognition consensus element [Wu et al., 1987], and several copies of a 9-bp-long GC-rich consensus sequence where sequence-specific binding has been shown to occur by gel-retardation studies [Salbaum et al., 1988]. In addition, the CpG:GpC ratio in this promoter region has been found to be 1:1, in contrast to the 1:5 ratio found in many eucaryotic DNAs [Razin and Riggs, 1980]; CpG dinucleotides are known to control gene expression via DNA methylation [Doerfler, 1983]. In addition, palindromic sequences capable of forming hairpin-like structures are found around the RNA start sites [La Fauci et al., 1989].

Recently, several groups of investigators have determined the consensus binding sequence (AT-rich decamer) for a number of different homeobox proteins

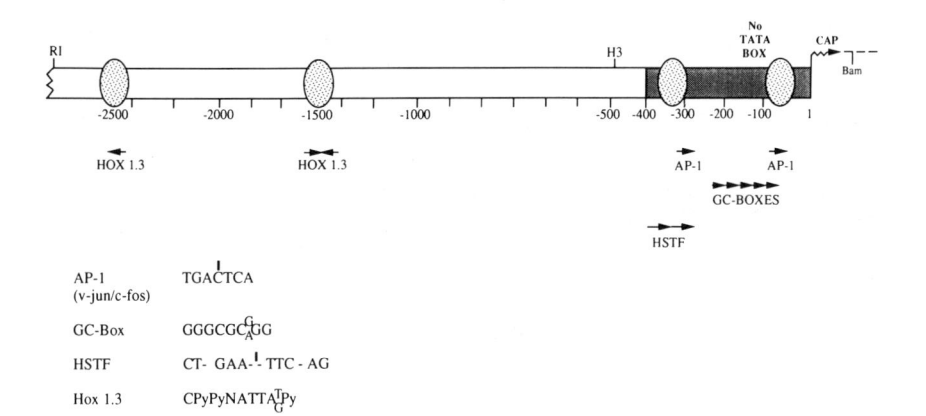

Fig. 4. The 5' regulatory region of the APP gene. The DNA segment 400 bp upstream (shaded segment) of the CAP sites contains typical promoter elements such as binding sites for the transcriptional factor AP-1, which is analogous to the product of the proto-oncogene v-jun and associated with the product of c-fos [Salbaum et al., 1988]. Multiple GC-boxes are located in this region, two of which participate in sequence-specific protein binding [Salbaum et al., 1988]. At least one consensus sequence for heat shock transcription factors (HSTF) is present in this region [Salbaum et al., 1988]. Consensus sequences for the homeobox protein Hox-1.3 [Odenwald et al., 1989] are located further upstream between positions −1400 bp and −2600 bp (see text, The APP Promoter).

[Desplan et al., 1988; Hoey and Levine, 1988; Ko et al., 1988; Odenwald et al., 1989], which most likely act as transcription factors in specific regions during embryogenesis [for review, see Gehring, 1987; Holland and Hogan, 1988]. As yet, target genes, developmentally regulated by homeobox proteins, have not been identified. It is apparent, however, that these genes will have an important role during embryogenesis and potentially throughout the life span. The APP gene promoter contains at least five of these homeobox binding sites upstream of the RNA start sites (Fig. 4). Preliminary experiments have shown that the homeobox protein Hox-1.3 [Odenwald et al., 1987,1989] binds at two of these sites [Goldgaber and Odenwald, personal communication]. Thus the APP gene, whose expression is developmentally regulated, appears to be a candidate gene for homeobox protein regulation. It remains to be determined whether any of these putative recognition consensus elements modulate the expression of the APP gene promoter.

TRANSGENIC ANIMAL TECHNOLOGY

The introduction of genes into the germline of animals is an extremely powerful technique for the generation of disease models that may lead to a better understanding of disease mechanisms [Cuthbertson and Klintworth, 1988; Jaenisch, 1988; Rosenfeld et al., 1988]. Cell culture in vitro systems can not duplicate the complex physiological interactions inherent in animal systems. Transgenic animals have been successfully generated from a number of species, including mice, sheep, and pig [Church, 1987; Clark et al., 1987]. Genes are microinjected directly into the pronuclei of one-cell embryos. A high percentage of reimplanted embryos develops normally, and, in a significant proportion of progeny, the transgene has integrated into chromosomal DNA. Usually, multiple copies of the transgene integrate as a head-to-tail array. Although mosaic animals can be generated, germline transmission of the transgene usually occurs [Hogan et al., 1986; DePamphilis et al., 1988].

Tissue-Specific Expression

Introduction of genes encoding various forms of the APP protein into transgenic mice may provide a powerful tool to elucidate the relationship between expression of the proteins and neuronal cell death. For these experiments it is important to direct expression to appropriate cell types. Transgenic animal technology has facilitated the characterization of regulatory elements that do confer appropriate expression patterns for a variety of genes. To date, the identification of essential regulatory elements for many genes has not been straightforward. A number of critical factors contribute to the complexity of this problem. First, gene promoters that exert cell-specific regulation in DNA transfection experiments do not necessarily confer cell and tissue specificity in transgenic animals. For example, transfection experiments revealed that important cell-specific regulatory elements reside within 400 bp upstream of the cap site of the rat albumin gene [Ott et al., 1984; Friedman et al., 1986]. However, an additional enhancer, located 10 kb upstream from the albumin promoter, was found to be necessary to direct liver-specific expression in transgenic mice [Pinkert et al., 1987]. While promoter sequences of the α-fetoprotein gene confer cell specificity in cell culture [Godbout et al., 1986; Muglia and Rothman-Denes, 1986; Widen and Papaconstantinou, 1986], additional enhancer elements located between −1 kb and −7 kb were found to be necessary for liver-specific expression in transgenic animals [Hammer et al., 1987]. Second, the organization of various genes differs considerably, and essential regulatory elements have been found in numerous positions. In some cases, the necessary regulatory elements are located within a compact regional proximal to the cap site. For example, sequences residing within −205 nt to +8 nt of the rat elastase I gene are sufficient to confer an appropriate expression pattern in transgenic mice [MacDonald et al., 1987]. A tightly defined regulatory region has also been identified in the human γ-crystalin gene [Goring et al., 1987]. The human β-globin gene, however, has at least four separate regulatory elements: a positive globin-specific promoter element, a negative regulatory element, and two gene enhancers, one located within the second intron and the other located 3′ of the structural gene [Behringer et al., 1987; Grosveld et al., 1987]. Third, in many cases, the site of integration exerts a strong influence on the level and pattern of expression of transgenes. Regions of several genes have been identified that overcome, at least in part, these position effects. DNase I hypersensitive sites located approximately 50 kb 5′ and 20 kb 3′ of the β-globin gene facilitate position-independent, high-level expression of a β-globin minigene in transgenic mice [Grosveld et al., 1987]. Furthermore, introns of the rat growth hormone and mouse metallothionein genes increase transcriptional efficiency of transgenes on average 10- to 100-fold [Brinster et al., 1988]. Rat growth hormone intronic sequences exerted a positive effect even on heterologous gene constructions utilizing either the metallothionein or elastase promoters. The effect of the introns was not related to an increased efficiency of RNA processing but was due to an actual increase in the rate of transcription [Brinster et al., 1988]. It is possible that introns and other genomic regions contain sequence elements that are recognized at particular stages of development or may contain elements that influence chromatin structure. In many cases, the inclusion of genomic elements that diminish position effects may be essential for a transgene to maintain an expression level sufficient to generate a phenotype. The identification of these elements in some cases may be a formidable task; for example, the APP gene locus encompasses at least 50 kb [Lemaire et al., 1989]. The identification of Dnase I hypersensitive sites [Grosveld et al., 1987] in and around the gene may facilitate the identification of essential regulatory elements.

ANIMAL MODELS FOR DISORDERS OF THE NERVOUS SYSTEM

A number of unique approaches have been applied to the development of transgenic models applicable to disorders of the nervous system. With the use of tissue-specific gene promoters, the function of dominant acting genes, such as oncogenes, mutant genes, and viral gene products, can be assessed in the nervous system of transgenic animals. The effects of copy number and overexpression of a specific gene can be determined by the introduction of a transgene. Levels of expression of an endogenous mouse gene can be modulated either by correcting an heritable mutation with a transgene expressing a normal gene product or possibly by expressing an antisense RNA moiety that is complementary to a normal cellular transcript.

Transgenic animals expressing viral gene products provide insight into the involvement of viruses in specific brain pathologies. Transgenic mice expressing SV40 T-antigen under the regulation of the viral enhancer/promoter develop papillomas of the choroid plexus [Brinster et al., 1984; Small et al., 1985; Messing et al., 1988; Marks et al., 1989]. A related human papovavirus, JC, has been implicated, in humans in progressive multifocal leukoencoephalopathy (PML), a chronic CNS demyelinating disease [ZuRhein, 1972; Johnson, 1983], and will cause astrocytomas and glioblastomas in primates [London et al., 1978]. Transgenic mice containing the JC virus early region in some cases develop neuroblastomas of the adrenal medulla and in other cases develop a dysmyelination of the CHS and exhibit a shaking disorder similar to that observed with *jimpy* or *quaking* mutant mice [Small et al., 1986; Trapp et al., 1988]. The JC virus-containing transgenic mice, therefore, are a good model for investigating the mechanisms of JC virus-induced demyelinating lesions in PML. Another interesting model has arisen from experiments in which the *tat* gene from human T-lymphotropic virus type 1 (HTLV-1), under the regulation of the viral long-terminal repeat, was introduced into transgenic mice. These transgenic mice developed tumors with properties that closely resemble human neurofibromatosis (von Recklinghausen's disease) [Hinrichs et al., 1987].

Transgenic animals have been generated that express altered levels of myelin basic protein (MBP). The animals should facilitate a better understanding of the myelination process in the CNS. A cosmid clone containing all seven exons of mouse MBP was introduced into the germline of shiverer (*shi*) mice [Readhead et al., 1987]. Shiverer mice are homozygote mutants that lack MBP and exhibit tremors, convulsions, and premature death. The *shi* phenotype was partially corrected in the MBP transgenic mice. Different lines of transgenic mice expressed graded levels of MBP [Popko et al., 1987]. In another approach, normal mice were converted to a shiverer phenotype by expressing mouse MPB antisense RNA [Katsuki et al., 1988]. Antisense RNA, being complementary to cellular mRNA, has been shown to repress the expression of a number of genes in cell culture [Kim and Wold, 1985; Weintraub et al., 1985; Trevor et al., 1987]. The MBP antisense construct contained the MBP promoter region (about 1.3 kb), and antisense RNA was expressed in a tissue-specific manner in transgenic animals. Again, graded phenotypes were obtained that were correlated with different levels of antisense RNA expression and endogenous mRNA levels. Given a strong tissue-specific promoter, the antisense RNA approach in transgenic animals may be utilized to modulate the expression levels of APP and other genes.

Neuropathological lesions such as neurofibrillary tangles, senile plaques, and neuronal cell loss are common in Down's patients reaching the third decade of life and are indistinguishable from Alzheimer disease pathology [Epstein, 1986a,b]. A mouse model for Down's syndrome may provide valuable insights into the process of amyloidosis. Mouse chromosome 16 carries a number of genes found on human chromosome 21, including: Cu/Zn superoxide dismutase (SOD-1), phosphoribosylglycinamine synthetase (PRGS), ETS-2, and APP [Reeves et al., 1986; Cheng et al., 1988]. Unfortunately, trisomy 16 mice are not fully viable and die before birth [Reeves et al., 1986]. As an alternative, transgenic mice overexpressing specific human chromosome 21 genes may develop Down's and/or Alzheimer pathology. Transgenic mice expressing human Zn/Cu superoxide dismutase have been generated [Epstein et al., 1987; Avraham et al., 1988]. The tongue neuromuscular junctions of the transgenic animals exhibited significant pathological changes, which were similar to those observed in aged mice and rats and in tongue muscles of patients with Down's syndrome. Overexpression of superoxide dismutase has been postulated to cause oxidative damage to biologically important molecules [Halliwell and Gutteridge, 1985]. In fact, mouse L cells overexpressing superoxide dismutase showed increased lipid peroxidation, and PC12 cells overexpressing the enzyme were impaired in neurotransmitter uptake [Elroy-Stein et al., 1986, 1988].

Another gene implicated in the neurologic abnormalities in DS encodes the β subunit of the S100 protein and maps to the distal half of the long arm of human chromosome 21 [Allore et al., 1989]. The S100 protein is a calcium-binding protein and participates in several calcium-dependent interactions with neuroleptic drugs and brain proteins [Zimmer and Van Eldik, 1986]. A transgenic mouse overexpressing the S100 protein alone or in conjunction with other human chromosome 21 genes such as APP and SOD-1 may generate Alzheimer pathology.

HUMAN AMYLOID PRECURSOR EXPRESSION IN TRANSGENIC MICE

The primary defect leading to AD is not yet known, and specific mutations that cause AD in humans have not been defined. With the exception of aged primates [Price et al., 1989], no laboratory animal model for AD exists. Because of these limitations, the generation of transgenic mouse models of AD appears to be a rational approach for defining the role APP plays in the pathology of AD. Transgenic mice carrying APP genes altered in their protein-coding sequences or in their expression levels may lead to dominant mutant phenotypes resembling those displayed in AD pathology. The choice of an appropriate gene promoter for the minigene is a critical step in the development of the transgenic mouse model. A gene promoter must be utilized that will facilitate the expression of recombinant genes with a cell and tissue specificity consistent with the formation of amyloid plaque and perhaps with the expression pattern of the endogenous mouse APP gene. As a first step toward this goal, the 5' end of the human APP gene can be cloned into minigene constructs (Fig. 5).

A reporter gene can be utilized to establish the cell and tissue specificity of the 5' end of the APP gene in transgenic mice. The *E. coli lac* Z gene (encoding β-galactosidase) functions as a convenient reporter gene in mammalian cells [Hall et al., 1983]. A histochemical reaction specifically stains any cell expressing the *lac* Z gene product [Lojda, 1970]. Background staining, resulting from endogenous β-galactosidase activity, is usually not observed. Recombinant retroviruses expressing the *lac* Z gene have been used to study

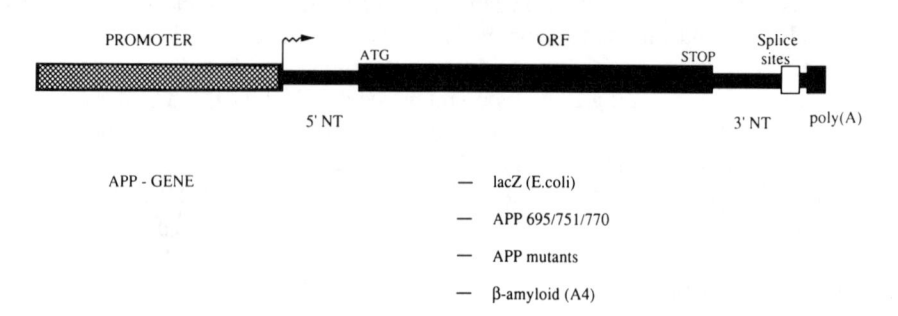

Fig. 5. Generic scheme of typical APP mini-gene constructions for expression in transgenic mice. The elements are: the APP gene promoter fused to a cDNA containing the CAP site and the 5' nontranslated region (5'NT), an open reading frame (ORF), and the 3' nontranslated region (3'NT) followed by donor and acceptor RNA splice sites and signals for polyadenylation of the transcript. Examples of alternative ORFs and their corresponding protein products are indicated.

postimplantation cell lineages in mouse embryos [Sanes et al., 1986]. The precise tissue specificity of the γ-crystallin promoter was demonstrated in transgenic mice using a γ-crystallin/*lac* Z minigene [Goring et al., 1987]. Furthermore, the *lac* Z gene has facilitated the identification of a 5' upstream element of the Hox-1.3 gene, which, in transgenic mice, directs spatially restricted expression in the embryonic central nervous system [Zakany et al., 1988]. Using the *lac* Z reporter gene (Fig. 5), we have initial results suggesting that the 5' end of the APP gene carries sufficient regulatory elements to direct cell and tissue-specific expression in transgenic mice. This expression pattern appears to be consistent with the expression of the endogenous mouse APP gene (Wirak et al., 1989).

The choice of an open reading frame for a minigene that would promote amyloidosis in transgenic animals is not immediately apparent. Based on available information regarding the molecular biology of the disease, one can, at this point, only test a number of hypotheses. First, the appearance of Alzheimer pathology in Down's patients suggests that gene dosage and/or gene expression levels may be a contributing factor in Alzheimer's disease. Expression of one or more of the alternatively spliced forms of the human APP gene products in transgenic mice will test whether overexpression or anomalous expression of one or more forms of the gene will lead to Alzheimer's pathology. To mimic the Alzheimer disease phenotype, it may be necessary to generate transgenic mice carrying multiple copies of several human chromosome 21 genes, including the APP, SOD-1, and S100 genes. Second, the accumulation of the A4 peptide, in amyloid plaques may be the result of anomalous proteolytic degradation of one or more forms of APP. Mutant APP gene products can be expressed in transgenic mice that may mimic the onset of proteolytic events, leading to amyloid plaque formation. Third, expression of the A4 peptide can evaluate the cytotoxicity of amyloid formation in transgenic mice. It is possible that the accumulation of amyloid plaques in mice will lead to additional neurodegenerative processes that may be analogous to those observed in Alzheimer and Down's patients. These and future approaches, utilizing transgenic animal technology, may lead to a better understanding of the biological function of APP and its role in the pathology of Alzheimer's disease.

FUTURE PROSPECTS

In the context of models for human disease, transgenic animals can be viewed as sophisticated assay systems in which the role of specific genes and proteins in complex physiological processes can be determined. Genes encoding proteins suspected of playing a role in disease processed can be introduced into animals, and any resulting pathology or phenotypes can be assayed. This approach is potentially very informative in elucidating disease mechanisms and in identifying appropriate targets for pharmacological intervention.

The transgenic animal approach to disease models, although powerful, presently suffers from four significant limitations. The first is that many diseases may have a polygenic etiology. In such a case, introduction of a single gene at best will result in only a partial pathology or phenotype. Introducing individual genes into mice and mating the offspring is a way of combining the effects of multiple genes, but such an approach still does not take into account relative expression levels and requires knowledge of many genes involved in the disease process. One possible way around such a problem, especially when many genes involved are genetically linked, such as in AD and Down's syndrome, is the introduction of large chromosomal regions into animals. Development of techniques to introduce large chromosomal regions will be a major contribution to the understanding of disease processes.

A second major limitation is that in order to elicit a phenotype, the introduced genes must act dominantly. Thus deficiencies that arise from mutant genes often can not be duplicated on the normal genetic background found in transgenic animals. One approach to overcome this problem is gene replacement, i.e., the replacement of the wild-type gene with a mutant version. Recently it has been demonstrated that embryonic stem cells can be maintained in culture [Evans and Kaufman, 1981; Martin, 1981], manipulated, and introduced into mouse embryos; they then participate in development, resulting in chimeric animals [Bradley et al., 1984; Robertson et al., 1986]. The cultured cells often participated in germ cell formation, thus resulting in F1 progeny derived from the cultured cells. The demonstration of gene replacement via homologous recombination in embryonic stem cells [Thomas and Capecchi, 1987; Doetschman et al., 1987] provided a route to gene replacement in intact animals. There is little doubt that this technology will develop into an extremely powerful tool for genetic analysis and generation of disease models.

A third limitation is a result of our incomplete understanding of gene regulation. Although many gene constructs introduced into transgenic animals are expressed in patterns similar to their endogenous counterparts, there are a number of examples (cited above) in which proper regulation in animals requires sequences in addition to those required in cell culture. It could be that additional sequences are required for fine-level gene control. For example, in focal diseases such as AD, if localized environmental factors contribute to the pathology via effects on gene expression, there is no assurance that the introduced constructs will contain the *cis*-acting responsive sequences, even if they are expressed in overall patterns similar to the endogenous genes.

A final caveat pertains to differences in physiology between the transgenic animals and humans. For example, mice made negative for HPRT had no phenotype, while HPRT deficiency is an extremely serious condition in humans [Hooper et al., 1987; Kuehn et al., 1987]. Additionally, species differences may interfere with intermolecular interactions, thus causing failure to observe pathology. For example, if the A4 peptide arises from proteolytic cleavage, there is no assurance that the murine enzyme will cleave the human precursor.

For now, studies of Alzheimer's disease in transgenic animals will consist mostly of introducing individual genes thought to be involved in the pathology. It is likely that significant information about the etiology of Alzheimer's disease will result from these studies. However, if the disease has a polygenic etiology, or if the known genes are not the ultimate cause, then the information will be limited and the resultant animals will be, at best, partial models. Isolation of the FAD locus may partially overcome this potential problem, but even then there is no assurance that this mutant gene will function appropriately in a transgenic mouse system or act in a dominant fashion.

The use of trangenic animals for studies of AD, and of other human diseases, is a field in the early stages of development. It is likely that the development of additional techniques for genetic manipulation will parallel, and act synergistically with, a more detailed understanding of the biochemistry of the disease processes. We can look forward to a time when truly informative and powerful disease models can be generated. There can be little doubt that such animals will be a major tool in the elucidation of the etiology of major human diseases.

REFERENCES

Allore R, O'Hanlon D, Price R, Neilson K, Willard HF, Cox DR, Marks A, Dunn RJ (1989): Gene encoding the β subunit of S100 protein is on chromosome 21: Implications for Down syndrome. Science 239:1311–1313.

Alzheimer A (1907): Uber eine eigenartige Erkrankung der Hirnrinde. Allg Z Psychiat Psych Gerichtl Med 64:146–148.

Avraham KB, Schickler M, Sapoznikov D, Yarom R, Groner Y (1988): Down's syndrome: Abnormal neuromuscular junction in tongue of transgenic mice with elevated levels of human Cu/Zn-superoxide dismutase. Cell 54:823–829.

Baymanyar S, Higgins GA, Goldgaber D, Lewis DA, Morrison JH, Wilson MC, Shankar SK, Gajdusek DC (1987): Localization of amyloid β protein messenger RNA in brains from patients with Alzheimer's disease. Science 237:77–79.

Behringer RR, Hammer RE, Brinster RL, Palmiter RD, Townes TM (1987): Two 3Ψ sequences direct adult erythroid-specific expression of human β-globin genes in transgenic mice. Proc Natl Acad Sci USA 84:7056–7060.

Blanquet V, Turleau C, Stehelin D, Creau-Goldberg N, Delabar JM, Sinet PM, Davous P, de Grouchy Y (1987): Cytogenet Cell Genet (abstract) 46:583.

Botstein D, White RL, Skolnick M, Davis RW (1980): Construction of a genetic linkage map in man using restriction fragment length polymorphisms. Am J Hum Genet 32:314–331.

Bradley A, Evans M, Kaufman MH, Robertson E (1984): Formation of germ-like chimaeras from embryo-derived teratocarcinoma cell lines. Nature 309:255–256.

Brinster RL, Chen HY, Messing A, van Dyke T, Levine AJ, Palmiter RD (1984): Transgenic mice harboring SV40 T-antigen genes develop characteristic brain tumors. Cell 37:367–379.

Brinster RL, Allen JM, Behringer RR, Gelinas RE, Palmiter RD (1988): Introns increase transcriptional efficiency in transgenic mice. Proc Natl Acad Sci USA 85:836–840.

Brion JP, Couck AM, Passareiro E, Flament-Durand J (1985): Neurofibrillary tangles of Alzheimer's disease: An immunohistochemical study. J Submicrosc Cytol 17:89–96.

Cheng SV, Nadeu JH, Tanzi RE, Watkins PC, Jagadesh J, Taylor BA, Haines JL, Sacchi N,

Gusella JF (1988): Comparative mapping of DNA markers from the familial Alzheimer disease and Down syndrome regions of human chromosome 21 to mouse chromosomes 17 and 17. Proc Natl Acad Sci USA 85:6032–6036.

Church RB (1987): Embryo manipulation and gene transfer in domestic animals. Trends Biotechnol 5:13–19.

Clark AJ, Simons P, Wilmut I, Lathe R (1987): Pharmaceuticals from transgenic livestock. Trends Biotechnol 5:20–24.

Cohen ML, Golde TE, Usiak MF, Younkin LH, Younkin SG (1988): In situ hybridization of nucleus basalis neurons shows increased β-amyloid mRNA in Alzheimer's disease. Proc Natl Acad Sci USA 85:1227–1231.

Crome L, Stern J (1972): "Pathology of Mental Retardation." Edinburgh: Churchill Livingstone.

Cuthbertson R, Klintworth GK (1988): Biology of disease: Transgenic mice—a gold mine for furthering knowledge in pathobiology. Lab Invest 58:484–501.

Delabar JM, Goldgaber D, Lamour Y, Nicole A, Huret JL, de Grouchy J, Brown P, Gajdusek DC, Sinet PM (1987): β-amyloid gene duplication in Alzheimer's disease and karyotypically normal Down's syndrome. Science 235:1390–1392.

DePamphilis ML, Herman SA, Martinez-Salas E, Chalifour LE, Wirak DO, Cupo DY, Miranda M (1988): Microinjecting DNA into mouse ova to study DNA replication and gene expression and to produce transgenic animals. BioTechniques 6:662–680.

Desplan C. Theis J, O'Farrel P (1988): The sequence specificity of homeodomain-DNA interaction. Cell 54:1081–1090.

Doerfler W (1983): DNA methylation and gene activity. Annu Rev Biochem 52:93–124.

Doetschman T, Gregg RG, Maeda N, Hooper ML, Melton DW, Thompson S, Smithies O (1987): Targetted correction of a mutant HPRT gene in mouse embryonic stem cells. Nature 330:576–578.

Dyrks T, Weidemann A, Multhaup G, Salbaum M, Lemaire HG, Kang J, Muller-Hill B, Masters CL, Beyreuther K (1988): Identification, transmembrane orientation and biogenesis of the amyloid A4 precursor of Alzheimer's disease. EMBO J 7:949–957.

Elroy-Stein O, Groner Y (1988): Impaired neurotransmitter uptake in PC12 cells overexpressing human Cu/Zn-superoxide dismutase—implication for gene dosage effects in Down syndrome. Cell 52:259–267.

Elroy-Stein O, Bernstein Y, Groner Y (1986): Overproduction of human Cu/Zn-superoxide disumutase in transfected cells: Extenuation of paraquat-mediated cytotoxicity and enhancement of lipid peroxidation. EMBO J 5:615–622.

Epstein CJ (1986a): The consequences of chromosome imbalance. In "Principles, Mechanisms, Models." New York: Cambridge University Press.

Epstein CJ (ed) (1986b): "The Neurobiology of Down Syndrome." New York: Raven Press.

Epstein CJ, Avraham KB, Lovett , Smith S, Elroy-Stein O, Rotman G, Bry C, Groner Y (1987): Transgenic mice with increased Cu/Zn-superoxide dismutase activity: Animal model of dosage effects in Down syndrome. Proc Natl Acad Sci USA 84:8044–8048.

Evans MJ, Kaufman MH (1981): Establishment in culture of pluripotential cells from mouse embryos. Nature 292:154–156.

Friedman JM, Babiss LE, Clayton DF, Darnell JE Jr (1986): Cellular promoter incorporated into the adenovirus genome: Cell specificity of albumin and immunoglobulin expression. Mol Cell Biol 6:3791–3797.

Gehring WJ (1987): Homeo boxes in the study of development. Science 236:1245–1252.

Glenner GG, Wong CW (1984a): Alzheimer's disease: Initial report of the purification and characterization of a novel cerebrovascular amyloid protein. Biochem Biophys Res Commun 120:885–890.

Glenner GG, Wong CW (1984b): Alzheimer's disease and Down's syndrome: Sharing of a unique cerebrovascular amyloid fibril protein. Biochem Biophys Res Commun 122:1131–1135.

Godbout R, Ingram R, Tilghman SM (1986): Multiple regulatory elements in the intergenic region between the α-fetoprotein and albumin genes. Mol Cell Biol 6:477–487.

Goedert M (1987): Neuronal localization of amyloid beta protein precursor mRNA in normal human brain and in Alzheimer's disease. EMBO J 6:3627–3632.

Goedert M, Wischik CM, Crowther RA, Walker JE, Klug A (1988): Cloning and sequencing of a cDNA encoding a core protein of the paired helical filament of Alzheimer's disease: Identification as the microtubule-associated protein tau. Proc Natl Acad Sci USA 85:4051–4055.

Goldgaber D (1988): The amyloid β-protein precursor gene encodes a family of secreted polyproteins: A hypothesis. In Finch CE, David P (eds): "The Molecular Biology of Alzheimer's Disease. Current Communications in Molecular Biology." New York: Cold Spring Harbor Press, pp 66–70.

Goldgaber D, Lerman MI, McBride OW, Saffiotti U, Gajdusek DC (1987): Characterization and chromosomal of a cDNA encoding brain amyloid of Alzheimer's disease. Science 235:877–880.

Goring DR, Rossant J, Clapoff S, Breitman ML, Tsui L-C (1987): In situ detection of β-galactosidase in lenses of transgenic mice with a γ-crystallin/*lac* Z gene. Science 235:456–458.

Grosveld F, van Assendelft GB, Greaves DR, Kolias G (1987): Position-independent, high-level expression of the human β-globin gene in transgenic mice. Cell 51:975–985.

Grundke-Iqbal I, Iqbal K, Quinlan M, Tung YC, Zaidi MS, et al (1986): Microtubule-associated protein tau: A component of Alzheimer paired helical filaments. J Biol Chem 261:6084–6089.

Hall CV, Jacob PE, Ringold GM, Lee F (1983): Expression and regulation of *Escherichia coli lacZ* gene fusions in mammalian cells. J Mol Appl Genet 2:101–109.

Halliwell B, Gutteridge JMC (1985): "Free Radicals in Biology and Medicine." Oxford: Clarendon.

Hammer RE, Krumlauf R, Camper SA, Brinsrer RL, Tilghman SM (1987): Diversity of α-fetoprotein gene expression in mice is generated by a combination of separate enhancer elements. Science 235:53–58.

Higgins GA, Lewis DA, Bahmanyar S, Goldgaber D, Gajdusek DC, Young WG, Morrison JH, Wilson MC (1988): Differential regulation of amyloid-β-protein mRNA expression within hippocampal neuronal subpopulations in Alzheimer's disease. Proc Natl Acad Sci USA 85:1297–1301.

Hinrichs SH, Nerenberg M, Reynolds RK, Khoury G, Jay G (1987): A transgenic mouse model for human neurofibromatosis. Science 237:1340–1343.

Hoey T, Levine M (1988): Divergent homeo box proteins recognize similar DNA sequences in *Drosophila*. Nature 332:858–861.

Hogan B, Constantini F, Lacy E (1986): "Manipulating the Mouse Embryo: A Laboratory Manual." Cold Spring Harbor: Cold Spring Harbor Press.

Holland PWH, Hogan BLM (1988): Expression of homeo box genes during mouse development: A review. Gene Dev 2:773–782.

Hooper M, Hardy K, Handyside A, Hunter S, Monk M (1987): HPRT-deficient (Lesch-Nyhan) mouse embryos derived from germline colonization by cultured cells. Nature 326:292–295.

Jaenisch R (1988): Transgenic animals. Science 240:1468–1474.

Johnson RT (1983): Evidence for polyomaviruses in human neurological diseases. In Sever JL, Madden DL (eds): "Polyomaviruses and Human Neurological Diseases." New York: Alan R. Liss, pp 183–190.

Johnson SA, Pasinetti GM, May PC, Ponte PA, Cordell B, Finch CE (1988): Selective reduction of mRNA for the β-amyloid precursor protein that lacks a Kunitz-type protease inhibitor motif in cortex from Alzheimer brains. Exp Neurol 102:264–268.

Kang J, Muller-Hill B (1989): The sequence of the two extra exons in rat preA4. Nucleic Acids Res 17:2130.
Kang J, Lemaire HG, Unterbeck A, Salbaum MJ, Masters CL, Grzeschik KH, Multhaup G, Beyreuther K, Muller-Hill B (1987): The precursor of Alzheimer's disease amyloid A4 protein resembles a cell-surface receptor. Nature 325:733–736.
Katsuki M, Sato M, Kimura M, Yokoyama M, Kobayashi K, Nomura T (1988): Conversion of normal behavior to shiverer by myelin basic protein antisense cDNA in transgenic mice. Science 241:593–595.
Katzman R (1986): Alzheimer's disease. N Engl J Med 314:964–973.
Kemper T (1984): Neuroanatomical and neuropathological changes in normal aging and in dementia. In Albert ML (ed.): "Clinical Neurology of Aging." New York: Oxford University Press, pp 9–52.
Kidd M (1963): Paired helical filaments in electron microscopy of Alzheimer's disease. Nature 197:192–193.
Kim SK, Wold BJ (1985): Stable reduction of thymidine kinase activity in cells expressing high levels of anti-sense RNA. Cell 42:129–138.
Kitaguchi N, Takahashi Y, Tokushima Y, Shiojiri S, Ito H (1988): Novel precursor of Alzheimer's disease amyloid protein shows protease inhibitory activity. Nature 331:530–532.
Ko HS, Fast P, McBride W, Staudt LM (1988): A human protein specific for the immunoglobulin octamer DNA motif contains a functional homeobox domain. Cell 55:135–144.
Kosik KS, Joachim CL, Selkoe DJ (1986): The microtubule-associated protein, tau, is the major antigenic component of paired helical filaments in Alzheimer's disease. Proc Natl Acad Sci USA 83:4044–4048.
Keuhn MR, Bradley A, Robertson EJ, Evans MJ (1987): A potential animal model for Lesch–Nyhan syndrome through introduction of HPRT mutations into mice. Nature 326:295–297.
La Fauci G, Lahiri DK, Salton SRJ, Robakis NK (1989): Characterization of the 5Ψ-end region and the first two exons of the β-protein precursor gene. Biochem Biophys Res Commun 159:297–304.
Laskowski M Jr, Kato I (1980): Protein inhibitors of proteinases. Annu Rev Biochem 49:593–626.
Lee W, Haslinger A, Karin M, Tjian R (1987): Activation of transcription by two factors that bind promoter and enhancer sequences of the human metallothionein gene and SV 40. Nature 325:369–372.
Lemaire HF, Salbaum JM, Multhaup G, Kang J, Bayney RM, Unterbeck A, Beyreuther K, Muller-Hill B (1989): The preA4695 precursor protein of Alzheimer's disease A4 amyloid is encoded by 16 exons. Nucleic Acids Res 17:517–522.
Lewis DA, Higgins GA, Young WG, Goldgaber D, Gajdusek DC, Wilson MC, Morrison JH (1988): Distribution of precursor amyloid β-protein messenger RNA in human cerebral cortex: Relationship to neurofibrillary tangles and neuritic plaques. Proc Natl Acad Sci USA 85:1691–1695.
Lojda Z (1970): Indigogenic methods for glycosidases. II. An improved method for beta-D-gactosidase and its application to localization studies of the enzymes in the intestine and other tissues. Histochemie 23:266–288.
London WT, Houff SA, Madden DL (1978): Brain tumors in owl monkeys inoculated with a human polyoma virus (JC virus). Science 201:1246–1249.
MacDonald RJ, Swift GH, Hammer RE, Ornitz DM, Davis BP, Brinster RL, Palmiter RD (1987): Targeted expression of cloned genes in transgenic mice. In Seil FJ, Herbert E, Carlson BM (eds): "Progress in Brain Research," Vol. 71. Amsterdam: Elsevier Science Publishers, pp 3–12.
Manning RW, Reid CM, Lampe RA, Davis LG (1988): Identification in rodents and other species of a mRNA homologous to the beta-amyloid precursor. Brain Res 427:293–297.

Marks JR, Lin J, Hinds P, Miller D, Levine AJ (1989): Cellular gene expression in papillomas of the choroid plexus from transgenic mice that express the simian virus 40 large T antigen. J Virol 63:790–797.

Martin GR (1981): Isolation of a pluripotent cell line from early mouse embryos cultured in medium conditioned by teratocarcinoma stem cells. Proc Natl Acad Sci USA 78:7634–7638.

Masters CL, Multhaup G, Simms G, Pottgiesser J, Martins RN, Beyreuther K (1985a): Neuronal origin of a cerebral amyloid: Neurofibrillary tangles of Alzheimer's disease contain the same protein as the amyloid of plaque cores and blood vessels. EMBO J 4:2757–2763.

Masters CL, Simms G, Weinman NA, Multhaup G, McDonald BL, Beyreuther K (1985b): Amyloid plaque core protein in Alzheimer's disease and Down's syndrome. Proc Natl Acad Sci USA 82:4245–4249.

McKhann G, Drachman D, Folstein M, Katzman R, Price D, Stadlan EM (1986): Clinical diagnosis of Alzheimer's disease: Report of the NINCDS–ADRDA Work Group under the auspices of Department of Health and Human Services Task Force on Alzheimer's disease. Neurology 34:939–944.

Messing A, Pinkert CA, Palmiter RD, Brinster RL (1988): Developmental study of SV40 large T antigen expression in transgenic mice with choroid plexus neoplasia. In ''Oncogene Research,'' Vol 3. Harwood Academic Publishers USA, pp 87–97.

Muller-Hill B, Beyreuther K (1989): Molecular biology of Alzheimer's disease. Annu Rev Biochem 58, in press.

Muglia L. Rothman-Denes LB (1986): Cell type-specific negative regulatory element in the control region of the rat α-fetoprotein gene. Proc Natl Acad Sci 83:7653–7657.

Neve RL, Finch EA, Dawes LR (1988): Expression of the Alzheimer amyloid precursor gene transcripts in the human brain. Neuron 1:669–677.

Odenwald WF, Taylor CF, Palmer-Hill FJ, Friedrich Jr V, Tani M, Lazzarini RA (1987): Expression of a homeo domain protein in noncontact-inhibited cultured cells and postmitotic neurons. Gene Dev 1:482–496.

Odenwald WF, Garbern J, Arnheiter H, Tournier-Lasserve E, Lazzarini RA (1989): The Hox-1.3 homeo box protein is a sequence-specific DNA-binding phosphoprotein. Gene Dev 3:158–172.

Ott MO, Sperling L, Herbomel P, Yaniv M, Weiss MC (1984): Tissue specific expression is conferred by a sequence from the 5' end of the rat albumin gene. EMBO J 3:2505–2510.

Palmert MR, Golde TE, Cohen ML, Kovacs DM, Tanzi RE, Gusella JF, Usiak MF, Younkin LH, Younkin SG (1988): Amyloid protein precursor messenger RNAs: Differential expression in Alzheimer's disease. Science 241:1080–1084.

Patterson D, Gardiner K, Kao FT, Tanzi R, Watkins P, and Gussella JF (1988): Mapping of the gene encoding the β-amyloid precursor protein and its relationship to the Down syndrome region on chromosome 21. Proc Natl Acad Sci USA 85:8266–8270.

Pinkert CA, Ornitz DM, Brinster RL, Palmiter RD (1987): An albumin enhancer located 10 kb upstream functions along with its promoter to direct efficient, liver-specific expression in transgenic mice. Gene Dev 1:268–276.

Podlisny MB, Lee G, Selkoe DJ (1987): Gene dosage of the amyloid β precursor protein in Alzheimer's disease. Science 238:669–671.

Ponte P, Gonzalez-DeWhitt P, Schilling J, Miller J, Hsu D, Greenberg B, Davis K, Wallace W, Lieberburg I, Fuller F, Cordell B (1988): A new A4 amyloid mRNA contains a domain homologous to serine proteinase inhibitors. Nature 331:525–527.

Popko B, Puckett C, Lai S, Shine HD, Readhead C, Takahashi N, Hunt III SW, Sidman RL, Hood L (1987): Myelin deficient mice: Expression of myelin basic protein and generation of mice with varying levels of myelin. Cell 48:713–721.

Price DL (1986): New perspectives on Alzheimer's disease. Annu Rev Neurosci 9:489–512.

Price DL, Koo EH, Unterbeck A (1989): Cellular and molecular biology of Alzheimer's disease. BioEssays 10:69–74.

Razin A, Riggs AD (1980): DNA methylation and gene function. Science 210:604–610.

Readhead C, Popko B, Takahashi N, Shine HD, Saavedra RA, Sidman RL, Hood L (1987): Expression of myelin basic protein gene in transgenic shiverer mice: Correction of the dysmyelinating phenotype. Cell 48:703–712.

Reeves RH, Oster-Granite ML, Gearhart JD (1986): The trisomy 16 mouse as a model of Down syndrome. ILAR News 29:4–9.

Robakis NK, Wisniewski HM, Jenkins EC, Devine-Gage E, Houck GE, Yao XL, Ramakrishna N, Wolfe G, Silverman WP, Brown WT (1987): Chromosome 21q21 sublocalization of gene encoding beta-amyloid peptide in cerebral vessels and neuritic (senile) plaques of people with Alzheimer's disease and Down's syndrome. Lancet 8529:384–385.

Robertson E, Bradley A, Keuhn M, Evans M (1986): Germ-line transmission of genes introduced into cultured pluripotential cells by retroviral vector. Nature 323:445–448.

Rosenfeld MG, Crenshaw III EB, Lira SA, Swanson L, Borrelli E, Heyman R, Evans RM (1988): Transgenic mice: Applications to the study of the nervous system. Annu Rev Neurosci 11:353–372.

Roses AD, Pericak-Vance MA, Haynes CS, Haines JL, Gaskell PA, Yamaoka LH, et al. (1988): Genetic linkage studies in Alzheimer's disease. 40th Annual Meeting of the American Academy of Neurology, Cincinnati, Ohio, April 17–23, 1988. Neurology 38:(3 Suppl.1):173.

Salbaum JM, Weidemann A, Lemaire HG, Masters C, Beyreuther K, (1988): The promoter of Alzheimer's disease amyloid A4 precursor gene. EMBO J 7:2807–2813.

Sanes JE, Rubenstein JLR, Nicolas JF (1986): Use of a recombinant retrovirus to study postimplantation cell lineage in mouse embryos. EMBO J 5:3133–3142.

Schellenberg GD, Bird TD, Wijsman EM, Moore DK, Boehnke M, Bryart EM, Lampe TH, Nocklin D, Sumi SD, Deeb SS, Beyreuther K, Martin GM (1988): Absence of linkage of chromosome 21q21 markers to familial Alzheimer's disease. Science 241:1507–1510.

Schmechel DE, Goldgaber D, Burkhart DS, Gilbert JR, Gajdusek DC, Roses AD (1988): Cellular localization of messenger RNA encoding amyloid-beta-protein in normal tissue and in Alzheimer disease. Alzheimer Dis Assoc Disord (US)2:96–111.

Schweber M, Tuson C, Shiloh R, Ben-Neriah Z (1987): Triplication of chromosome 21 material in Alzheimer's disease. 39th Annual Meeting of the American Academy of Neurology, New York, NY, April 5–11, 1987. Neurology 37:(3 Suppl.1):222.

Selkoe DJ (1989): Biochemistry of altered brain proteins in Alzheimer's disease. Annu Rev Neurosci 12:463–490.

Selkoe DJ, Ihara Y, Salazar FJ (1982): Alzheimer's disease: Insolubility of partially-purified paired helical filaments in sodium dodecyl sulfate and urea. Science 215:1243–1245.

Shivers BD, Hilbich C, Multhaup G, Salbaum M, Beyreuther K, Seeburg PH (1988): Alzheimer's disease amyloidogenic glycoprotein: Expression pattern in rat brain suggests a role in cell contact. EMBO J 7:1365–1370.

Small JA, Blair DG, Showalter SD, Scangos GA (1985): Analysis of a transgenic mouse containing simian virus 40 and v-myc sequences. Mol Cell Biol 5:642–648.

Small JA, Scangos GA, Cork L, Jay G, Khoury G (1986): The early region of human papovavirus JC induces dysmyelination in transgenic mice. Cell 46:13–18.

St. George-Hyslop PH, Tanzi RE, Polinsky RJ, Haines JL, Nee L, Watkins PC, Myers RH, Feldman RG, Pollen D, Drachman D, Growdon J, Bruni A, Foncin JF, Salmon D, Frommelt P, Amaducci , Sorbi S, Piacentini S, Stewart GD, Hobbs WJ, Conneally M, Gusella JF (1987a): The genetic defect causing familial Alzheimer's disease maps on chromosome 21. Science 235:885–890.

St. George-Hyslop PH, Tanzi RE, Polinsky RJ, Neve RL, Pollen D, et al. (1987b): Absence of

duplication of chromosome 21 genes in familial and sporadic Alzheimer's disease. Science 238:664–666.
Summit RL (1981): Specific segments that cause the phenotype of Down's syndrome. In de la Cruz FF, Gerald PS (eds): "Trisomy 21 (Downs syndrome): Research Perspectives." Baltimore: University Park Press, pp 225–235.
Tanzi R, Gusella JF, Watkins PC, Bruns GAP, St. George-Hyslop P, Van Keuren ML, Patterson D, Pegan S, Kurnit DM, Neve RL (1987a): Amyloid β protein gene: cDNA,mRNA distribution, and genetic linkage near the Alzheimer locus. Science 235:880–884.
Tanzi RE, St. George-Hyslop PH, Haines JL, Polinsky RJ, Nee L, et al., (1987b): The genetic defect in familiar Alzheimer's disease is not tightly linked to the amyloid β-protein gene. Nature 329:156–157.
Tanzi RE, Bird ED, Latt SA, Neve RL (1987c): The amyloid β protein gene is not duplicated in brains from patients with Alzheimer's disease. Science 238:666–669.
Tanzi RE, McClatchey AI, Lamperti ED, Villa-Komaroff L, Gusella JF, Neve RL (1988): Protease inhibitor domain encoded by an amyloid protein precursos mRNA associated with Alzheimer's disease. Nature 331:528–530.
Thomas KR, Capecchi MR (1987): Site-directed mutagenesis by gene targeting in mouse embryo-derived stem cells. Cell 51:503–512.
Trapp BD, Small JA, Pulley M, Khoury G, Scangos GA (1988): Dysmyelination in transgenic mice containing JC virus early region. Ann Neurol 23:38–48.
Trevor K, Linney E, Oshima RG (1987): Suppression of endo B cytokeratin by its antisense RNA inhibits the normal coexpression of endo A cytokeratin. Proc Natl Acad Sci USA 84:1040–1044.
Unterbeck A (1986): Dissertation, University of Cologne, FRG.
Van Broeckhoven C, Genthe AM, Vandenberghe A, Horsthemke B, Backhovens H, Raeymaekers P, Van Hul W, Wehnert A, et al. (1987): Failure of familial Alzheimer's disease to segregate with the A4-amyloid gene in several European families. Nature 329:153–155.
Vitek MP, Rasool CG, de Sauvage F, Vitek SM, Bartus RT, Beer B, Ashton RA, Macq AF, Maloteaux JM, Blume AJ, Octave JN (1988): Absence of mutation in the β-amyloid cDNAs cloned from the brains of three patients with sporadic Alzheimer's disease. Mol Brain Res 4:121–131.
Warren AC, Robakis NK, Ramakrishna N, Koo EH, Ross CA, Robb AS, Folstein MF, Price DL, Antonarakis SE (1987): β-amyloid gene is not present in three copies in autopsy-validated Alzheimer's disease. Genomics 1:307–312.
Weintraub H, Izant JG, Harland RM (1985): Anti-sense RNA as a molecular tool for genetic analysis. In "Trends in Genetics." Amsterdam: Elsevier Science Publishers, pp 22–25.
Widen SG, Papaconstantinou J (1986): Liver-specific expression of the mouse α-fetoprotein gene is mediated by *cis*-acting DNA elements. Proc Natl Acad Sci USA 83:8196–8200.
Wirak DO, Unterbeck AJ, Bayney RM, Trapp BD, Koo EH, Price DL, Scangos G (1989): Expression of the human β-amyloid precursor gene in transgenic mice. J Cell Biochem [Suppl] 13B:185
Wischik CM, Novak M, Edwards PC, Klug A, Tichelaar W, et al. (1988a): Structural characterization of the core of the paired helical filament of Alzheimer's disease. Proc Natl Acad Sci USA 85:4884–4888.
Wischik CM, Novak M, Thogersen HC, Edwards PC, Runswick MJ, Eet al. (1988b): Isolation of a fragment of tau derived from the core of the paired helical filament of Alzheimer's disease. Proc Natl Acad Sci USA 85:4506–4510.
Wisniewski HM, Narang MK, Terry TD (1976): Neurofibrillary tangles of paired helical fragments. J Neurol Sci 27:173–181.

Wolozin B, Davies P (1987): Alzheimer related neuronal protein A68: Specificity and distribution. Ann Neurol 22:521–526.
Wolozin BL, Pruchnicki A, Dickson DW, Davies P (1986): A neuronal antigen in the brains of Alzheimer patients. Science 232:648–650.
Wolozin B, Scicutella A, Davies P (1988): Reexpression of a developmentally regulated antigen in Down syndrome and Alzheimer disease. Proc Natl Acad Sci USA 85:6202–6206.
Wong CW, Quaranta V, Glenner GG (1985): Neuritic plaques and cerebrovascular amyloid in Alzheimer's disease are antigenically related. Proc Natl Acad Sci USA 82:8729–8732.
Wood JG, Mirra SS, Pollock NJ, Binder LI (1986): Neurofibrillary tangles of Alzheimer's disease share antigenic determinants with the axonal microtubule-associated protein tau. Proc Natl Acad Sci USA 83:4040–4043.
Wu C, Wilson S, Walker B, David I, Paisley T, Zimarino V, Ueda H (1987): Purification and properties of Drosophila heat shock activator protein. Science 238:1247–1253.
Wurtman RJ (1985): Alzheimer's disease. Sci Am 252:62–74.
Yamada T, Sasaki H, Furuya H, Miyata T, Goto I, Sakaki Y (1987): Complementary DNA for the mouse homolog of the human amyloid beta protein precursor. Biochem Biophys Res Commun 149:665–671.
Yamada T, Sasaki H, Dohura K, Goto I, Sakaki Y (1989): Structure and expression of the alternatively spliced forms of mRNA for the mouse homolog of Alzheimer's disease amyloid beta protein precursor. Biochem Biophys Res Commun 158:906–9012.
Zain SB, Salim M, Chou WG, Sajdel-Sulkowska EM, Majocha RE, Marotta CA (1988): Molecular cloning of amyloid cDNA derived from mRNA of the Alzheimer disease brain: Coding and noncoding regions of the fetal precursor mRNA are expressed in the cortex. Proc Natl Acad Sci USA 85:929–933.
Zakany J, Tuggle CK, Patel MD, Nguyen-Huu MC (1988): Spatial regulation of homeobox gene fusions in the embryonic central nervous system of transgenic mice. Neuron 1:679–691.
Zimmer DB, Van Eldik LJ (1986): Identification of a molecular target for the calcium-modulated protein S100. J Biol Chem 261:11424–11428.
ZuRhein GM (1972): Virions in progressive multifocal leukoencephalopathy. In Minkler J (ed): "Pathology of the Nervous System," Vol. 3. New York: McGraw-Hill, pp 2893–2912.

Amyloid Precursor Protein (APP) mRNA in Normal and Alzheimer's Disease Brain

Steven A. Johnson

INTRODUCTION

One of the neuropathologic features of Alzheimer's disease (AD) is the abnormal accumulation of a specific type of amyloid, the amyloid β-peptide (AP), which is found in the brain, but not elsewhere. I note at the outset that other polypeptides, including cystatin C [Ghiso et al., 1986] and prion PrP 27-30 [Prusiner, 1987], have the structural (β-pleated sheet), biophysical (non-branching, aggregating fibrils), and tinctorial (thioflavin S staining and congo red birefringence) properties common to all amyloids, but are not associated with AD. AP is deposited as a highly insoluble proteinaceous mass in extracellular neuritic plaques and in cerebral and meningeal blood vessels [Glenner, 1988]. Whether AP is also a major constituent of intracellular neurofibrillary tangles in AD [Masters et al., 1985] is quite controversial [Klug et al., 1988], but, if true, would implicate amyloid in the two major pathologies associated with the disease. Plaque and cerebrovascular (CV) AP deposition occurs to a similar extent in older Down's syndrome (DS) individuals, who generally have AD symptomology [Wisniewski et al., 1985], but is also seen at a much lesser extent in the brains of normal individuals 60 years of age or older (reviewed by Ogomori et al., this volume). AP deposition in AD, DS, and normal aging brain is primarily in the neocortex, hippocampus, and amygdala.

A fundamental unanswered question in AD concerns the etiology of AP deposition and formation of neuritic plaques. Morphologically, plaques are spherical clusters of degenerating axons and dendrites, which often surround a core of insoluble proteinaceous filaments. Isolated AP from either CV amyloid or plaque amyloid cores has an identical amino acid sequence [Glenner and Wong, 1984; Masters et al., 1985]. Furthermore, these data and recent immunochemical data show that AD amyloid differs from other brain amyloids,

---

Andrus Gerontology Center and Department of Neurobiology, University of Southern California, Los Angeles, California 90089-0191

such as the prion protein-containing amyloid associated with Scrapie and Creutzfeldt–Jacob disease [Kitamoto et al., 1986].

Several groups have used the cerebrovascular AP peptide sequence to clone cDNA's coding for the β-amyloid precursor protein (APP) from human cDNA libraries [Goldgaber et al., 1987; Robakis et al., 1987; Kang et al., 1987]. Based on the cDNA nucleotide sequence, APP is a 695 amino acid membrane-associated glycoprotein, with the 42 amino acid AP located near the C terminus, overlapping a presumptive membrane-spanning domain (Fig. 1) [Kang et al., 1987; Dyrks et al., 1988]. The 3' untranslated region of human APP mRNA has three polyadenylation signals, two of which are clustered and separated from the third by 242 nucleotides (NT) [Kang et al., 1987].

The normal function of APP and the presumably abnormal processing that results in AP accumulation in plaques and cerebral vessels are presently unknown. Selkoe [1988] has recently written an excellent review covering recent research on the APP protein from normal and AD brain. The APP gene has 18 exons spanning at least 50 kb of DNA [Lemaire et al., 1989] and is located on chromosome 21 [Goldgaber et al., 1987; Kang et al., 1987; Robakis et al., 1987; Tanzi et al., 1987]. Studies on the APP gene 5' transcriptional control region using transgenic rats are reviewed by Unterbeck et al. in this volume. Accordingly, the remainder of this review addresses the current status of work on the APP mRNA family.

APP mRNAs

Further cloning studies recently revealed additional APP mRNAs that include an extra 56 amino acid (APP-751) or 75 amino acid (APP-770) of coding sequence with a significant similarity (50%) to the Kunitz family of serine protease inhibitors [Ponte et al., 1988; Tanzi et al,. 1988; Kitaguchi et al., 1988]. These additional transcripts presumably arise by alternate splicing of a single nuclear RNA transcribed from the single APP gene, including one (168 NT) or two (168 NT + 57 NT) separate exons in the APP-695 coding sequence (Fig. 1). Lysates from APP-770 transfected COS cells had increased trypsin inhibitor activity, suggesting that the Kunitz protease inhibitor motif may function as a protease inhibitor in vivo [Kitaguchi et al., 1988].

Tissue distribution studies by RNA gel blot analysis have shown that the APP-751 transcript is present in most human tissues including brain [Tanzi et al., 1988; Neve et al., 1988], whereas APP-695 mRNA appears to be brain-specific [Ponte et al., 1988; Neve et al., 1988]. In adult controls, APP-695 mRNA prevalence is higher in association cortex, i.e., the region of the neocortex that is generally most affected in AD, while APP-751 transcript prevalence shows little difference between any brain region or nonbrain tissue studied [Neve et al., 1988]. APP-770 mRNA is slightly less abundant than APP-695

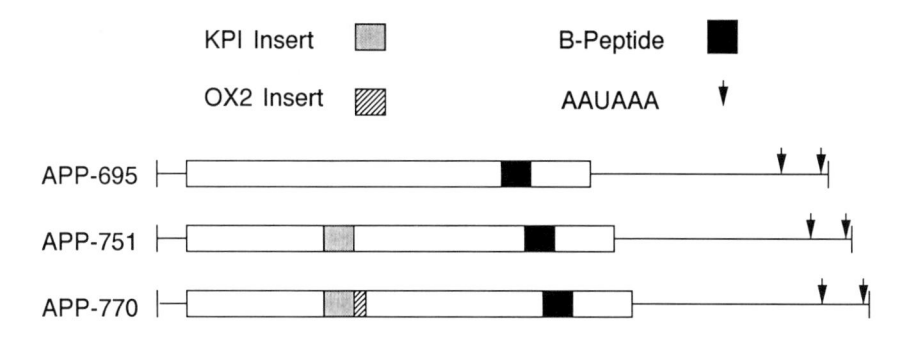

Fig. 1. Structure of APP mRNA forms. Lines represent noncoding sequence. Boxed areas are APP coding sequences. Filled boxes are indicated above. Arrowheads point to AAUAAA polyadenylation sites.

mRNA in fetal brain [Kitaguchi et al., 1988]. However, in normal aged adult cortex it represents only 5% of APP-695 or APP-751 mRNA prevalence [Johnson et al., 1989]. Thus, APP-695 and APP-751 mRNAs dominate the APP mRNA pool in the aged human brain.

Little is known about changes in the steady-state prevalence of each APP mRNA form during development or with aging. Limited data indicates that total APP mRNA prevalence is similar in a 19 week fetal brain and a normal, aged adult [Tanzi et al., 1987]. However, an additional study indicates an apparent shift in the APP-751/APP-695 mRNA ratio; APP-695 mRNA is more prevalent in fetal brain, while APP-751 mRNA is more prevalent in adult [Neve et al., 1988]. More work is necessary to establish the temporal expression of each APP mRNA form.

A number of in situ hybridization studies indicate that APP mRNA is localized primarily in neurons, with pyramidal cells having the highest levels [Goedert, 1987; Bahmanyar et al., 1987; Neve et al., 1988; Schmechel et al., 1988]. The laminar distribution of APP mRNA in normal brain cortex varied with the cortical area in one study [Neve et al., 1988], but not in another [Lewis et al., 1988]. In a single AD study there was also no difference in laminar distribution of APP mRNA between cortical regions [Neve et al., 1988]. Whether APP mRNA is present in various glial cell types or endothelial cells lining cerebral blood vessels is somewhat controversial [Schmechel et al., 1988], although most studies cannot detect APP mRNA in non-neuronal cells [Goedert, 1987; Higgins et al., 1988; Neve et al., 1988; Palmert et al., 1988]. Thus, if APP mRNA is present in glial or endothelial cells, its prevalence is much lower than in most neurons. Note that these in situ hybridization studies employed 3′ APP mRNA probes, which do not discriminate APP-695 from APP-751 or APP-770 mRNAs. One recent study using selective probes did not find any obvious cell type difference in localization of APP-695 mRNA vs APP-751 mRNA [Neve et al., 1988]. However, a remaining question concerns whether both APP-695 and APP-751 mRNAs are present in the same neuron.

APP coding sequence is evolutionarily well conserved throughout the mammalian radiation [Manning et al., 1988]. For example, both mouse and rat APP-695 mRNA have 97% aa sequence similarity to the human cognate [Yamada et al., 1987; Shivers et al., 1988]. Furthermore, from Northern blot studies, both APP-695 and APP-751 mRNA species are the same size in mouse as in human [Johnson et al., 1988], suggesting that the overall structure of both transcripts has been maintained since the ancestral lines of mouse and human diverged about 80 million years ago.

APP mRNA IN AD AND CONTROL BRAIN

Since plaque and CV amyloid neuropathology is much greater in affected AD brain regions, a number of laboratories have compared APP mRNA between AD and control (CTL) tissues. The APP gene has been mapped to chromosome 21 [Goldgaber et al., 1987; Tanzi et al., 1987], which is triplicated in DS; moreover, older DS individuals show clinical manifestations of Alzheimer dementia with abundant amyloid deposition. However, despite an early report of APP gene triplication in AD [Delabar et al., 1987], other work shows that APP gene triplication does not occur in AD [Podlisny et al., 1987; Van Broeckhoven et al., 1987] and thus is probably not a common feature of AD. Vitek et al. [1988] and Zain et al. [1988] have compared nearly 85% of the APP mRNA sequence between AD and fetal control and found no differences. While the 5' 15% of the sequence remains to be compared, it is clear that the AP sequence, the potential N-linked glycosylation sites, and the presumptive membrane-spanning domain sequences are unchanged.

APP mRNA prevalence has also been examined in AD vs. CTL tissues. Earlier studies used APP probes that detected all APP mRNA forms. Higgins et al., [1988] found little overall change in prevalence in AD hippocampus by in situ hybridization. However, they did show a 2.5-fold total APP mRNA increase in AD parasubicular neurons compared with CA3 pyramidal neurons, suggesting that cell-specific regulation of APP mRNA expression may be altered in AD. On the other hand, Schmechel et al. [1988] found an overall increase in neuronal APP mRNA prevalence in AD hippocampus and entorhinal cortex. Similarly, Cohen et al. [1988] showed that remaining neurons of nucleus basalis, a region that often suffers severe cell loss, have a two-fold increase of APP mRNA in AD. Northern blot analysis of total APP mRNA prevalence has generally failed to show any significant changes in AD cortex [Robakis et al., 1987; Goedert, 1987]. Thus, while specific changes in APP mRNA prevalence have been shown for certain brain regions, the large increase in plaque and CV amyloid in AD is not due to an overall increase in total APP mRNA prevalence.

However, using probes that discriminate APP-695 and APP-751 mRNA species separately on Northern blots with five AD and five CTL specimens, we found a selective loss of the brain-specific APP-695 transcript in two heavily affected brain regions, frontal cortex and hippocampus, which results in a 2.5-fold increase in the APP-751/APP-695 transcript ratio in AD. However, in cerebellum, which rarely contains plaques, the APP-751/APP-695 transcript ratio was much lower (6× lower) than in AD cortex or AD hippocampus and, importantly, was not different between AD and CTL [Johnson et al., 1988, 1989]. Preliminary data by Tanzi et al. [1988], deduced from one AD and one control specimen but without direct assay of APP-695 mRNA, support this finding.

Palmert et al. [1988] measured total APP transcript prevalence and APP-751-specific transcript prevalence in individual neurons of the nucleus basalis of Meynert (NbM), locus ceruleus (LC), basis pontis, and subiculum in several AD and CTL individuals by in situ hybridization. These data indirectly show that APP-751 mRNA is unchanged, whereas APP-695 mRNA increases two-fold in surviving neurons of NbM and LC, but does not change in basis pontis or subiculum in AD. How this finding relates to the decreased APP-695 mRNA prevalence in AD cortex and hippocampus, as determined by Northern blot analysis, is presently unclear. The apparent contradiction may be due to different methodology; in situ hybridization measures cellular RNA prevalence, while Northern blot hybridization measures RNA prevalence in tissue. However, there is no evidence for massive AD-related loss of neurons that *selectively* express APP-695.

IMPLICATIONS AND FUTURE DIRECTIONS

Numerous studies that compared APP mRNA nucleotide sequence, gene copy number, cell and tissue localization, and total APP mRNA prevalence all failed to show any significant differences between AD and CTL. However, when each major adult APP transcript was compared in AD and CTL brain regions that contain (cortex, hippocampus) or lack (cerebellum) abundant amyloid pathology, the APP-751/APP-695 mRNA ratio was increased two- to three-fold in AD cortex and hippocampus, but was six-fold lower and unchanged in AD cerebellum as compared with controls. The APP-751/APP-695 ratio is also low in mouse cortex [Johnson et al., 1988] and cerebellum [Johnson et al., 1989] and in fetal brain [Neve et al., 1988] (all tissues lacking amyloid pathology). Interestingly, the APP-751/APP-695 mRNA ratio is also quite low in fetal DS brain, which has no amyloid pathology at this age. However, in hippocampus and neocortex of a 37-year-old DS patient, which should have abundant amyloid pathology [Wisniewski et al., 1985], the APP-751/APP-695 mRNA

TABLE I. APP-751/APP-695 mRNA Prevalence Ratio in RNA From Various Brain Regions of Human Control (CTL), AD, and DS Individuals

Condition	Brain region	APP-751/APP-695 mRNA ratio	Reference
Adult AD	Hippocampus[a]	3.8	Johnson et al., 1989
Adult AD	Cortex[a]	2.8	Johnson et al., 1988
37 yr DS	Hippocampus[a]	2–3	Neve et al., 1988
37 yr DS	Cortex[a]	2–3	Neve et al., 1988
Adult CTl	Hippocampus	2.2	Johnson et al., 1989
Adult CTl	Cortex	1.2	Johnson et al., 1989
37 yr DS	Cerebellum	<1.0	Neve et al., 1988
Adult CTL	Cerebellum	0.6	Johnson et al., 1989
Adult AD	Cerebellum	0.6	Johnson et al., 1989
Fetal CTL	Whole brain	<0.5	Neve et al., 1988
Fetal DS	Whole brain	<0.5	Neve et al., 1988

[a]Region typically has numerous amyloid plaques.

ratio is similar to that of AD cortex and AD hippocampus [Neve et al., 1988]. Thus, there may be a correlation between a relatively high APP-751/APP-695 mRNA ratio and amyloid deposition in AD as well as DS (Table I).

Most of the data discussed above suggest that the prevalence of APP-751 mRNA, found in most tissues, is relatively unaffected in AD, while the brain-specific APP-695 transcript is changed. In future work it will continue to be important to examine the neuronal prevalence of each major APP mRNA by quantitative in situ hybridization [Palmert et al., 1988]. One possible explanation for the specific decrease of APP-695 mRNA in AD would be its selective presence in a subset of neurons that degenerate in affected brain regions. Other explanations include an alteration in the differential splicing of the APP nuclear transcript in AD, or the possibility that APP-695 mRNA is less stable in AD. Thus, it is necessary to determine: 1) the number of neurons expressing each APP mRNA form, and 2) whether both APP-751 and APP-695 mRNAs are coexpressed in the same neuron. Also, while these questions are pursued, it will be important to assess the cause/effect relationship between changes in specific APP mRNA prevalence and the presence of amyloid pathology. Finally, more information concerning the normal function of each APP protein type will undoubtedly be necessary before the etiology of amyloid deposition in AD can be elucidated.

REFERENCES

Bahmanyar S, Higgins GA, Goldgaber D, Lewis DA, Morrison JH, Wilson MC, Shankar SK, Gajdusek DC (1987): Localization of amyloid beta protein messenger RNA in brains from patients with Alzheimer's disease. Science 237:77–80.

Cohen ML, Golde TE, Usiak MF, Younkin LH, Younkin SG (1988): In situ hybridization of nucleus basalis neurons shows increased β-amyloid mRNA in Alzheimer disease. Proc Natl Acad Sci USA 85:1227–1231.

Delabar J-M, Goldgaber D, Lamour Y, Nicole A, Huret J-L, de Grouchy J, Brown P, Gajdusek DC, Sinet P-M (1987): β amyloid gene duplication in Alzheimer's disease and karyotypically normal Down syndrome. Science 235:1388–1392.

Dyrks T, Weidemann A, Multhrup G, Salbaum JM, Lemaire H, Kang J, Muller-Hill B, Masters CL, Beyreuther K (1988): Identification, transmembrane orientation and biogenesis of the amyloid A4 precursor of Alzheimer's disease. EMBO J 7:949–957.

Ghiso J, Jensson O, Frangione B (1986): Amyloid fibrils in hereditary cerebral hemorrhage with amyloidosis of Icelandic type is a variant of γ-trace basic protein (cystatin C). Proc Natl Acad Sci USA 83:2974–2978.

Glenner GG (1988): Alzheimer's disease: Its proteins and genes. Cell 52:307–308.

Glenner GG, Wong CW (1984): Initial report of the purification and characterization of a novel cerebrovascular amyloid protein. Biochem Biophys Res Commun 120:885–890.

Goedert M (1987): Neuronal localization of amyloid beta protein precursor mRNA in normal human brain and in Alzheimer's disease. EMBO J 6:3627–3632.

Goldgaber D, Lerman MI, McBride OW, Saffiotti U, Gajdusek DC (1987): Characterization and chromosomal localization of a cDNA encoding brain amyloid of Alzheimer's disease. Science 235:877–880.

Higgins GA, Lewis DA, Bahmanyar S, Goldgaber D, Gajdusek DC, Young WG, Morrison JH, Wilson MC (1988): Differential regulation of amyloid-β-protein mRNA expression within hippocampal neuronal subpopulations in Alzheimer disease. Proc Natl Acad Sci USA 85:1297–1301.

Johnson SA, Pasinetti GM, May PC, Ponte PA, Cordell B, Finch CE (1988): Selective reduction of mRNA for the β-amyloid precursor protein that lacks a Kunitz-type protease inhibitor motif in cortex from Alzheimer brains. Exp Neurol 102:264–268.

Johnson SA, Rogers J, Finch CE (1989): APP-695 transcript prevalence is selectively reduced during Alzheimer's disease in cortex and hippocampus but not in cerebellum. Neurobiol Aging 10:267–272.

Kang J, Lemaire H, Unterbeck A, Salbaum JM, Masters CL, Grzeschik K, Multhaup G, Beyreuther K, Muller-Hill B (1987): The precursor of Alzheimer's disease amyloid A4 protein resembles a cell surface receptor. Nature 325:733–736.

Kitaguchi N, Takahashi Y, Tokushima Y, Shiojiri S, Ito H (1988): Novel precursor of Alzheimer's disease amyloid protein shows protease inhibitory activity. Nature 331:530–532.

Kitamoto T, Tateishi J, Tashima T, Takeshita I, Barry RA, DeArmond SJ, Prusiner SB (1986): Amyloid plaques in Creutzfeld-Jacob disease stain with prion protein antibodies. Ann Neurol 20:204–208.

Klug A, Goedert M, Wischik CM (1988): Summary. In "The Molecular Biology of Alzheimer's Disease," "Current Communications in Molecular Biology." Finch CE, Davies P (eds): Cold Spring Harbor: Cold Spring Harbor Laboratory, pp 187–197.

Lemaire HG, Salbaum JM, Multhaup G, Kang J, Bayney RM, Unterbeck A, Beyreuther K, Muller-Hill B (1989): The pre-A4$_{695}$ precursor protein of Alzheimer's disease A4 amyloid is encoded by 16 exons. Nucleic Acids Res 17:517.

Lewis DA, Higgins GA, Young WG, Goldgaber D, Gajdusek DC, Wilson MC, Morrison JH (1988): Distribution of precursor amyloid-β-protein messenger RNA in human cerebral cortex: Relationship to neurofibrillary tangles and neuritic plaques. Proc Natl Acad Sci 85:1691–1695.

Manning RW, Reid CM, Lampe RA, Davis LG (1988): Identification in rodents and other species of an mRNA homologous to the human β-amyloid precursor. Mol Brain Res 3:293–298.

Masters CL, Multhaup G, Simms G, Pottgeisser J, Martins RN, Beyreuther K (1985): Neuronal origin of a cerebral amyloid: Neurofibrillary tangles of Alzheimer's disease contain the same protein as the amyloid of plaque cores and blood vessels. EMBO J 4:2757–2763.

Neve RL, Finch EA, Dawes LR (1988): Expression of the Alzheimer amyloid prescursor gene transcripts in the human brain. Nature 1:669–677.

Palmert MR, Golde TE, Cohen ML, Kovacs DM, Tanzi RE, Gusella JF, Usiak LH, Younkin LH, Younkin SG (1988): Amyloid protein precursor messenger RNAs: Differential expression in Alzheimer's disease. Science 241:1080–1084.

Podlisny MB, Lee G, Selkoe DJ (1987): Gene dosage of the amyloid β precursor protein in Alzheimer's disease. Science 238:669–671.

Ponte P, Gonzalez-DeWhitt P, Schilling J, Miller J, Hsu D, Greenberg B, Davis K, Wallace W, Lieberburg I, Fuller F, Cordell B (1988): A new A4 amyloid mRNA contains a domain homologous to serine proteinase inhibitors. Nature 331:525–527.

Prusiner SB (1987): Prions and neurodegenerative diseases. N Engl J Med 317:1571.

Robakis NK, Ramakrishna N, Wolfe G, Wisniewski HM (1987): Molecular cloning and characterization of a cDNA encoding the cerebrovascular and the neuritic plaque amyloid peptides. Proc Natl Acad Sci 84:4190–4194.

Schmechel DE, Goldgaber D, Burkhart DS, Gilbert JR, Gajdusek DC, Roses AD (1988): Cellular localization of messenger RNA encoding amyloid-beta-protein in normal tissue and in Alzheimer disease. Alzheimer Dis Assoc Dis 2:96–111.

Selkoe DJ (1988): Biochemistry of altered brain proteins in Alzheimer's disease. In Cowan WM (ed): "Annual Review of Neuroscience," Vol. 12. Palo Alto, CA: Annual Reviews, Inc., in press.

Shivers BD, Hilbich C, Multhaup G, Salbaum M, Beyreuther K, Seeburg PH (1988): Alzheimer's disease amyloidogenic glycoprotein: Expression pattern in rat brain suggests a role in cell contact. EMBO J 7:1365–1370.

Tanzi RE, Gusella JF, Watkins PC, Bruns GAP, St. George-Hyslop P, Van Keuren ML, Patterson D, Pagen S, Kurnit DM, Neve RL (1987): Amyloid β protein gene: cDNA, mRNA distribution, and genetic linkage near the Alzheimer locus. Science 235:880–884.

Tanzi RE, McClatchey AI, Lamperti ED, Villa-Komaroff L, Gusella JF, Neve RL (1988): Protease inhibitor domain encoded by an amyloid protein precursor mRNA associated with Alzheimer's disease. Nature 331:528–530.

Van Broeckhoven, et al. (1987): Failure of familial Alzheimer's disease to segregate with the A4-amyloid gene in several European families. Nature 329:153–155.

Vitek MP, Rasool CG, de Sauvage F, Vitek SM, Bartus RT, Beer B, Ashton RA, Macq AF, Maloteaux JM, Blume AJ, Octave JN (1988): Absence of mutation in the β-amyloid cDNAs cloned from the brains of three patients with sporadic Alzheimer's disease. Mol Brain Res 4:121–131.

Wisniewski KE, Dalton AJ, McLachlan C, Wen GY, Wisniewski HM (1985): Alzheimer's disease in Down's syndrome: Clinicopathologic studies. Neurology 35:957–961.

Yamada T, Sasaki H, Furuya H, Miyata T, Goto I, Sakaki Y (1987): Complementary DNA for the mouse homolog of the human beta protein precursor. Biochem Biophys Res Commun 149:665–671.

Zain SB, Salim M, Chou WG, Sajdel-Sulkowska EM, Majocha RE, Marotta CA (1988): Molecular cloning of amyloid cDNA derived from mRNA of the Alzheimer disease brain: Coding and noncoding regions of the fetal precursor mRNA are expressed in the cortex. Proc Natl Acad Sci USA 85:929–933.

Brain Amyloid in Aging and Alzheimer's Disease

Koji Ogomori, Tetsuyuki Kitamoto, and Jun Tateishi

INTRODUCTION

"Amyloid" is a general term for the abnormal fibrillar proteinaceous substance deposited in tissues. Regardless of the biochemical composition, all types of amyloid fibrils share the following properties [Glenner, 1980]: 1) green birefringence under polarized light after Congo red staining; 2) nonbranching, aggregated fibrils under electron microscope; and 3) β-pleated sheet structure, determined by X-ray diffraction. Moreover, the fibrils share the physical property of being highly insoluble in physiologic solvents and resistant to mild proteolytic digestion.

Classification of amyloid depends on its constituent fibril protein. Three types of amyloid proteins have to date been identified in the brain. The first is β or A4 protein [Glenner and Wong, 1984a; Masters et al., 1985], found in senile plaques and amyloid angiopathy in patients with Alzheimer's disease (AD), Down's syndrome (DS) [Glenner and Wong, 1984b], and normal aging [Coria et al., 1987]. The second is prion [Prusiner, 1982] or SAF (scrapie-associated fibrils) [Merz et al., 1981] protein, which is found in kuru plaques in cases of Creutzfeldt–Jakob disease (CJD) [Kitamoto et al., 1986], Gerstmann-Sträussler syndrome (GSS), kuru, and scrapie, all belonging to transmissible spongiform encephalopathy. The third is cystatin C [Ghiso et al., 1986], which is found in amyloid angiopathy of hereditary cerebral hemorrhage with amyloidosis of Icelandic type (HCHWA-I). This report is directed mainly to β-protein amyloid in normal aging and in those patients with Alzheimer's disease. Alzheimer's neurofibrillary tangle (NFT), one of the main pathological features of AD, also exhibits green birefringence after Congo red staining; however, it has an unique ultrastructure known as paired helical filaments. These have been classified separately from amyloid. Here, we exclude NFT from the term "amyloid."

Department of Neuropathology, Neurological Institute, Faculty of Medicine, Kyushu University, Fukuoka 812, Japan

β-PROTEIN

The neuropathology of AD (including senile dementia of Alzheimer type) is marked by amyloid deposits in the senile plaques and in walls of the cerebral vessels (amyloid angiopathy). The same type of amyloid deposits occurs in the brains of DS patients [Glenner and Wong, 1984a; Jervis and Thiells, 1948] and, to a lesser degree, in association with the normal aging process. Glenner and Wong [1984a] isolated a 4,200 dalton protein from cerebrovascular amyloid of AD and elucidated the amino acid sequence of N-terminal 24 residues of the protein, termed β-protein. Masters et al. [1985] reported the amino acid sequence of the amyloid (senile) plaque core protein and termed it A4 protein. This protein was all but identical to that of β-protein. With the knowledge of the N-terminal sequence of β or A4 protein, several groups of workers have cloned the cDNA encoding it [Kang et al., 1987; Goldgaber et al., 1987; Robakis et al., 1987]. The β-protein precursor was first reported to be composed of 695 amino acids. Recently, a cDNA-encoding alternate form of the β-protein precursors that contains a protease inhibitor domain and is composed of 751 or 770 amino acids was reported [Kitaguchi et al., 1988; Tanzi et al., 1988; Ponte et al., 1988]. While the mechanism of the conversion of the precursor protein to β-protein amyloid remains unclear, clarification of the pathogenesis of AD will not be feasible. Molecular studies of β-protein precursor are advancing rapidly and are reviewed by Dr. A. Unterbeck in this volume.

AGING AND BRAIN AMYLOID

There are reports on the relationship between aging and brain amyloid [Stam et al., 1986; Störkel et al., 1983; Wright et al., 1969; Yamada et al., 1987; Masuda et al., 1988], determined using former histochemical methods such as Congo red staining. All noted that senile plaques and amyloid angiopathy were rarely seen under 60 years of age, but the incidence increased over 60 in association with aging, to 70% or more (mainly senile plaques) in subjects over 80 years of age.

Recently, antibodies against synthetic peptide of β-protein have been used for the immunohistochemical detection of β-protein amyloid (senile plaques and amyloid angiopathy) [Wong et al., 1985; Allsop et al., 1986; Bobin et al., 1987; Koike et al., 1988; Dickson et al., 1988]. With the formic acid pretreatment [Kitamoto et al., 1987], immunostaining using anti-β protein antiserum revealed that all the senile plaques contained some amyloid [Ogomuri et al., 1988]. This method, used by many investigators, led to a clear detection of brain amyloid, more specifically and more sensitively than the heretofore employed histochemical methods [Ogomuri et al., 1988; Davies et al., 1988; Yamaguchi et al., 1988; Vinters et al., 1988; Tagliavini et al., 1988].

As reported [Kitamoto et al., 1986], age-related cerebral amyloid can be efficiently extracted using the SAF purification method [Hilmert and Diringer, 1984]. We investigated tissues from the frontal lobe of 66 nondemented subjects of all age groups in three ways: the histochemical method (Congo red staining), immunostaining using anti-β protein antiserum [Ogomori et al., 1988], and the amyloid extraction method [Kitamoto et al., 1986]. The positivity was compared among them [Ogomori et al., 1988], as shown in Figure 1. Amyloid was extracted initially from 67% of the brains of 9 persons in their 50s and from 97% of the brains of 35 persons over 60 years of age. These data sug-

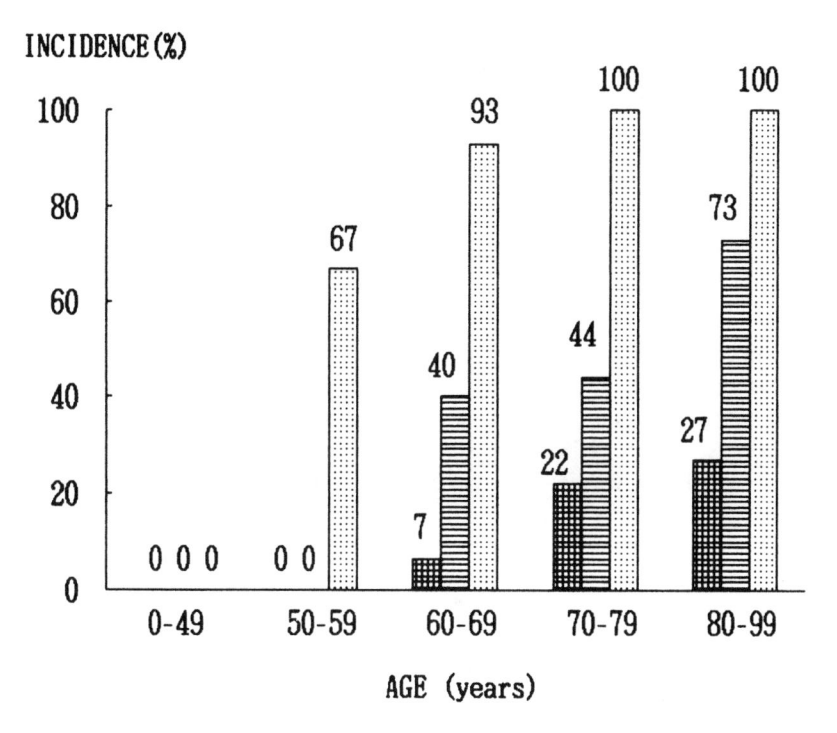

Fig. 1. Incidence of amyloid deposits in the human brain of nondemented subjects according to age, determined by three different methods.

gest that amyloid deposition in the brains of nondemented subjects begins at least from their 50s and is an inevitable part of the aging process. For chemical analysis of the extracted amyloid, we boiled the amyloid fraction in SDS buffer and solubilized SDS-insoluble amyloid fraction with 100% formic acid for analysis of amino acid composition and sequencing. The amino acid composition matched well with that of β-protein [Glenner and Wong, 1984a], and two main components showed the following N-terminal sequences: A, Asp-Ala-Glu-Phe-Arg-His-Asp-Ser-Gly-Tyr-Glx-Val-His-; B, Phe-Arg-His-Asp-Ser-Gly-Tyr-Glx-Val-His-His-Gln-Lys-. "A" matches with the sequence of β-protein [Glenner and Wong, 1984a] and "B" corresopnds with the sequence from the fourth amino acid of β-protein. Heterogeneity of the N-terminal sequence was also reported by Masters et al. [1985]. In our extraction method, proteinase K digestion is performed; hence, related effects must be kept in mind during data evaluations. Quantification of the β-protein amyloid will provide additional knowledge on the relationship of amyloid to aging.

AMYLOID AND DIAGNOSIS OF AD

The pathologic diagnosis of AD is difficult in older aged persons, for the following reasons: 1) in AD, pathologic severity (especially counts of senile plaques) is inversely proportional to age [Hansen et al., 1988], while in nondemented subjects, senile plaques increase with aging; 2) in older onset AD (over 75 years of age), nerve cell loss is often not prominent and often lacks neocortical NFT [Terry et al., 1987]. Many investigators reported that AD differs from normal aging by the density of senile plaques in the cerebral cortex [Blessed et al., 1968; Tomlinson et al., 1968, 1970]. Recent clinicopathologic studies have shown that in older age there are subjects with numerous senile plaques, yet a pre mortem diagnosis of dementia would not be given [Katzman et al., 1988; Crystal et al., 1988]. Their frequency was reported to be up to one-third [Katzman et al., 1988] or two-thirds [Crystal et al., 1988] of nondemented old-age subjects. Katzman et al. [1988] reported that these subjects had better functional and cognitive performance than the average of normal subjects. On the other hand, Crystal et al. [1988] reported that these subjects did not meet criteria for dementia yet did show a consistent decline on neuropsychological test scores. Whether these subjects are "preclinical" AD or would never develop clinical dementia remains to be determined. Further clinicopathologic studies using sensitive histological methods are needed. At least, there may be no cut-off point for diagnosing AD in subjects over 75 years, based only on the density of senile plaques in the cerebral cortex. Other parameters, for example, cell loss, distribution of amyloid (mentioned in the next paragraph), tau protein immunoreactivity, etc., may be useful in distinguishing AD from nondemented subjects, histopathologically.

DISTRIBUTION OF AMYLOID IN THE CENTRAL NERVOUS SYSTEM

As for the distribution and density of the senile plaques (amyloid deposits), it is generally accepted that the cerebral cortex, amygdala, and hippocampus are the most affected regions; other regions of the brain were much less affected, for example, brainstem and cerebellum were rarely affected [Tomlinson and Corsellis, 1984]. This pattern was found both in AD and nondemented subjects [Mann et al., 1987]. We applied the sensitive immunostaining method using anti-β-protein antiserum to observe amyloid deposits of AD (onset under 65 years of age) and nondemented old subjects [Ogomori et al., 1989]. β-protein immunostaining with formic acid enhancement [Kitamoto et al., 1987] revealed numerous faintly stained plaques (described as amorphous spherical deposit [Davies et al., 1988], diffuse plaque [Yamaguchi et al., 1988], or preamyloid deposit [Tagliavini et al., 1988]), perivascular fibrillary deposits, and subpial and subependymal deposits, which were not marked either by Congo red staining or by silver impregnation. Whether or not these deposits contain amyloid fibrils should be determined electron microscopically. If amyloid fibrils are not formed in these deposits, then they represent an abnormal focal accumulation of β-protein or β-protein precursor. The positivity of amyloid deposits (of any type) was calculated for each region of the central nervous system, both in AD and nondemented aged subjects, as shown in Table I. Amyloid deposits were found mainly in the cerebral cortex in nondemented subjects, while in AD they were distributed widely in the regions that were not affected in the nondemented subjects. This is evidence of a wider and denser distribution of amyloid deposits (senile plaques) in AD than heretofore reported. An assessment of the distribution of amyloid deposits should prove to be useful for the histopathologic diagnosis of AD. According to our results, amyloid

TABLE 1. Positivity (%) of β-Protein Amyloid Deposits in Patients With Alzheimer's Disease and Nondemented Aged Subjects for Each Area of the Central Nervous System

Area	Alzheimer's disease	Nondemented aged subjects
Cerebral cortex	100	60
Amygdala	100	42
Caudate nucleus	100	22
Thalamus	100	11
Putamen	86	10
Midbrain	92	5
Cerebellum	100	0
Pons	100	0
Medulla	64	0
Globus pallidus	43	0
Spinal cord	33	0

deposit in the cerebellum is a specific and constant finding in AD (early onset) but not in nondemented aged subjects [Ogomori et al., 1989].

Senile plaques can distribute in any part of the cerebral cortex. Some workers postulated that in the neocortex, these lesions were more numerous in associative regions than in primary sensory and motor areas [Rogers and Morrison, 1985; Pearson et al., 1985], while another study showed that their number did not vary between primary and associative visual cortical regions [Lewis et al., 1987]. Regional distribution of amyloid in the cerebral cortex will need further study, using highly sensitive methods.

Laminar distribution of amyloid in the cerebral cortex of AD and normal aging was also studied. Using classical techniques, abundant senile plaques were observed in layers 2 and 3 in AD, [Rogers and Morrison, 1985; Pearson et al., 1985; Duyckaerts, et al., 1986]; the same finding was confirmed by anti-β protein immunostaining [Majocha et al., 1988]. In our observations using β-protein immunostaining [Ogomori et al., 1989], small (granular) deposits were numerous in layers 1, 3, and 4, large deposits were numerous in layers 3 and 4, and core deposits were numerous in layers 5 and 6. Rafalowska et al. [1988] reported that in nondemented subjects senile plaques were present in deeper layers of the cortex than in those with AD, determined using classical techniques. Laminar distribution of amyloid also requires a more detailed study, using β-protein immunostaining.

β-protein amyloid presumably plays an important role in the pathogenesis of AD and aging of the brain. To clarify the role of brain amyloid in aging and AD, advanced molecular studies and clinicopathologic studies should be done.

ACKNOWLEDGMENTS

We thank Drs. K. Yamaguchi, T. Hirata, and T. Hara, the laboratories of Kyowa Hakko Co. Ltd. for collaboration on the amino acid analysis and sequencing of amyloid protein, and M. Ohara for comments.

REFERENCES

Allsop D, Landon M, Kidd M, Lowe JS, Reynolds GP, Gardner A (1986): Monoclonal antibodies raised against a sequence of senile plaque core protein react with plaque cores, plaque periphery and cerebrovascular amyloid in Alzheimer's disease. Neurosci Lett 68:252–256.

Blessed G, Tomlinson BE, Roth M (1968): The association between quantitative measures of dementia and senile change in the cerebral grey matter of elderly subjects. Br J Psychiatry 114:797–811.

Bobin SA, Currie JR, Merz PA, Miller DL, Styles J, Walker WA, Wen GY, Wisniewski HM (1987): The comparative immunoreactivities of brain amyloids in Alzheimer's disease and scrapie. Acta Neuropathol 74:313–323.

Coria F, Castaño E, Frangione B (1987): Brain amyloid in normal aging and cerebral amyloid angiopathy is antigenically related to Alzheimer's disease β-protein. Am J Pathol 129:422–428.

Crystal H, Dickson D, Fuld P, Masur D, Scott R, Mehler M, Masdeu J, Kawas C, Aronson M, Wolfson L (1988): Clinico-pathologic studies in dementia: Nondemented subjects with pathologically confirmed Alzheimer's disease. Neurology 38:1682–1687.

Davies L, Wolska B, Hilbich C, Multhaup G, Martins R, Simms G, Beyreuther K, Masters CL (1988): A4 amyloid protein deposition and the diagnosis of Alzheimer's disease: Prevalence in aged brains determined by immunocytochemistry compared with conventional neuropathologic techniques. Neurology 38:1688–1693.

Dickson DW, Farlo J, Davies P, Crystal H, Fuld P, Yen SC (1988): Alzheimer's disease: A double-labeling immunohistochemical study of senile plaques. Am J Pathol 132:86–101.

Duyckaerts C, Hauw JJ, Bastenaire F, Piette F, Poulain C, Rainsard V, Javoy-Agid F, Berthaux P (1986): Laminar distribution of neocortical senile plaques in senile dementia of the Alzheimer type. Acta Neuropathol 70:249–256.

Ghiso J, Jensson O, Frangione B (1986): Amyloid fibrils in hereditary cerebral hemorrhage with amyloidosis of Icelandic type is a variant of γ-trace basic protein (cystatin C). Proc Natl Acad Sci USA 83:2974–2978.

Glenner GG (1980): Amyloid deposits and amyloidosis: The β-fibrilloses. New Engl J Med 302:1283–1292.

Glenner GG, Wong CW (1984a): Alzheimer's disease: Initial report on the purification and characterization of a novel cerebrovascular amyloid protein. Biochem Biophys Res Commun 120:885–890.

Glenner GG, Wong CW (1984b): Alzheimer's disease and Down's syndrome: Sharing of a unique cerebrovascular amyloid fibril protein. Biochem Biophys Res Commun 122:1131–1135.

Goldgaber D, Lerman MI, McBride W, Saffiotti U, Gajdusek DC (1897): Characterization and chromosomal localization of a cDNA encoding brain amyloid of Alzheimer's disease. Science 235:877–880.

Hansen LA, DeTeresa R, Davies P, Terry RD (1988): Neocortical morphometry, lesion counts, and choline acetyltransferase levels in the age spectrum of Alzheimer's disease. Neurology 38:48–54.

Hilmert H, Diringer H (1984): A rapid and efficient method to enrich SAF-protein from scrapie brains of hamsters. Biosci Rep 4:165–170.

Jervis GA, Thiells NY (1948): Early senile dementia in mongoloid idiocy. Am J Psychiatry 105:102–106.

Kang J, Lamaire HG, Unterbeck A, Salbaum JM, Masters CL, Grzeschik KH, Multhaup G, Müller-Hill B (1987): The precursor of Alzheimer's disease amyloid A4 protein resembles cell-surface receptor. Nature 325:733–736.

Katzman R, Terry R, DeTeresa R, Brown T, Davies P, Fuld P, Renbing X, Peck A (1988): Clinical, pathological, and neurochemical changes in dementia: A subgroup with preserved mental status and numerous neocortical plaques. Ann Neurol 23:138–144.

Kitaguchi N, Takahashi Y, Tokushima Y, Shiojiri S, Ito H (1988): Novel precursor of Alzheimer's disease amyloid protein shows protease inhibitory activity. Nature 331:530–532.

Kitamoto T, Tateishi J, Tashima T, Takeshita I, Barry RA, DeArmond SJ, Prusiner SB (1986a): Amyloid plaques in Creutzfeldt–Jakob disease stain with prion protein antibodies. Ann Neurol 20:204–208.

Kitamoto T, Hikita K, Tashima T, Tateishi J, Sato Y (1986b): Scrapie-associated fibrils (SAF) purification method yields amyloid proteins from systemic and cerebral amyloidosis. Biosci Rep 6:459–465.

Kitamoto T, Ogomori K, Tateishi J, Prusiner SB (1987): Formic acid pretreatment enhances immunostaining of cerebral and systemic amyloids. Lab Invest 57:230–236.

Koike F, Kunishita T, Nakayama H, Tabira T (1988): Immunohistochemical study of Alzheimer's disease using antibodies to synthetic amyloid and fibronectin. J Neurol Sci 85:9–15.

Lewis DA, Campbell MJ, Terry RD, Morrison JH (1987): Laminar and regional distribution of neurofibrillary tangles and neuritic plaques in Alzheimer's disease: A quantitative study of visual and auditory cortex. J Neurosci 7:1799–1808.

Majocha RE, Benes FM, Reifel JL, Rodenrys AM, Marotta CA (1988): Laminar-specific distribution and infrastructural detail of amyloid in the Alzheimer disease cortex visualized by computer-enhanced imaging of epitopes recognized by monoclonal antibodies. Proc Natl Acad Sci USA 85:6182–6186.

Mann DMA, Tucker CM, Yates PO (1987): The topographic distribution of senile plaques and neurofibrillary tangles in the brains of non-demented persons of different ages. Neuropathol Appl Neurobiol 13:123–139.

Masters CL, Simms G, Weinman NA, Multhaup G, McDonald BL, Beyreuther K (1985): Amyloid plaque core protein in Alzheimer disease and Down syndrome. Proc Natl Acad Sci USA 82:4245–4249.

Masuda J, Tanaka K, Ueda K, Omae T (1988): Autopsy study of incidence and distribution of cerebral amyloid angiopathy in Hisayama, Japan. Stroke 19:205–210.

Merz PA, Somerville RA, Wisniewski HM, Iqbal K (1981): Abnormal fibrils from scrapie-infected brain. Acta Neuropathol 54:63–74.

Ogomori K, Kitamoto T, Tateishi J, Sato Y, Tashima T (1988): Aging and cerebral amyloid: Early detection of amyloid in the human brain using biochemical extraction and immunostain. J Gerontol 43:B157–B162.

Ogomori K, Kitamoto T, Tateishi J, Sato Y, Suetsugu M, Abe M (1989): β-protein amyloid is widely distributed in the central nervous system of patients with Alzheimer's disease. Am J Pathol 134:243–251.

Pearson RCA, Esiri MM, Hiorns RW, Wilcock GK, Powell TPS (1985): Anatomical correlates of the distribution of the pathological changes in the neocortex in Alzheimer disease. Proc Natl Acad Sci USA 82:4531–4534.

Ponte P, Gonzalez-DeWhitt P, Schilling J, Miller J, Hsu D, Greenberg B, Davis K, Wallace W, Lieberburg I, Fuller F, Cordell B (1988): A new A4 amyloid mRNA contains a domain homologous to serine protease inhibitors. Nature 331:525–527.

Prusiner SB (1982): Novel infectious particles cause scrapie. Science 216:136–144.

Rafalowska J, Barcikowska M, Wen GY, Wisniewski HM (1988): Laminar distribution of neuritic plaques in normal aging, Alzheimer's disease and Down's syndrome. Acta Neuropathol 77:21–25.

Robakis NK, Ramakrishna N, Wolfe G, Wisniewski HM (1987): Molecular cloning and characterization of cDNA encoding the cerebrovascular and neuritic plaque amyloid peptides. Proc Natl Acad Sci USA 84:4190–4194.

Rogers J, Morrison JH (1985): Quantitative morphology and regional distributions of senile plaques in Alzheimer's disease. J Neurosci 5:2801–2808.

Stam FC, Wigboldus Jm, Smeulder WM (1986): Age incidence of senile brain amyloidosis. Pathol Res Pract 181:558–562.

Störkel S, Bohl J, Schneider HM (1983): Senile amyloidosis: Principles of location in a heterogeneous form of amyloidosis. Virchows Arch (Cell Pathol) 44:145–161.

Tagliavini F, Giaccone G, Frangione B, Bugiani O (1988): Preamyloid deposits in the cerebral cortex of patients with Alzheimer's disease and nondemented individuals. Neurosci Lett 93:191–196.

Tanzi RE, McClatchey AI, Lamberti ED, Villa-Komaroff L, Gusella JF, Neve RL (1988): Protease inhibitor domain encoded by an amyloid protein precursor mRNA associated with Alzheimer's disease. Nature 331:528–530.

Terry RD, Hansen LA, DeTeresa R, Davies P, Tobias H, Katzman R (1987): Senile dementia of the Alzheimer type without neocortical tangles. J Neuropathol Exp Neurol 46:262–268.

Tomlinson BE, Blessed G, Roth M (1968): Observation on the brains of non-demented old people. J Neurol Sci 7:331–356.

Tomlinson BE, Blessed G, Roth M (1970): Observation on the brains of demented old people. J Neurol Sci 11:205–242.

Tomlinson BE, Corsellis JAN (1984): Aging and dementias. In: Adams JH, Corsellis JAN, Duchen LW (eds): "Greenfield's Neuropathology," 4th ed. London: Edward Arnold pp 951–1025.

Vinters H, Pardridge WM, Secor DL, Ishii N (1988): Immunohistochemical study of cerebral amyloid angiopathy. II. Enhancement of immunostaining using formic acid pretreatment of tissue sections. Am J Pathol 133:150–162.

Wong CW, Quaranta V, Glenner GG (1985): Neuritic plaques and cerebrovascular amyloid in Alzheimer's disease are antigenically related. Proc Natl Acad Sci USA 82:8729–8732.

Wright JR, Calkins E, Breen WJ, Stolte G, Schutz RT (1969): Relationship of amyloid to aging. Review of the literature and systemic study of 83 patients derived from a general hospital population. Medicine 48:39–60.

Yamada M, Tsukagoshi H, Otomo E, Hayakawa M (1987): Cerebral amyloid angiopathy in the aged. J Neurol 234:371–376.

Yamaguchi H, Hirai S, Morimatsu M, Shoji M, Ihara Y (1988): A variety of cerebral amyloid deposits in the brains of the Alzheimer-type dementia demonstrated by β-protein immunostaining. Acta Neuropathol 76:541–549.

The Striatum, A Microcosm for the Examination of Age-Related Alterations in the CNS: A Selected Review

J.A. Joseph, G.S. Roth, and R. Strong

INTRODUCTION

The objective of this chapter was to review literature from 1986 to the present concerned with alterations in central neuronal functioning in senescence. After assembling and reviewing the relevant papers, three concerns became readily apparent: 1) space considerations would not permit an adequate discussion of the myriad of age-related structural and synaptic alterations in the CNS: 2) if such a feat were attempted the end result would be a "telephone directory" of all the changes in the aging CNS and their numerous directions; and 3) such a presentation would leave little space for discussion of the possible mechanisms of these changes or their functional correlates.

To address these concerns, we chose instead to confine this review to the striatum, an area that shows profound, consistent alterations in aged organisms and that traditionally has been associated with the mediation of such functions as motor behavior [e.g., see Carp and Anderson, 1979; Marshall, 1979; Sabol et al., 1985] and the programming of complex behavioral patterns associated with feeding [Ungerstedt, 1971] and arousal [Marshall et al. 1980]. The striatum is also involved in mediating the ordering and sequencing of behaviors that are not directed by exteroceptive stimuli [Cools, 1980; Jaspers et al., 1984; Vrijmoed-De Vries and Cools, 1986].

Since many of these same behaviors decline with age [see Joseph and Roth, 1988a,b, for review], one can easily postulate that age-related alterations in this "central processing area" (the striatum) underlie observed decrements in performance. Therefore, the striatum can be employed as a model for determining how specific, age-related alterations in various aspects of neurotransmission (e.g., synthesis, release, modulation of receptor function, etc.) can

---

Gerontology Research Center/NIA, Francis Scott Key Medical Center, Baltimore, Maryland 21224 (J.A.J., G.S.R.) and VA Medical Center, GRECC, St. Louis, Missouri 63125 (R.S.)

have important consequences for behavioral control in senescence. The elaboration and specification of these various neuronal parameters are the subjects of the present review. We have confined the review to studies conducted between 1986 and 1989. However, in many cases we found it necessary to cite earlier work to make specific points in the discussion.

THE STRIATAL DOPAMINE SYSTEM

Numerous experiments have demonstrated that interference with striatal function by any one of several methods can retard motor performance in young animals. These procedures have generally utilized techniques that alter striatal dopamine (DA) activity [see Joseph and Roth, 1988a,b, for review]. Because of this relationship, extensive research in the field of the neurobiology of aging has been directed toward determining the various aspects of DA synaptic activity exhibiting age-related changes. Facets of this work have been reviewed previously [e.g., Joseph and Roth, 1988a,b; Morgan, 1987; Morgan et al., 1987a,b; Severson, 1987]. Generally, the studies cited in these reviews have indicated consistent age-related declines in this system in rodents and humans that include: 1) decreases in striatal DA and tyrosine hydroxylase levels; 2) loss of striatal DA receptors of both the D_1 and D_2 subtypes (see below, however); 3) shifts in the proportion of high-to-low affinity striatal D_2 sites; 4) decreases in the synthesis rates of both D_1 and D_2 receptor subtypes; and 5) decreases in DA uptake.

Since these reviews, studies concerned with assessing possible age-induced alterations on the striatal system have generally supported and extended these previous findings. For example, Watanabe [1987] demonstrated that the rate of DA synthesis, as assessed via DOPA accumulation, was slower in striatum and nucleus accumbens obtained from aged rats than mature rats. However, no age differences were observed in the olfactory tubercles, indicating a selective vulnerability of the DA neurons in the subcortical motor control areas in senescence. Similarly, findings from Missale et al. [1986] indicated age-dependent decreases in the binding of [^{3}H]-cocaine to striatal (but not frontal cortex) membranes that was accompanied by specific increases in the Km for DA uptake into striatal slices. These studies are supported by those cited in reviews mentioned above and by additional studies that have indicated reduced density of striatal DA uptake sites in aging [e.g., Henry et al., 1986; Morgan et al., 1987b; Henry et al., 1987; Rinne, 1987; Zelnik et al., 1986]. However, while general agreement exists that DA binding sites are lost from the striatum with age, the pattern of such loss, especially of the D_1 subtype, remains controversial and is possibly species- and strain-dependent. D_2 receptor concentrations show remarkably consistent, progressive age-related declines in numerous species [e.g., rat, Henry et al., 1986; Hyttel, 1987; Petkov et al.,

1988; monkey, Lai et al., 1987; human, Morgan et al., 1987b; mouse, Severson and Finch, 1980; rabbit, Thal et al., 1980] under a variety of experimental conditions. The actual percentage of this decline depends upon the assay conditions and the particular DA-sensitive ligand utilized. For example, findings from studies using [^{3}H]-spiperone, for example, which is selective for the D_2 subtype, indicate consistent decreases in D_2 receptor concentrations of 36–66% over the life span of the animal.

If age-related changes in concentration or affinity of the striatal D_1 ("adenylate cyclase"-linked) receptor are assessed, results are more variable. Findings from at least one laboratory show parallel losses between D_1 receptor concentration and DA-stimulated adenylate cyclase activity that are greatest between 3 and 14 months of age in Wistar [Henry et al., 1986; Hyttel, 1987] or Sprague-Dawley [Giorgi et al., 1987] rats. However, results from other laboratories using Sprague-Dawley rats [O'Boyle and Waddington, 1984] or C57BL/6J mice [Finch and Morgan, 1984] indicate no change in striatal D_1 receptor levels throughout the life span. Examination of human post mortem material suggests that the density of D_1 receptors increases progressively in the caudate and putamen, while the D_2 receptor concentrations show consistent decreases in the caudate nucleus. The D_1/D_2 ratio increases in both regions from approximately 1 at age 20 to 2 by age 75 [Morgan et al., 1987a]. These authors suggest that D_1 receptors may be up-regulated in the human as a result of decreases in DA content in the basal ganglia during aging.

As indicated above, strain or species differences can possibly account for some of the discrepancies observed with respect to age-related D_1 receptor changes between laboratories. Support for this suggestion can be provided from extrapolations of a recent report of Ebel et al. [1987] in which they assessed several parameters of DA functioning between C57BL and BALB/c mouse strains. Results indicated that striatal DOPA and 3MT levels declined selectively in the BALB/c but not in the C57BL strain, while HVA and DOPAC declined in both strains as a function of age. Additionally, differences in DA receptor concentrations occur between different strains of young mice. As one extreme example, the weaver mutant mouse, which is homozygous for an autosomal recessive gene (weaver), shows decreased concentrations of both DA receptor subtypes in the striatum at 30–60 days of age [Pullara and Marshall, 1989].

However, it is difficult to account for the discrepancies among the various laboratories that have been reported (with respect to the pattern of striatal D_1 receptor loss) in the Sprague-Dawley rat. Giorgi et al. [1987] report decreased D_1 receptor binding in the striatum with age in this rat strain, while O'Boyle and Waddington [1984] do not. Note that O'Boyle and Waddington also failed to find progressive declines in striatal D_2 receptor concentrations after 11 months of age in their rats. Nearly all other laboratories (see above) report significant, progressive, age-related declines in this receptor subtype. Additionally, early experiments by Schmidt and Thornbury [1978], as well as later studies [Henry et al., 1986] indicate substantial declines in DA-stimulated adenylate cyclase activity by 12 months of age, declines that parallel the observed D_1 receptor loss [Henry et al., 1986]. These differences are difficult to explain, and they have not been adequately addressed. Possibly they reflect a specific characteristic of rats from a particular source [e.g., Wolfson Institute of Gerontology—O'Boyle and Waddington, 1986; Charles River, Como Italy—Giorgi et al., 1987].

It is also possible that differences in dissection procedures among the various laboratories could account for these discrepancies. Previous experiments have indicated that the striatum is not a homogeneous structure and that the receptor distribution is not uniform (see below). Very few examinations have been undertaken to determine the regional intra-striatal age differences in the concentrations of these receptors, and these have focused exclusively on the D_2 subtype. Strong et al. [1982] reported that the decline in DA content of the senescent (Sprague-Dawley) rat occurred primarily in the caudal portion of the striatum. Later studies by Marshall and his colleagues, using quantitative autoradiographic procedures, [Joyce and Marshall, 1987; Joyce et al., 1986; Marshall and Joyce, 1988] extended these findings to include the mediolateral and dorsoventral striatal axes. Their results indicated that the density of [^{3}H]-spiperone-labeled sites (putative D_2 sites) was much greater in the dorsolateral than in the medial striatum. This distribution corresponds to innervation by the motor/somatosensory cortex; thus, it might be surmised that the D_2 sites might be located on the axons of these corticostriatal projections. However, ablations that virtually isolated the striatum from the cortex produced no loss of D_2 receptor binding, while injections of the selective (for interneurons) neurotoxin quinolinic acid reduced binding by 90 to 95%, indicating primary localization of D_2 sites on intrinsic striatal neurons. This percentage, when compared with that found in earlier experiments, which employed radioreceptor binding techniques [Garau et al., 1978; Govoni et al., 1978; Schwarcz et al., 1978; Spano et al., 1978] appear to be rather high (40% on intrinsic neurons and 60% on corticostriatal afferents).

The reasons for this discrepancy ($< 90\%$ to $> 50\%$) are not clear. One explanation [Memo et al., 1986; Missale, et al., 1987; Onali et al., 1985] suggested that there may be more than one population of D_2 sites with those located on intrinsic neurons associated with inhibition of adenylate cyclase and those on extrinsic neurons not connected to the cAMP generating system. Missale et al. [1987] have also indicated that the latter type of D_2 receptor is lost in aging. The cyclase-inhibiting D_2 receptor remains relatively intact. Given the differences in techniques, it is difficult to reconcile the disparate findings among these various localization studies, especially since Marshall and Joyce [1988] have also shown that aged rats exhibited a heterogenous loss of D_2 sites, with the declines being most extensive in the regions that showed the highest D_2 density in young animals [e.g., anterior dorsolateral CP, Joyce et al., 1986].

These findings indicate a selective loss of control over the sensorimotor portion of the striatum in senescence, but fail to specify precise localization.

A second aspect of these age-related changes in receptor density to consider is that concerned with the mechanism(s) involved in this decline. Various approaches, including manipulation of striatal membrane fluidity [Cimino et al., 1984; Henry and Roth, 1986a] and solubilization of the receptors [Henry and Roth, 1986b] have yielded no support for hypotheses that suggest possible masking or sequestration of DA receptors in senescence. Instead, it appears that losses in receptor containing neurons or dendritic components occur with increasing age. Unfortunately, none of the studies cited above attempted to correlate cell numbers and D_2 receptor density as a function of age. Recently, however, these determinations have been carried out [Han et al., in press]. The findings indicated that up to one-half the decline of the striatal D_2 receptors may result from a variable loss of neurons from different striatal regions. The other one-half of the decrements in D_2 concentration appear to be independent of neuronal loss and may result from impaired regulation of receptor synthesis and/or degradation [Henry and Roth, 1984; Henry et al., 1987; Leff et al., 1984].

These same observations need to be carried out with respect to the D_1 receptor, but it seems clear that at least some of the D_1 and D_2 receptors may be colocalized on the same interneurons. However, the failure to detect D_1 receptor loss in several other species including man [Morgan et al., 1987a] may suggest that some portion of this receptor population may be located on corticostriatal or nigrostriatal neurons. To make these determinations, it is necessary to couple quantitative visualization techniques with morphometric analyses.

In view of the findings cited above [e.g., Morgan and Finch, 1986; Morgan et al., 1987a], it is also possible that loss of cyclase stimulation with aging may be independent of D_1 receptor cell loss. It is the delineation of possible functional deficits in poststimulation events that is of paramount importance in altered DA transmission in aging. Govoni et al. [1988] assessed cyclic AMP-dependent phosphorylation in particulate and in cytosolic fractions of striatum and nucleus accumbens of 3- and 24-month-old Sprague-Dawley rats. Results indicated a reduced cAMP-stimulated ^{32}P incorporation in several protein bands in both brain areas with increased age. One of these, DARPP-32, is specific for dopaminoceptive neurons containing D_1 receptor sites. Govoni et al. [1988] point out that through phosphorylation DARPP-32 is converted to an active phosphatase inhibitor. This conversion allows DA to modulate its own action or the action of other first messengers. Unfortunately, these investigators did not examine a middle-aged group; thus it is difficult to determine if these deficits are only seen in senescent animals or whether they occur in a developmental time frame similar to that seen for adenylate cyclase activation.

Nevertheless, one important fact emerges from these studies: in the senescent rat D_2 (possibly D_1) receptor concentrations appear to be reduced, and this reduction is coupled with a decreased modulatory capability of the D_1 system. Over the past few years, the use of pharmacologic agents specific for D_1 and D_2 has allowed the determination of the particular motor behavioral patterns controlled by the D_1 or D_2 subtype [e.g., see Molloy et al., 1986; Robertson and Robertson, 1986]. Using these pharmacologic probes, future examinations should be able to determine the impact of declines in concentration of each DA receptor subtype on specific motor behaviors reduced during aging. Also helpful will be the employment of cDNA probes for each receptor subtype to determine specific patterns of loss throughout the brain. Recently, the D_2 cDNA has been developed [Bunzow et al., 1988]. Hopefully the D_1 cDNA will follow in the near future.

The importance of these two subtypes of receptors cannot be overestimated, since both experimental down-regulation or up-regulation of their concentrations (of D_1 and D_2) can produce, respectively, deficits or improvements in motor behavioral performance [see Joseph and Roth, 1988a,b, for review].

DIFFERENTIAL CHANGES IN SYNAPTIC PARAMETERS OF CHOLINERGIC FUNCTION

Acetylcholine (ACh) is synthesized from choline (taken up from presynaptic terminals by a sodium-dependent, high-affinity choline uptake (HACU) system) and acetyl coenzyme A (synthesized by the mitochondria). The reaction is catalyzed by choline acetyl-transferase activity (ChAT). As was seen with respect to the striatal DA system and aging, findings assessing central ACh changes in senescence tend to vary when the rather static measurements related to ACh levels, ChAT activity, HACU (under nondepolarizing conditions; see below), and acetylcholinesterase activity (AChE) are made.

ChAT, HAUC, and ACHE

ChAT as well as HAUC have been used as indices for ACh function in the senescent striatum, and results generally have been equivocal. Examinations of ChAT activity have indicated decreases [McGeer et al., 1971; Meek et al., 1977; Strong et al., 1982] in the rodent striatum by some investigators. Others have reported no change [e.g., Morin and Wasterlain, 1980; Perry et al., 1977; Strong et al., 1986]. No consistent age-related findings in this parameter have been seen in the cortex or hippocampus either [e.g., see Table I in Decker, 1987 and Sirvio et al., 1988]. Generally, changes in ChAT in areas such as the hippocampus and cortex that occur as a function of normal aging are less reliable than those seen in Alzheimer's disease (AD), but, as pointed out in an excellent review by Decker [1987], there are no reports of increases of ChAT activity as a function of age in rats or humans (there is one report of increases in mice [Waller et al., 1983]).

The reasons for these inconsistent findings are numerous and could include assay procedures [Waller et al., 1983; Waller and London, 1983] and sex [Luine et al., 1986] and strain/species differences [Strong et al., 1980; Gilad and Gilad, 1987]. One other important consideration is the precise localization of the neural area utilized for the assay. There appears to be intraregional specificity for a number of biomarkers, such as ChAT, within the striatum. The activity of this enzyme, for example, is highest in the rostral (relative to the caudal) neostriatum, and age-associated decreases occur primarily in the caudal regions [Strong et al., 1982, in Sprague-Dawley rats]. However, a later study by Strong et al. [1986], which specifically examined regional intrastriatal ChAT activity in three age groups of Fischer rats, failed to find any decreases in the striatum. Thus strain differences may, in some cases, supercede regional specificity. Note that this strain of rats also failed to show age-related alterations in ChAT activity in the cortex [Norman et al., 1986] and showed either zero [Lippa et al., 1985; Luine et al., 1986; Sherman et al., 1981] or minimal declines [Dravid, 1983] in the hippocampus.

Intra- and interregional differences within and among the striatum, cortex, and hippocampus become even more apparent when HACU is assessed. While several studies have reported no age-related differences in HACU from the cortex [Consolo et al., 1986; Meyer et al., 1984; Saito et al., 1986] or hippocampus (synaptosomes preincubated in a medium containing either physiological or depolarizing concentrations of K^+ [Sherman et al., 1981], a later study [Strong et al., 1986] using 6-, 18-, and 30-month-old Fischer rats indicated significant intrastriatal differences in HACU. Decrements were observed in this index in the lateral caudal and medial rostral striatal regions. Interestingly, this is the same study in which no age-related differences in ChAT activity were observed.

The results with respect to AChE have been even more contradictory than those for ChAT. Various studies have indicated that declines in this parameter range from 0% to 26% in the rodent [e.g., Morin and Waterlain, 1980; Santos-Benito and Gonzalez, 1985] and are less than 20% in the aged human. In one recent study, Sirvio et al. [1988] found that the ratio of detergent-soluble to salt-soluble AChE was lower in 30- than in 4-month-old rats. Again (as in the determination of age-related loss in DA parameters), it is very difficult to determine the developmental period when the changes actually occurred. Hollander and Barrows [1968], for example, have reported that the largest declines in AChE occurred between 3 and 6 months of age. Unfortunately, these determinations have primarily focused on the cortical and hippocampal regions and little, if any, work has been conducted in the striatum.

Muscarinic Receptor Binding

If this parameter is considered, there is better general agreement that there are decreases in the number of binding sites in several brain areas, including the striatum, as a function of age. Using the ligand [^{3}H]-quinuclidinyl benzilate (^{3}H-QNB), these studies have generally shown changes in the concentration of muscarinic receptors (mAChR) of about 11–36% in the hippocampus [Kubanis et al., 1982; Lippa et al., 1985; Waller and London, 1983, see also Nordberg and Winblad, 1981] and cortex [Kubanis et al., 1982; Norman et al., 1986; Pedigo and Polk, 1985; Strong et al., 1980]. More recent studies have generally supported these findings [e.g., cortex and hippocampus, Bigeon et al., 1988; Gurwitz et al., 1987] and extended them to the striatum, where reductions similar to those observed in the cortex and hippocampus were observed [10–20%, Bigeon et al., 1988; 19%, Gurwitz et al., 1987].

Distribution of these receptors within the striatum was similar to that seen for the D_2 receptors, being highest in the rostral region and lowest in the caudal portion of the striatum. mAChR were lost as a function of age primarily from the medial and caudal portions of the striatum [Strong et al., 1982]. However, in a later study using Fischer rather than Sprague-Dawley rats by this group [Strong et al., 1986], significant mAChR loss of about 20% throughout the striatum was observed. Strain differences appear again to be important in this regard.

There have been no attempts to determine the pattern of loss of mAChR subtypes (see below) (M_1 and M_2) as a function of normal aging. Bigeon et al. [1988] utilized 100 μM carbachol in the incubation medium [see Mash and Potter, 1986] for preliminary autoradiographic analysis of M_1 receptor distribution, but no significant differences as a function of age were seen. These findings support a previous experiment that had employed membrane binding using the putative M_1 antagonist pirenzepine [Norman et al., 1986]. Detailed analyses of alterations in muscarinic receptor subtypes in senescence are likely to be facilitated by the use of AF-DX 116, a selective M_2 antagonist, and the development of specific cDNA probes, which, when used with in situ hybridization, can determine age differences in regional expression of four muscarinic receptor subtypes (m_1–m_4). At present pharmacologic agents exist that distinguish only between the M_1 (may consist of three gene products (m_1, m_3, m_4) [Brann et al., 1988] and M_2 mAChR subtypes.

Autoradiographic analysis by Bigeon et al. [1988] also revealed that the pattern of mAChR loss throughout the brain was highest in areas such as the striatum, which contain high concentrations of cholinergic perikaryon. Thus a selective loss of cholinergic neurons could be a contributing factor to the decline in mAChR concentration. Studies by Albanese et al. [1985] and Gozzo et al. [1986] have indicated, using AChE staining techniques, a loss of forebrain cholinergic neurons as a function of age. Armstrong et al. [1988], using immunocytochemical staining for ChAT, observed swollen cholinergic profiles and "grapelike clusters of immunoreactivity" in the neocortex and cingulate of the aged Fischer rat. Similar findings have been reported in aged human and nonhuman primates [Armstrong et al., 1986; Kitt et al., 1984;

Struble et al., 1982; Tago et al., 1987]. In support of these latter findings, Adams [1987] reported decreases in the number of synapses and increases in the lengths of the postsynaptic contact zone in the human precentral motor cortex. Additional experiments by Altavista et al. [1988] using AChE staining extended these findings to the striatum to show that the size of the neostriatum and the number of putative, cholinergic, AChE-positive perikarya are significantly reduced as of 24 months in the rat. Cell density decreased by approximately 16%. The loss is most pronounced in the dorsal, lateral, and ventromedial areas of the striatum. Unfortunately, this study employed only four animals from two age groups (3 and 24 months) to make these determinations, so the results are inconclusive. Nevertheless, when taken together with previous studies cited above, they suggest that nearly all of the mAChR loss in the aged striatum can be accounted for by neuronal loss. In this respect, these findings differ from the striatal D_1 and D_2 results wherein only 50% of the age-related decline could be accounted for by neuronal loss.

In summary, the studies cited thus far in this section indicate that, generally, there are variable changes in these rather static indices. Such small changes (which are species- and strain-dependent) provide little insight into the nature of the rather ubiquitous decline in memory and motor function seen in senescence. Numerous studies have demonstrated relationships between central cholinergic function and memory and motor performance [e.g., Bartus et al., 1985]. Thus an important question becomes: are there indices of cholinergic function that show consistent change with age associated with impaired behavioral performance? So far, the answer appears to be yes. When more dynamic properties of cholinergic function are assessed, consistent age-related decrements are observed. There are at least two sources of evidence suggesting such decrements. These sources are derived from experiments that have assessed: 1) synthesis and release; and 2) responsiveness of central mAChR to agonist stimulation.

ACh synthesis and release. Several studies have observed age-related decrements in two dynamic indices of neuronal functioning—synthesis and release. Declines in ACh synthesis are seen as early as 13 months of age in rat striatum, hippocampus, and cortex in both basal and high K^+ conditions [see Decker, 1987]. Unfortunately, there are insufficient data to determine how this parameter may vary as a function of the variables described above (e.g., species, strain, or inter- and intrastriatal regions). Available findings suggest that these factors may be less important with respect to synthesis than with respect to ChAT activity, etc. Moreover, synthesis decline may be a maturational occurrence rather than one specific to senescence.

Considerable work concerned with ACh release has been conducted in the cortex and hippocampus. While age differences in basal ACh release may [e.g., Sastry et al., 1983; cortex] or may not [e.g., Pedata et al., 1983, cortex] be observed, nearly all of the studies thus far have shown decrements in stimulated ACh release [e.g., Pedata et al., 1983 cortex; Consolo et al., 1986, hippocampus; Meyer et al., 1984, Pedata et al., 1983, cortex]. Thompson et al. [1984] extended these findings to include the striatum and observed similar decreases in ACh release.

These investigators [Thompson et al., 1984] also attempted to determine whether the well-known reciprocal inhibitory control between striatal DA and ACh systems might be altered during aging by assessing the effects of the DA agonist apomorphine on [^{3}H]ACh release from perifused striatal slices. Thompson et al. [1984] reported that apomorphine was effective in inhibiting [^{3}H]ACH release from striatal slices from young (7 months) and middle-aged (12 months) Wistar rats but not from old (24 months) (35% inhibition young and middle; 0% in old).

mAChR responsivity. More recent experiments have extended and supported these findings and have suggested that profound age-related declines in the ability of muscarinic agonists to alter DA responses involved inhibition of DA-stimulated adenylate cyclase from striatal broken cell preparations. Coupet et al. [1985] showed that the muscarinic agonists oxotremorine and carbachol were effective in young (6 months) and middle-aged (12 months) animals in inhibiting DA-stimulated adenylate cyclase but not in old (24 months). A similar inability of muscarinic agonists to regulate striatal DA function was also observed by Joseph et al. [1988a].

It has been known for several years that striatal DA release is under the control of a group of striatal D_2 autoreceptors [Dwoskin and Zahniser, 1986]. If these autoreceptors are enhanced or inhibited, K^+-evoked release of DA will be respectively enhanced or inhibited [see Joseph et al., 1988a,b, for methods and reviews]. This control is mediated through inhibitory presynaptic cholinergic heteroreceptors presumably located on the same terminals as the autoreceptors [Raiteri et al., 1984]. Cholinergic agonist stimulation of perifused striatal slices normally results in enhancement of K^+-evoked release of DA. However, Joseph et al. [1988a] demonstrated that even though no age-related differences were observed in levels of striatal DA or in K^+-evoked release of DA from perifused striatal slices, significant reductions occurred in the efficacy of several muscarinic agonists (e.g., oxotremorine, carbachol) to enhance K^+-evoked release of DA. Application of some agonists (e.g., carbachol) resulted in diminished responding as early as 12 months of age.

These deficits in mAChR responsivity to agonist stimulation appear to extend to the hippocampus as well. In this latter region, both extracellular [Lippa et al., 1985] and intracellular [Segal, 1982] recording studies have revealed a highly selective reduction in the ability of iontophoretically applied ACh to increase the excitability of hippocampal pyramidal cells. Moreover, Lippa et al. [1985] also suggested that this diminished responsivity began at about 15–16 months of age, at which time reductions in the burst firing rate of these cells to applied ACh were observed. Further reductions were observed in the senescent (24 months) group.

Thus these studies, as well as those cited above, suggest that there are profound, consistent, age-related declines in the sensitivity of mAChR to agonist stimulation beginning around middle age in the rodent. Given findings such as these, it is not surprising that previous work has indicated only minimal success in improving performance on tasks that assess memory function in senescent animals or humans through use of agents that enhance cholinergic functioning [e.g., Bartus et al., 1985].

An important question becomes: what alterations occur in the functioning of the muscarinic receptor to blunt its responsivity to agonists? The answer(s) to this question has important implications for geroneurobiology directed toward determining the role of the cholinergic system in memory loss in normal aging and in dementing diseases, such as Alzheimer's disease. It is becoming increasingly clear that deficits in cholinergic function may not be sufficient to explain the cognitive deficits that accompany aging and age-related diseases. There are three sources of evidence for this contention: 1) not all dementing diseases are associated with cholinergic deficits [Rinne et al., 1988]; 2) other pathological disorders may be associated with cholinergic deficits as profound as those in Alzheimer's disease [Kish et al., 1988]; and 3) deficiencies occur in other transmitter systems in patients who die of Alzheimer's disease. Nevertheless, it is possible that improvements in cholinergic functioning could result in enhanced performance on memory and perhaps motor tasks.

It would be tempting to answer the above question by holding the small decline in central muscarinic receptors responsible for the reductions in agonist/receptor efficacy. However, it was emphasized previously that significant age-related reductions occur in both D_1 and D_2, but few indications are found in the literature to suggest loss of responsivity to agonist stimulation even in the face of this significant ($>$ 30%) loss. Joseph et al. [1988a,b] observed no reductions in K_+-evoked release of DA in striatal tissue from senescent animals. These findings support those of earlier work [Thompson et al., 1981] in the striatal slice and at least partially support a study by Rose et al. [1986] in which DA release was assessed in the striatal slice using in vivo voltometry. Very preliminary findings from our laboratory, using microdialysis, indicate no decrements in striatal DA release following direct cocaine application to this area. Moreover, Joseph et al. [1988b] have shown that if DA autoreceptors are directly inhibited by haloperidol, or inhibited via activation of nicotinic heteroreceptors, no age-related decrements are observed in the enhancement of K^+-evoked release of DA. Therefore, simple explanations implying that receptor loss is sufficient to explain functional loss in a striatal cholinergic system are not adequate. Rather, it may be necessary to examine possible age-related reductions in signal transduction parameters to explain decrements in responsivity in this system.

An enormous amount of data has accumulated in recent years showing that ligand activation of mAChR results in Ca^{2+} mobilization brought about through a signal transduction process involving ligand-induced increases in the turnover of phosphatidylinositol (PI) (through the activation of G-proteins), a product that may be further phosphorylated to form phosphatidylinositol 4,5-bisphosphate (PIP_2). Phospholipase C, activated by the ligand–receptor complex, cleaves PIP_2. One product resulting from this cleavage is 1,4,5-inositol trisphosphate (IP_3), which diffuses into the cytoplasm, releases Ca^{2+} from storage, and ultimately induces a physiological response [Fisher and Agranoff, 1987]. As of this writing, few experiments have attempted to make these determinations in central mAChR systems. However, findings from at least one study suggest that if the mAChR is bypassed and Ca^{2+} mobilized directly with the Ca^{2+} ionophore A23187 or IP_3, no age-related deficits in enhancement of K^+-evoked release are observed [Joseph et al., 1988b]. These observations suggest that the deficits may occur at the receptor/ligand interface or very early in the signal transduction process. Indeed, results from an earlier report suggest that is much more mAChR receptor heterogeneity exists in old animals, with a greater number of receptors being in desensitized conformational/orientational states [Lippa et al., 1985]. This finding has not been supported by additional research [e.g., Waller and London, 1983; Avissar et al., 1981; Consolo et al., 1986].

There are also indications from other sources suggesting that there may be age-related decrements in signal transduction pathways mediated by G-proteins in several peripheral neurotransmitter systems including noradrenergic (beta receptors) and mAChR [see reviews by Severson, 1987; Scarpace and Abrass, 1987]. Taken together, these studies indicate that the deficits in mAChR responsivity may occur early in the signal transduction process and may lie either at the receptor/ligand interface or with initial receptor G-protein coupling.

CONCLUSIONS AND FUTURE DIRECTIONS

We have attempted to indicate in this review that research in the area concerned with striatal age-related changes in DA or ACh functioning that focuses upon static measurements of receptor loss or alterations in various synthetic catabolic enzymes will yield little additional information. Yet, even as of 1988 there are still papers appearing in the literature that measure, for example, mAChR or DA receptor loss without any attempt to determine the possible intraregional patterns of declines or mechanisms involved in their loss. No attempt is made even to relate decrements in concentrations to neuronal loss. Numerous critical determinations needed in this area are not being pursued. Some of these are listed below.

1. It should be clear that since a significant portion of the age-related decrements in receptor concentration may be accounted for by neuronal loss or perhaps by alterations in cell surface membranes, it is necessary to determine the mechanism of such changes. For example, one possibility is that these changes result from increased peroxidative damage to membranes in aged organisms [Schroeder, 1984]. Catecholamines are well-known cytotoxins. They are rapidly autoxidized to free radical-forming quinone by-products, and DA is known to be toxic to neuroblastoma, leukemia, and melanoma cell lines [Cadet et al., 1986]. It is also important to note in this regard (as we mentioned in the introduction) that the nigrostriatal system is one of the most sensitive to the ravages of time. Other findings demonstrate: 1) increased sensitivity to the neurotoxin 1-methyl-4-phenyl-1,2,3,6 tetrahydropyridine (MPTP) (which may produce damage through free radical generation and is more neurotoxic in aged animals [Ricaurte et al., 1987ab]); and 2) correlated motor behavioral deficits and blunting of muscarinic enhancement of K^+-evoked release of DA from striatal slices of young animals at 8 and 14 days after low doses ($<$ 25 rads) of ^{56}Fe irradiation [Joseph et al., in preparation]. All the factors cited just above illustrate the necessity for examining possible free radical damage as the responsible mechanism for the age-related changes observed in this system. Clearly, more mechanistic research using various models that have construct validity with neuronal systems needs to be carried out.

2. Future research efforts should also be directed toward elucidating the steps in signal transduction that may be altered in aging, beginning with the receptor/ligand interface. As of this writing, almost no work on this subject has been performed in the aging brain. Here again, some of the membrane changes may be the result of lipid peroxidation and cross-linking. As pointed out by Schroeder [1984], alterations in membrane structural properties may be intimately involved in the loss of receptor function seen here and in other systems [e.g., D_1-adenylate cyclase, Nomura et al., 1984, 1987].

One direction for this work would be to examine transmembrane segments of the various forms [m_1–m_4, Bonner et al., 1987; Brann et al., 1988] of the mAChR. Also important is the determination of factors that may alter receptor G-protein interactions in aging [e.g., see Green and Johnson, 1989]. A second approach would be to examine possible impaired Ca^{2+} mobilization in this system [see Roth, in press]. Perhaps once these events have been studied, it may be possible to design specific agents that could enhance cholinergic function and improve behavioral performance in senescence.

3. Finally, it is important to mention that none of these neurotransmitter systems exist in a vacuum. There is a great deal of cross-talk among the various systems to the effect that alterations in one system (e.g., as a function of aging) can profoundly affect another. The studies reviewed here clearly show loss in reciprocal inhibitory control between striatal ACh and DA in senes-

cence. However, these interactions are only a small part of the enormous modulatory control that exists between these systems. As pointed out in a recent review by Hill and Kendall [1989], stimulatory or inhibitory effects on one set of neurotransmitters and second messengers can be greatly influenced by simultaneous activation of other neurotransmitter–receptor systems, which can influence cyclic nucleotide, and IP$_3$ formation as well as the opening and closing of ion channels. At present, these determinations have not been made with respect to aging. Their further specification is critical in future research concerned with brain behavior relationships in aging.

REFERENCES

Adams I (1987): Plasticity of the synaptic contact zone following loss of synapses in the cerebral cortex of aging humans. Brain Res 424:343–351.

Albanese A, Gozzo S, Iacopino C, Altavista MC (1985): Strain-dependent variations in the number of forebrain cholinergic neurones. Brain Res 334:380–384.

Altavista MC, Bentivoglio AR, Crociani P, Rossi P, Albanese A (1988): Age-dependent loss of cholinergic neurones in basal ganglia of rats. Brain Res 455:177–181.

Armstrong DM, Bruce G, Gersh LB, Terry RD (1986): Choline acetyltransferase immunoreactivity in neuritic plaques of Alzheimer brain. Neurosci Lett 71:229–234.

Armstrong DM, Hersh LB, Gage FH (1988): Morphologic alterations of cholinergic processes in the neocortex of aged rats. Neurobiol Aging 9:199–205.

Avissar S, Egozi Y, Skolovsky M (1981): Aging process decreases the density of muscarinic receptors in rat adenohypophysis. FEBS Lett 133:275–278.

Bartus RT, Dean RL, Pontecorvo MJ, Flicker C (1985): The cholinergic hypothesis: A historical overview, current perspectives and future directions. Ann NY Acad Sci 444:332–358.

Bigeon A, Duvdevani R, Greenberger V, Segal M. (1988): Aging and brain cholinergic muscarinic receptors: An autoradiographic study in the rat. J Neurochem 51:1381–1385.

Bonner TI, Buckley NJ, Young AC, Brann MR (1987): Identification of a family of muscarinic acetylcholine receptor genes. Science 237:527–532.

Brann MR, Buckley NJ, Bonner TI (1988): The striatum and cerebral cortex express different muscarinic receptor mRNAs. FEBS Lett 230:90–94.

Bunzow JR, HHM Van Tol, Grandy DK, Albert P, Salon J, Christie M, Machida CA, Neve K, Civelli O (1988): Cloning and expression of a rat D$_2$ dopamine receptor cDNA. Nature 336:783–787.

Cadet JL, Lohr JB, Jeste DV (1986): Free radicals and tardive dyskinesia. Science 9:107–108.

Carp JS, Anderson RJ (1979): Sensorimotor deficits produced by phenytoin and chlorpromazine in unanesthetized cats. Pharmacol Biochem Behav 10:513–520.

Cimino M, Vantini G, Algeri S, Curatola G, Pezzoli C, Stramentinoli G (1984): Age-related modification of dopaminergic and B-adrenergic receptor systems: Restoration to normal activity by modifying membrane fluidity with S-adenosyl-methionine. Life Sci 34:2029–2039.

Consolo S, Wang J-W, Fiorentini F, Vezzani A, Ladinsky H (1986): In vivo and in vitro studies on the regulation of cholinergic neurotransmission in striatum, hippocampus and cortex of aged rats. Brain Res 374:212–218.

Cools AR (1980): Role of neostriatal dopaminergic activity in sequencing and selecting behavioral strategies: Facilitation of processes involved in selecting the best strategy in a stressful situation. Behav Brian Res 1:361–378.

Coupet J, Rauh CE, Joseph JA (1985): Age-related loss in the capacity of muscarinic agonists to inhibit the activation of adenylate cyclase in the striatum. Soc Neurosci Abs 11:573.

Decker MJ (1987): The effects of aging on hippocampal and cortical projections of the forebrain cholinergic system. Brain Res Rev 12:423–438.

Dravid AR (1983): Deficits in cholinergic enzymes and muscarinic receptors in the hippocampus and striatum of senescent rats: Effects of chronic hydergine treatment. Arch Int Pharmacodyn Ther 264:195–202.

Dwoskin LP, Zahniser NR (1986): Robust modulation of [^{3}H]dopamine release from rat striatal slices by D_2 dopamine receptors. J Pharmacol Exp Ther 239:442–454.

Ebel A, Strosser MT, Kempf E (1987): Genotypic differences in central neurotransmitter responses to aging in mice. Neurobiol Aging 8:417–427.

Finch CE, Morgan DG (1984): Serotonin (S-2) binding sites decrease in old mice, but are downregulated to the same extent by subchronic amitriptyline at all ages. Soc Neurosci Abs 10:16.

Freeman GB, Gibson GE (1987): Selective alteration of mouse brain neurotransmitter release with age. Neurobiol Aging 8:147–152.

Garau L, Govoni S, Stefanini E, Trabucchi M, Spano PF (1978): Dopamine receptors: pharmacological and anatomical evidences indicate that two distinct dopamine receptor population are present in rat striatum. Life Sci 23:1745–1750.

Gilad GM, Gilad VH (1987): Age-related reductions in brain cholinergic and dopaminergic indices in two rat strains differing in longevity. Brain Research 408:247–250.

Giorgi O, De Montis G, Porceddu ML, Mele S, Calderini G, Toffano G, Biggio G (1987): Developmental and age-related changes in D_1-dopamine receptors and dopamine content in the rat striatum. Dev Brain Res 35: 283–290.

Govoni S, Olgiati VR, Trabucchi M, Garau L, Stefanini E, Spano PF (1978): [^{3}H]-haloperidol and [^{3}H]-spiroperidol receptor binding after injection of kainic acid. 8:207–210.

Govoni S, Rius RA, Battaini F, Trabucchi M (1988): Reduced cAMP-dependent phosphorylation in striatum and nucleus accumbens of aged rats: Evidence of an altered functioning of D_1 dopaminoceptive neurons. J Gerontol 43:B93–97.

Gozzo S, Iacopino C, Altavista MC, Albanese A (1986): Early senescence of forebrain cholinergic systems: A genetic animal model. In Biggio G, Spano PF, Toffano G, Gl Gessa (eds): "Modulation of Central and Peripheral Transmitter Function. Symposia in Neuroscience," Vol 3 Paudua: Liviana, pp 299–303.

Green A, Johnson JL (1989)): Evidence for altered expression of the GTP-dependent regulatory proteins G_s and G_i in adipocytes from aged rats. Biochem J 258:607–610.

Gurwitz D, Egozi Y, Henis YI, Kloog Y, Sokolovsky M (1987): Agonist and antagonist binding to rat brain muscarinic receptors: Influence of aging. Neurobiol Aging 8:115–122.

Han Z, Kuyatt BJ, Kochman KA, De Souza EB, Roth GS (in press): Effect of aging on the concentration of D_2-receptor containing neurons in the rat striatum. Brain Res.

Henry JM, Roth GS (1984): Effect of aging on recovery of striatal dopamine receptors following N-ethoxycarbonyl-2-ethoxy-1, 2-dihydroquinoline (EEDQ) blockade. 35:899–904.

Henry JM, Roth GS (1986a): Modulation of rat striatal membrane fluidity: Effects on age-related differences in dopamine receptor concentrations. Life Sci 39:1223–1229.

Henry JM, Roth GS (1986b): Solubilization of striatal D_2 dopamine receptors: Evidence that apparent loss during aging is not due to membrane sequestration. J Gerontol 41:129–135.

Henry JM, Filburn CR, Joseph JA, Roth GS (1986): Effect of aging on striatal dopamine receptor subtypes in Wistar rats. Neurobiol Aging 7:357–361.

Henry JM, Joseph JA, Kochman K, Roth GS (1987): Effect of aging on striatal dopamine receptor subtype recovery following N-ethoxycarbonyl-2-ethoxy-1,2 dihydroquinoline blockade and relation to motor function in Wistar rats. Brain Res 418:334–342.

Hill SJ, Kendall DA (1989): Cross-talk between different receptor-effector systems in the mammalian CNS. Cell Signal 1:135–141.

Hollander J, Barrows CH (1968): Enzymatic studies in senescent rodent brains. J Gerontol 23:174–179.

Hyttel J (1987): Age related decrease in the density of dopamine D_1 and D_2 receptors in corpus striatum of rats. Pharmcol and Toxicol 61:126–129.

Jaspers R, Schwarz M, Sontag KH, Cools AR (1984): Caudate nucleus and programming behavior in cats: Role of dopamine in switching motor patterns. Behav Brain Res 14:17–28.

Joseph JA, Dalton TK, Hunt WA (1988a): Age-related decrements in the muscarinic enhancement of K^+-evoked release of endogenous striatal dopamine: An indicator of altered cholinergic–dopaminergic reciprocal inhibitory control in senescence. Brain Res 454:140–148.

Joseph JA, Dalton TK, Roth GS, Hunt WA (1988b): Alterations in muscarinic control of striatal dopamine autoreceptors in senescence: A deficit at the ligand–muscarinic interface? Brain Res 454:149–155.

Joseph JA, Roth GS (1988a): Upregulation of striatal dopamine receptors and improvement of motor performance in senescence. In Joseph JA (ed): "Central Determinants of Age Related Declines in Motor Function," Vol. 515. New York: New York Academy of Sciences, pp 355–366.

Joseph JA, Roth GS (1988b): Altered striatal dopaminergic and cholinergic reciprocal inhibitory control and motor behavioral decrements in senescence. In Pierpaoli W (ed): "Proceedings of the First International Conference on Neuroimmunomodulation," Vol. 521. New York: New York Academy of Sciences, pp 110–122.

Joyce JN, Marshall JF (1987): Quantitative autoradiography of dopamine D_2 sites in rat caudate–putamen: Localization to intrinsic neurons and not to neocortical afferents. Neuroscience 20:773–795.

Joyce JN, Sapp DW, Marshall JF (1986): Human striatal dopamine receptors are organized in compartments. Proc Natl Acad Sci USA 83:8002-8006.

Kubanis P, Zornetzer SF, Freund G (198): Memory and post-synaptic cholinergic receptors in aging mice. Pharmacol Biochem Behav 17:313–322.

Kish SJ, El-Awar M, Schut L, Leach L, Oscar-Berman M, Freedman M (1988): Cognitive deficits in olivopontocerebellar atrophy: Implications for the cholinergic hypothesis of Alzheimer's dementia. Ann Neurol 24:200–206.

Kitt CA, Price DL, Struble RG, Cork LC, Wainer BH, Becher MW, Mobley WC (1984): Evidence for cholinergic neurites in senile plaques. Science 226:1443–1444.

Lai H, Bowden DM, Horita A (1987): Age-related decreases in dopamine receptors in the caudate nucleus and putamen of the Rhesus monkey (*Macaca mulatta*). Neurobiol Aging 8:45–49.

Leff SE, Gariano R, Creese I (1984): Dopamine receptor turnover rates in rat striatum are age-dependent. Proc Natl Acad Sci USA 81:3910–3914.

Lippa AS, Loullis CC, Rotrosen J, Cordasco DM, Critchett DJ, Joseph JA (1985): Conformational changes in muscarinic receptors may produce diminished cholinergic neurotransmission and memory deficits in aged rats. Neurobiol Aging 6:317–325.

Luine VN, Renner KJ, Heady S, Jones KL (1986): Age and Sex-dependent decreases in ChAT in basal forebrain nuclei. Neurobiol Aging 7:193–198.

Marshall JF (1979): Somatosensory inattention after dopamine-depleting intracerebral 6-OHDA injections: Spontaneous recovery and pharmacological control. Brain Res 17:311–324.

Marshall J, Levitan D, Stricker E (1980): Activation-induced restoration of sensorimotor fuctions in rats with dopamine-depleting brain lesions. J Comp Physiol Psychol 90:536–546.

Marshall JF, Joyce JN (1988): Basal ganglia dopamine receptor autoradiography and age-related movement disorders. Ann NY Acad Sci 515:215–225.

Mash DC, Potter LT (1986): Autoradiographic localization of M_1 and M_2 receptors in rat brain. Neuroscience 19:551–564.

McGeer EG, Fibiger HC, McGeer PL, Wickson V (1971): Aging and brain enzymes. Exp Gerontol 6:391–396.

Meek JL, Bertilsson L, Cheney DL, Zsilla G, Costa E (1977): Aging-induced changes in acetylcholine and serotonin content of discrete brain nuclei. J Gerontol 32:129–131.

Memo M, Missale C, Carruba MO, Spano PF (1986): D_2 dopamine receptors associated with inhibition of dopamine release from rat neostriatum are independent of cyclic AMP. Neurosci Lett 71:192–196.

Meyer EM, Crews FT, Otero DH, Larson K (1984): Aging decreases the sensitivity of rat cortical synaptosomes to calcium ionophore-induced acetylcholine release. J Neurochem 47:1244–1246.

Missale C, Govoni S, Pasinetti G, Assini C, Spano PF, Battaini F, Trabucchi M (1986): Age-dependent changes in the mechanisms regulating dopamine uptake in the central nervous system. J Gerontol 41:136–139.

Missale C, Catelletti L, Pizzi M, Memo M, Carruba MO, Spano PF (1987): Striatal adenylate cyclase-inhibiting dopamine D_2 receptors are not affected by the aging process. Neurosci Lett 75:38–42.

Molloy AG, O'Boyle KM, Pugh MT, Waddington JL (1986): Locomotor behaviors in response to new selective D_1 and D_2 dopamine receptor agonists and the influence of selective antagonists. Pharmacol Biochem Behav 25:249–253.

Morgan DG (1987): The dopamine and serotonin systems during aging in human and rodent brain. A brief review. Prog Neuropsychopharmacol Biol Psychiatry 11:153–157.

Morgan DG, Finch CE (1986): [^{3}H]-fluphenazine binding to brain membranes and simultaneous measure of D_1 and D_2 receptor sites. J Neurochem 46:1623–1631.

Morgan DG, Marcusson JO, Nyberg P, Wester P, Winblad B, Gordon MN, Finch CE (1987a): Divergent changes in D_1 and D_2 dopamine binding sites in human brain during aging. Neurobiol Aging 8:195–201.

Morgan DG, May PC, Finch CE (1987b): Dopamine and serotonin systems in human and rodent brain: Effects of age and neurodegenerative disease. J Am Geriatr Soc 35:334–345.

Morin AM, Wasterlain CG (1980) Aging and rat brain muscarinic receptors as measured by quinuclidinyl benzilate binding. Neurochem Res 5:301–308.

Nomura Y, Makihata J, Segawa T (1984): Activation of adenylate cyclase by dopamine, GTP, naF and forskolin in striatal membranes of neonatal, adult, and senescent rats. 106:437–440.

Nomura Y, Aarima T Swgawa T (1987): Influences of pertussis toxin, guanine nucleotides and forskolin on adenylate cyclase in striatal membranes of infant, adult and senescent rats. 5:271–279.

Nordberg A, Winblad B (1981): Cholinergic receptors in human hippocampus—regional distribution and variance with age. Life Sci 29:1937–1944.

Norman AB, Blaker SN, Thal L, Creese I (1986): Effects of aging and cholinergic deafferentation on putative muscarinic cholinergic receptor subtypes in rat cerebral cortex. Neurosci Lett 70:289–294.

O'Boyle KM, Waddington JL (1984): Loss of rat striatal dopamine receptors with aging is selective for D_2 but not D_1 sites: Association with increased non-specific binding of the D_1 ligand [^{3}H]piflutixol. Eur J Pharmacol 105:171–174.

O'Boyle KM, Waddington JL (1986): A re-evaluation of changes in rat striatal D_2 dopamine receptors during development and aging. Neurobiol Aging 7:265–267.

Onali PL, Olianas MC, Gessa Gl (1985): Characterization of dopamine receptors mediating inhibition of adenylate cyclase activity in rat striatum. Mol Pharmacol 28:138–145.

Pedata F, Slavikova J, Kotas A, Pepeu G (1983): Acetylcholine release from rat cortical slices during postnatal development and aging. Neurobiol Aging 4:31–35.

Pedigo NW Jr, Polk DM (1985): Reduced muscarinic receptor plasticity in frontal cortex of aged rats after chronic administration of cholinergic drugs. Life Sci 37:1443–1449.

Perry EK, Perry RH, Gibson PH, Blessed G, Tomlinson BE (1977): A cholinergic connection between normal aging and senile dementia in human hippocampus. Neurosci Lett 6:85–89.

Petkov VD, Petkov VV, Sancheva SL (1988): Age-related changes in brain neurotransmission. Gerontology 34:14–21.

Pullara JM, Marshall JF (1989): Striatal dopamine innervation and receptor density: Regional effects of the weaver mutation. Brain Res 480:225–233.

Raiteri M, Riccardo L, Marchi M (1984): Heterogeneity of presynaptic muscarinic receptors regulating neurotransmitter release in the rat brain. J Pharmacol Exp Ther 228:209–215.

Ricaurte GA, Delanny LE, Irwin I, Langston JW (1987a): Older dopaminergic neurons do not recover from the effects of MPTP. Neuropharmacology 26:97–99.

Ricaurte GA, Irwin I, Forno LS, DeLanney LE, Langston E, Langston JW (1987b): Aging and 1-methyl-4-phenyl-1,2,3,6-tetrahydropyridine-induced degeneration of dopaminergic neurons in the substantia nigra. Brain Res 403:43–51.

Rinne JO (1987): Muscarinic and dopaminergic receptors in the aging human brain. Brain Res 404:162–168.

Rinne JO, Sako E, Paljarvi L, Molsa PK, Rinne UK (1988): A comparison of brain choline acetyltransferase activity in Alzheimer's disease, multi-infarct dementia and combined dementia. J Neural Transm 73:123–128.

Robertson GS, Robertson HA (1986): Synergistic effects of D_1 and D_2 dopamine agonists on turning behavior in rats. Brain Res 384:387–390.

Rose GM, Gerhardt GL, Hoffer BJ (1986): Age-related alterations in monoamine release from rat striatum: An in vivo electrochemical study. Neurobiol Aging 7:77–82.

Roth GS (in press) Changes in hormone action with age: Altered calcium mobilization and/or responsiveness impairs signal transduction. In Armbrecht HJ (ed) "Endocrine Function and Aging" New York: Springer-Verlag.

Roth GS (1988): Age changes in adrenergic and dopaminergic signal transduction mechanisms: Parallels and contrasts. Neurobiol Aging 9:63–64.

Sabol KE, Neill DB, Wages SA, Church WH, Justice JB (1985): Dopamine depletion in a striatal subregion disrupts performance of a skilled motor task in the rat. Brain Res 335:33–43.

Saito M, Kindel G, Karczmar AG, Rosenberg A (1986): Metabolism of choline in brain of the aged CBF-1 mouse. J Neurosci Res 15:197–204.

Santos-Benito FF, Gonzalez JL (1985): Decrease of choline acetyltransferase activity of rat cortex capillaries with aging. J Neurochem 45:633–636.

Sastry BVR, Janson VE, Jaiswal N, Tayeb OS (1983): Changes in enzymes of the cholinergic system and acetylcholine release in the cerebral of aging male Fischer rats. Pharmacology 26:61–72.

Scarpace PJ, Abrass IB (1988): Alpha- and beta-adrenergic receptor function in the brain during senescence. Neurobiol Aging 9:53–58.

Schmidt MJ, Thornbury JF (1978): Cyclic AMP and cyclic GMP accumulation in vitro in brain regions of young, old and aged rat. Brain Res 139:169–177.

Schroeder F (1984): Role of membrane lipid asymmetry in aging. Neurobiol Aging 5:323–333.

Schwarcz R, Creese I, Coyle JT, Snyder SH (1978): Dopamine receptors localized on cerebral cortical afferents of rat corpus striatum. Nature 271:766–768.

Segal M (1982): Changes in neurotransmitter actions in the aged rat hippocampus. Neurobiol Aging 3:121–132.

Severson JA (1987): Synaptic Regulation of Neurotransmitter Function in Aging. In Rothstein M (ed): "Review of Biological Research in Aging," Vol. 3. New York: Alan R. Liss, Inc., pp 191–206.

Severson JA, Finch CE (1980): Reduced dopaminergic binding during aging in the rodent striatum. Brain Res 192:147–162.

Sherman KA, Kuster JE, Dean RL, Bartus RT, Friedman E (1981): Presynaptic cholinergic mechanisms in brain of aged rats with memory impairments. Neurobiol Aging 2:99–104.

Sirvio J, Hervonen A, Valjakka A, Riekkinen PJ (1988): Pre- and postsynaptic markers of cholinergic neurons in the cerebral cortex of rats and different ages. Exp Gerontol 23:473–479.

Spano PF, Govoni S, Trabucchi M (1978): Studies on the pharmacological properties of dopamine receptors in various areas of the central nervous system. Adv Biochem Pshychopharmacol 19:155–165.

Strong R, Hicks P, Hsu L, Bartus RT, Enna SJ (1980): Age-related changes in the rodent brain cholinergic system and behavior. Neurobiol Aging 1:59–63.

Strong R, Samorjski T, Gottesfeld Z (1982): Regional mapping of neostriatal neurotransmitter systems as a function of aging. J Neurochem 39:831–836.

Strong R, Rehwaldt C, Wood GW (1986): Intra-regional variations in the effect of aging on high affinity choline uptake, choline acetyltransferase and muscarinic cholinergic receptors in rat neostriatum. Exp Gerontol 21:177–186.

Struble RG, Cork LC, Whitehouse PJ, Price DL (1982): Cholinergic innervation in neuritic plaques. Science 216:413–415.

Tago H, McGeer PL, McGeer EG (1987): Acetylcholinesterase fibers and the development of senile plaques. Brain Res 406:363–369.

Thal LJ, Horowitz SG, Dvorkin B, Makman (1980): Evidence for loss of brain [^{3}H]-ADTN binding sites in rabbit brain with aging. Brain Res 192:185–194.

Thompson JM, Whitaker J, Joseph JA (1981): [^{3}H]-dopamine release from striatal slices following amphetamine or KCl administration in young, mature and senescent rats. Brain Res 224:436–440.

Thompson JM, Makino CL, Whitaker JR, Joseph JA (1984): Age-related decrease in apomorphine modulation of acetylcholine release from rat striatal slices. Brain Res 299:169–173.

Ungerstedt U (1971): Adipsia and aphagia after 6-hydroxydopamine induced degeneration of the nigrostriatal dopamine system. Acta Physiol Scand [Suppl] 367:95–122.

Vrijmoed-DeVries MC, Cools AR (1986): Differential effects of striatal injections of dopaminergic cholinergic and GABAergic drugs upon swimming behavior in rats. Brain Res 364:77–90.

Waller S, London ED (1983): Age-differences in choline acetyltransferase activities and muscarinic receptor binding in brain regions of C57BL/6J mice. Exp Gerontol 18:419–425.

Waller SB, Ingram DK, Reynolds MA, London ED (1983): Age and strain of comparisons of neurotransmitter synthetic enzyme activities in the mouse. J Neurochem 41:1421–1428.

Watanabe H. (1987): Differential decrease in the rate of dopamine synthesis in several dopaminergic neurons of aged rat brain. Exp Gerontol 22:17–25.

Zelnik N, Angel I, Paul SM, Kleinman JE (1986): Decreased density of human striatal dopamine uptake sites with age. Eur J Pharmacol 126:175–176.

Neurotoxic Effects of Estrogen, Glucose, and Glucocorticoids: Neurohumoral Hysteresis and Its Pathological Consequences During Aging

Charles V. Mobbs

INTRODUCTION

The neuroendocrine control of some humoral substances, including certain hormones and glucose, progressively deteriorates with age; this deterioration, associated with a loss of feedback inhibition, may have pathological consequences [Dilman, 1971; Finch, 1976]. Some neuroendocrine deterioration may be caused by toxic effects of humoral substances on the very inhibitory neurons that control those substances. For example, estrogen, glucose, and glucocorticoid levels are normally regulated, through feedback inhibition, by neurons in the ventromedial hypothalamus (VMH, including ventromedial nucleus and arcuate nucleus) that regulate ovarian, pancreatic, and adrenal secretion; hippocampal neurons also regulate glucocorticoids. These neurons are impaired, and the neuroendocrine regulation of these substances deteriorates, with age. Elevations of estrogen, glucose, and glucocorticoids accelerate the corresponding age-correlated neuroendocrine impairments, and reductions of these substances will delay the impairments. Therefore some age-correlated neuroendocrine impairments may occur due to a destructive positive feedback cycle, in which the toxic effects of certain humoral substances on inhibitory neurons leads to a loss of neuroendocrine feedback inhibition, leading to relatively higher levels of the substances, leading to more impairments, etc. This general pattern of progressive damage of a neural regulatory (inhibitory) locus by the substance it regulates has been called "neurohumoral hysteresis" [Mobbs, 1989], since hysteresis refers to the influence of the previous history or treatment of a body on its subsequent response to a given force or changed condition. In the context of a cyclic (humoral) challenge, hysteresis constitutes both a memory and a clock: since each humoral cycle adds further damage to the neuroendo-

Rockefeller University, New York, New York 10021

crine substrate, the degree of impairment reflects the number of humoral cycles, which is usually correlated with the passage of time (Fig. 1).

The minimal criteria [Mobbs, 1989] for determining if neurohumoral hysteresis is a mechanism in the senescence of a physiological system are:

1. The regulation of the relevant humoral substance should exhibit age-correlated impairments, especially impairments in feedback inhibition and associated glandular hypersecretion, often followed by glandular exhaustion (hyposecretion)
2. Neurons that regulate the substance should exhibit age-correlated impairments
3. Damage to these neurons should induce age-like impairments in the regulation of the corresponding substance
4. Elevations of the substance should induce specific and irreversible age-like impairments in the corresponding regulatory neurons
5. Reduction of the substance during aging should attenuate age-correlated impairments in the corresponding regulatory neurons as well as concomitant neuroendocrine impairments
6. Ideally, food restriction should reduce levels of the humoral substance and concomitantly delay corresponding age-correlated neuroendocrine impairments.

HYSTERESIS—

The Influence of the Previous History or Treatment of a Body on its Subsequent Response to a Given Force or Changed Condition.

NEUROHUMORAL HYSTERESIS

ST negative feedback, LT damage

⟶

REGULATED SUBSTANCE   **REGULATORY LOCUS**

⟵

Normally suppressive

Estrogen   Hypothalamus
Glucose   Hippocampus
Glucocorticoids

Fig. 1. Essential features of neurohumoral hysteresis.

Although these criteria do not serve to prove that hysteresis causes the senescence of a given physiological system, the criteria do form a useful framework for critical assessment. By the first five criteria, age-correlated impairments in the regulation of estrogen, glucose, and glucocorticoids may be due in part to neurohumoral hysteresis. In addition, some effects of food restriction may be due to a decrease in estrogen and glucose.

Neurohumoral hysteresis seems to entail distinct phases, an early phase I of hypersecretion and relatively reversible neuroendocrine impairments, followed by a phase II of hypersecretion and irreversible impairments, often followed by a phase III of hyposecretion, due to glandular exhaustion. Different individuals enter these phases at different ages, so it is the phase, rather than the chronological age, which is most relevant in assessing the state of the individual. Therefore the average state of the neuroendocrine system in a population of a given age is only a crude indication of the dynamics of the age-correlated changes.

In theory, even normally fluctuating youthful levels of humoral substances could gradually impair neuronal feedback inhibition, leading to higher levels of the corresponding substance, causing more neuronal impairments, etc. Such a mechanism could constitute a primary mechanism of aging, since no other age-related mechanism need be invoked. It is clear that elevated levels of humoral substances can cause further (age-correlated) damage. However, it is difficult to determine if the neural damage occurs before (and thus causes) the elevated levels, since glandular (ovarian, pancreatic, and adrenal) impairments develop concomitantly with neuroendocrine impairments. As indicated below, the problem of primary impairment can at present be addressed only indirectly.

Neurohumoral hysteresis is a general mechanism some of whose forms may apply to humans, but different forms may apply to different species. As discussed below, estrogenic hysteresis may apply to female rats and mice, but the evidence that neuroendocrine impairments are associated with menopause is minimal. Hyperglycemic hysteresis may apply to different rodent species as well as to humans and may mediate some effects of food restriction on aging. Glucocorticoid hysteresis may apply to male rats, but the evidence that glucocorticoid levels increase with age in humans or other species is minimal, except in certain pathological situations. Thus neurohumoral hysteresis should be evaluated with regard to its applicability to different species, including humans, as well as with regard to the criteria outlined above.

ESTROGENIC HYSTERESIS

Toxic effects of estrogen on neurons in the VMH, which mediate effects of estrogen on reproductive hormones, may cause some neuroendocrine impairments associated with female reproductive senescence [Finch et al., 1984].

Neuroendocrine and ovarian impairments can develop independently and simultaneously contribute to female reproductive senescence in rodents [Mobbs et al., 1984a; Nelson et al., 1987]; this review will focus on neuroendocrine impairments apparently caused by estrogen. Although reproductive senescence in rats and mice may differ in certain details, such as the relative contributions of ovarian and neurendocrine impairments and the role of hyperprolactinemia, the general statements made below apply to both rats and mice, at least to the extent that comparable studies have been made in both species.

Estrogenic hysteresis appears to contribute to reproductive senescence in both rats and mice by the following criteria:

1. Reproductive senescence is marked by numerous neuroendocrine impairments, including impairments in feedback inhibition of luteinizing hormone (LH) by estrogen, and hypersecretion of prolactin and follicle-stimulating hormone. Estrogen is relatively elevated in the early phases of reproductive senescence, and the late phase (III) is marked by hyposecretion of estrogen.

2. Neurons in the VMH, including the ventromedial nucleus and arcuate nucleus, which contain estrogen receptors and regulate LH and prolactin, decrease with age and exhibit many other age-correlated impairments including a decrease in estrogen receptors and impaired regulation of catecholamines by estrogen.

3. Damage to neurons in the VMH causes numerous age-like neuroendocrine reproductive impairments, including impairments in the regulation of LH by estrogen and hypersecretion of prolactin.

4. Elevated estrogen causes numerous irreversible age-like neuroendocrine reproductive impairments, including morphological and neurochemical indications of VMH damage as well as impairments in the regulation of LH by estrogen and hypersecretion of prolaction.

5. Reduction of estrogen by ovariectomy when young attenuates numerous age-correlated neuroendocrine reproductive impairments, including damage in the VMH and impairments in the regulation of LH and prolactin by estrogen.

6. Food restriction and progesterone both decrease estrogen secretion and delay reproductive neuroendocrine impairments.

Neuroendocrine Impairments During Reproductive Senescence

Phase I of reproductive senescence in rodents begins before the initial transition, at around 7–9 months, from 4-day regular to longer irregular cycles and continues until cycles cease. Even before the transition to longer cycles, the regular 4-day cycles of rats are characterized by hypersecretion of estrogen [Nass et al., 1984], possibly due to stimulation by elevated follicle-stimulating hormone [Depaolo and Chappel, 1986], whose hypersecretion in turn suggests a failure in neuroendocrine–ovarian negative feedback. Decreased feedback inhibition of LH by estradiol in middle-aged cycling as well as acyclic females was directly demonstrated in rats [Gray and Wexler, 1980] and mice [Mobbs et al., 1985]. Rats [reviewed in Lu, 1983; Wise, 1987] and mice [Mobbs et al., 1984a, 1985] in phase I exhibit impairments in LH and prolactin regulation. Phase II begins at the transition from irregular cycles to an acyclic state at around 12 to 15 months of age, characterized by persistent vaginal cornification and relatively elevated and constant estrogen [Lu et al., 1981; Nelson et al., 1981], with more severe neuroendocrine impairments that become gradually less reversible by ovariectomy [Lu et al., 1981; Mobbs et al., 1984a, 1985]. Rats, and very occasionally mice, may also exhibit another pattern, characterized by a repetitive pseudopregnancy vaginal pattern, which can develop when prolactin is sufficiently elevated [Aschheim, 1983]. Phase III, common in mice after about 18 months of age and in humans after menopause but less common in rats, begins at the transition to persistent leukocytic vaginal cells, and is characterized by almost complete ovarian exhaustion and hyposecretion of estrogen [Nelson et al., 1981; Gosden et al., 1983; Gee et al., 1983]. Menopause in women is characterized by accelerated follicular depletion [Richardson et al., 1987], and in postmenopausal women phase III is characterized by some loss of feedback inhibition by estrogen [Mills and Mahesh, 1978] and hypersecretion of prolactin [Cocchi et al., 1986]. A relatively normal LH surge can be induced in postmenopausal women [Odell and Swedloff, 1968], in contrast to rodents [Lu et al., 1981; Mobbs et al., 1984a]. However, in humans the age-correlated hypersecretion of prolactin may be clinically important since elevated prolactin is associated with increased risk of breast cancer in both pre- and postmenopausal women [Meyer et al., 1986; Rose and Pruitt, 1981] and hyperprolactinemia is associated with decreased immune function [Gerli et al., 1986].

Neuronal Impairments in the VMH During Aging

Neurons in the VMH regulate reproductive function and are progressively impaired with age. (As mentioned above, the VMH contains both the ventromedial nucleus and the arcuate nucleus. The two nuclei are physically close and are often impossible to dissociate functionally, and relevant functional neurons are not necessarily found within the visible nuclear boundaries, so these nuclei will often be treated together as the VMH unit.) Sabel and Stein [1981] reported a progressive age-correlated loss of neurons from rat ventromedial nucleus of 30-month-old rats compared with the number in 3-month-old rats; in contrast, neurons in the nearby lateral amygdala were not significantly reduced [Sabel and Stein, 1981]. Similarly, in perhaps the most extensive analysis of hypothalamic neuron number during aging published to date, Sartin and Lamperti [1985] reported an age-correlated decrease in neuron number in both the ventromedial nucleus and arcuate nucleus counted in 3-, 12-, and 24 month-old male rats. In this study neurons in the VMH decreased 30% by 24 months, whereas six other hypothalamic nuclei, including the medial preoptic area, exhibited no significant age-correlated neuronal loss [Sartin and Lamperti, 1985]. Hsu and Peng [1978] reported a 50% loss in neurons in the arcuate nucleus of female rats by 24 months compared with the number in rats 4 months old, although they did not replicate this loss in male rats [Peng and Hsu, 1982]. Smith-West and Garris [1983] reported a 20% reduction in the ventromedial nuclear area from 3 months to 15 months, as well as a reduction in nuclear density, in the Chinese hamster. Schipper et al. [1981] reported progressive increased glial hyperactivity, an indication of neuronal degeneration, in the arcuate nucleus of female rats and mice during reproductive senescence, and this gliosis was not reversed by ovariectomy 6 weeks before sacrifice. Wise et al. [1984] found that by 10–11 months cytoplasmic estradiol receptors, specifically in the ventromedial nucleus, were reduced by about 50%, compared with levels in rats 3–4 months old; this decrease was apparently related to a loss of neurally mediated biochemical and behavioral responses to estradiol [Wise et al., 1984; Wise, 1987]. The loss of VMH estrogen receptors was corroborated by a different study, which indicated a 30% loss of cytoplasmic estrogen receptors by 12 months in the entire medial basal hypothalamus [Wise and Parsons, 1984]. Using steroid autoradiography Blaha and Lamperti [1983] demonstrated a decrease of estrogen-concentrating neurons in the arcuate nucleus of hamsters during reproductive senescence. Age-correlated impairments in the regulation of catecholamines and steroid receptors in the VMH of middle-aged female rats have been reported [Wise et al., 1984]. Furthermore, transplantation of newborn hypothalamic tissue into aged female rats facilitates recovery of some reproductive functions [Matsumoto et al., 1984].

Damage to the VMH Causes Age-Like Reproductive Impairments

Lesions in the VMH induce in several species numerous age-like reproductive impairments including acyclicity, impaired postovariectomy LH rise, increased prolactin, impaired negative feedback of estrogen on LH secretion, and impaired sexual behavior [Joseph and Knigge, 1968; Sridaran and Blake, 1978; for reviews, see Bleier and Byne, 1985; Pfaff, 1980]. It should be pointed out that seldom is it possible to lesion only the arcuate or the ventromedial nucleus without damaging the other. Even monosodium glutamate, often used to damage the arcuate nucleus specifically, can damage the ventromedial nucleus as well [Komeda et al., 1980].

Elevated Estrogen Causes Age-Like VMH Damage and Neuroendocrine Impairments

Elevated levels of estrogen in young rodents cause both reversible and irreversible reproductive impairments similar in many respects to those exhibited during reproductive senescence. As few as three injections of 2 mg estradiol valerate (EV) will, after 21 days, induce glial hyperactivity, an indication of neuronal degeneration, in *both* the arcuate nucleus and ventromedial nucleus, but not in the nearby dorsomedial nucleus or lateral hypothalamus, nor, importantly, in the preoptic area [Brawer and Sonneschein, 1975]. Estrogen-induced impairments in the ventromedial nucleus are also suggested by the reduction in lordosis that follows injection of estrogen over many weeks [Clark et al., 1986]. A single 2 mg EV injection can eventually lead to glial hyperactivity, but this effect requires intact ovaries [Brawer et al., 1978, 1980]. The induction of glial hyperactivity is specific to estrogen, since androgens are not effective [Brawer et al., 1983]. However, hyperprolactinemia and pituitary tumors can be induced reproducibly in male rats by prolonged exposure to the estrogenic compound diethylstilbesterol [Morgan et al., 1985]. The estrogen-induced hyperprolactinemia appears to be due to irreversible damage to hypothalamic dopaminergic neurons since hypothalamic dopamine impairments persist in ovariectomized rats 8 months after removal of estrogen implants, and the hyperprolactinemia can be reversed by dopamine agonists [Gottschall and Meites, 1986]. Furthermore, estrogen-induced hyperprolactinemia can be reversed by intracerebral transplantation of hypothalamic tissue containing dopaminergic tissue [Arendash and Leung, 1986]. A single 2 mg injection of EV or estradiol benzoate causes immediate and permanent loss of estrous cycles and leads to age-like impairments in the regulation of estrous cycles, LH, and prolactin in female rats [Brown-Grant, 1975; Brawer and Sonnenschein, 1975; Brawer et al., 1978; Simard et al., 1987] and in the regulation of estrous cycles and LH in female mice [Mobbs et al., 1984b]. Estradiol implants producing moderately sustained physiological (but not cyclic) levels of estrogen, as well as daily injections of very low levels of estradiol, will also result in premature loss of estrous cycles [Mobbs et al., 1984b; Kawashima, 1960].

Reductions of Estrogen by Ovariectomy Can Delay or Partially Reverse Some Reproductive Neuroendocrine Impairments

Several age-correlated neuroendocrine impairments are attenuated if rodents are ovariectomized when young, as originally noted by Aschheim [reviewed in Aschheim, 1983]. Age-correlated signs of neuronal damage in the arcuate nucleus of both rats and mice is attenuated by ovariectomy when young [Schipper et al., 1981]. Ovariectomy of young rats attenuates age-correlated impairments in estrous cycles with young ovarian grafts [Aschheim, 1983] and in LH secretion [Blake et al., 1983]. Ovariectomy of young mice attenuates age-correlated impairments in LH secretion [Gee et al., 1983], estrous cycles with ovarian grafts [Felicio et al., 1983; Mobbs et al., 1984a], the steroid-induced LH surge [Mobbs et al, 1984a], pituitary glucose-6-phosphate dehydrogenase [Gordon et al., 1988], and dopamine [Telford et al., 1986]. In particular, ovariectomy when young attenuates the age-correlated decrease in the feedback inhibition of LH by estrogen [Mobbs et al., 1985]. If rodents exhibiting persistent vaginal cornification are ovariectomized at least 1 month before assessment, neuroendocrine function is significantly improved, but not as much as if the rodents were ovariectomized when young [Lu et al., 1981; Mobbs et al., 1984b]. These results suggest that ovarian steroids impair neuroendocrine reproductive functions and that there are short-term reversible and long-term irreversible components to these impairments.

Food Restriction and Progesterone Reduce Estrogen and Can Delay or Reverse Some Reproductive Neuroendocrine Impairments

Food restriction delays many aspects of aging and can be used as a powerful tool to examine fundamental mechanisms of the aging process [Masoro, 1988]. Life-long food restriction delays impairments associated with reproductive senescence [Merry and Holehan, 1979; Nelson et al., 1985]. Similarly, a 10-week period of food restriction in rats already exhibiting acyclicity results, after restoration of a normal diet, in temporary restoration of cycles [Quigley et al., 1987]. Food restriction causes an average decrease in estrogen [Holehan and Merry, 1985] and prolactin [Quigley et al., 1987] levels compared with ad lib fed controls. Similarly, progesterone implants can delay reproductive neuroendocrine impairments [Lapolt et al., 1986; DePaolo and Rowlands, 1986]. These progesterone treatments decrease estrogen levels [Lapolt et al., 1986], and the effect of progesterone in delaying reproductive impairments was prevented by moderately increasing estrogen levels wiith implants [Lapolt et al., 1988]. Therefore food restriction and progesterone decrease estrogen, and the effect of these treatments to delay and partially reverse age-correlated reproductive neuroendocrine impairments could be due in part to a concomitant attenuation of estrogenic hysteresis.

HYPERGLYCEMIC HYSTERESIS

Toxic effects of glucose on neurons in the VMH, which mediate neuroendocrine regulation of insulin, may contribute to age-correlated impairments in glucose regulation [Mobbs, 1989].

1. In humans, rodents, and other species, plasma glucose regulation (glucose tolerance) is progressively impaired with age. This impairment is characterized by hypersecretion of insulin and glucagon, insulin resistance, and elevated plasma glucose (hyperglycemia).
2. VMH neurons regulate glucose and are impaired during aging.
3. Damage to VMH neurons impairs glucose regulation, as indicated by

hypersecretion of pancreatic insulin and glucagon, insulin resistance, and hyperglycemia.

4. Elevated plasma glucose causes irreversible and specific damage to VMH neurons, impairments in glucose regulation, and numerous other age-correlated impairments.

5. Food restriction decreases glucose levels and delays impairments in glucose regulation and numerous other age-correlated impairments.

Therefore, progressive age-correlated impairments in glucose regulation may be caused by a destructive feedback cycle between VMH neurons and the pancreas, in which mild (postprandial) hyperglycemia episodes when young induce mild suppression or damage to the VMH, leading to a decrease in glucose feedback inhibition, causing hypersecretion of pancreatic insulin and glucagon (phase I), leading to insulin resistance and more hyperglycemia (phase II), leading to further damage, etc. (Fig. 2). Eventually hypersecretion can be followed by pancreatic impairments and associated severe hyperglycemia (phase III). Age-correlated diabetes, cardiovascular complications, and other diseases could be precipitated by this pancreatic hypersecretion, insulin resistance, and hyperglycemia. The effect of food restriction to delay age-correlated impairments may be due in part to a reduction in glucose.

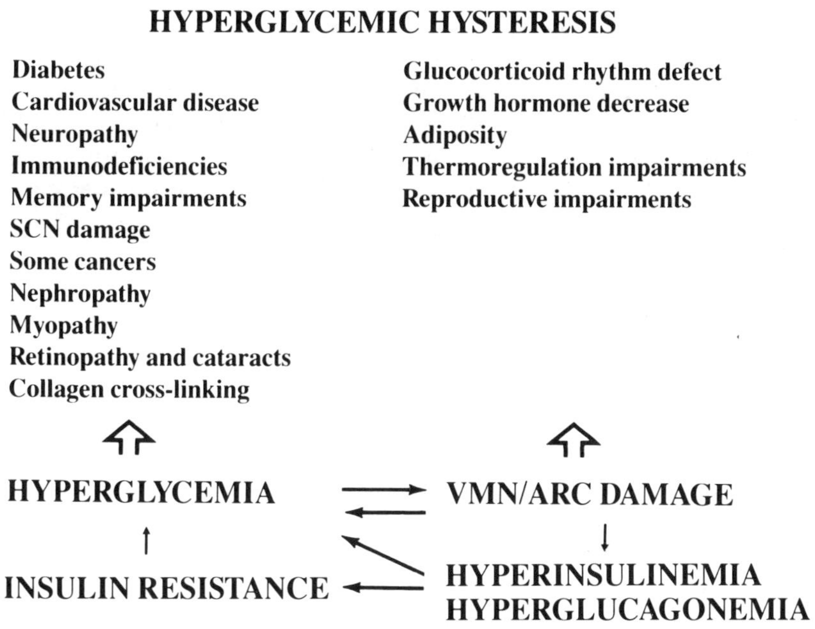

HYPERGLYCEMIC HYSTERESIS

Diabetes
Cardiovascular disease
Neuropathy
Immunodeficiencies
Memory impairments
SCN damage
Some cancers
Nephropathy
Myopathy
Retinopathy and cataracts
Collagen cross-linking

Glucocorticoid rhythm defect
Growth hormone decrease
Adiposity
Thermoregulation impairments
Reproductive impairments

⇑   ⇑
HYPERGLYCEMIA ⇄ **VMN/ARC DAMAGE**
↑   ↓
INSULIN RESISTANCE ⇐ **HYPERINSULINEMIA HYPERGLUCAGONEMIA**

Fig. 2. Essential features of hyperglycemic hysteresis.

Impairments of Glucose Regulation During Aging in Humans and Other Species

Glucose regulation in humans, rodents, and other species is progressively impaired with age [for reviews see DeFronzo, 1982; Davidson, 1979; Zimmet and Whitehouse, 1979]. The prevalence of diabetes in the United States at ages 20–44, 45–54, 55–64, and 65–74 years is 2%, 8.5%, 13.4% and 18.7%, respectively; prediabetic impaired glucose tolerance is even more prevalent [Harris et al., 1987], and, as indicated below, hypersecretion of insulin is more prevalent yet. Several studies have demonstrated that age-correlated glucose intolerance in humans is due primarily to a loss of insulin sensitivity, using glucose clamping [Defronzo et al., 1979] or insulin clamping [DeFronzo, 1979; Fink et al., 1984a; Rowe et al., 1983; Chen et al., 1985] techniques. Insulin levels in response to *oral* glucose are more than adequate with advancing age [DeFronzo, 1982], although individual pancreatic cells may be impaired [Reaven et al., 1979; Kithara and Adelman, 1979]. Despite pancreatic cellular impairments, the insulin output in response to oral glucose stays high because the pancreas hypertrophies with age [Reaven et al., 1979; Kithara and Adelman, 1979]. Therefore, insulin resistance precedes and presumably causes age-correlated impairments in glucose regulation.

Insulin resistance and subsequent hyperglycemia may be caused by pancreatic hypersecretion, rather than vice versa, since numerous studies suggest that during aging hyperinsulinemia *precedes* age-correlated insulin resistance and hyperglycemia. In middle age hyperinsulinemia occurs in the presence of *hypo*glycemia in rats, humans, and mice [Brancho-Romero and Reaven, 1977; Chlouverakis et al., 1967; Leiter et al., 1988]; only later does hyperglycemia develop. The pancreas secretes more insulin with increasing age after stimulation by oral glucose in both rats [Hayashi, 1982; Brancho-Romero and Reaven, 1977] and humans [Hayashi, 1980; Sandberg et al., 1973]. In vitro studies with perfused whole pancreas also demonstrate age-correlated increased total secretion, when the secretion is monitored for several hours [Gold et al., 1976]. Hyperinsulinemia is often detected in nonobese older individuals who do not show hyperglycemia [Hayashi, 1980; Chlouverakis et al., 1967]. Statistical analysis by multiple regression reveals that insulin secretion increases with age independent of adiposity, weight, height, or blood sugar [Welborn et al., 1969], although these factors also independently contribute to plasma insulin levels. While age-correlated glucose intolerance may depend on increased adiposity [Reaven and Reaven, 1985], clearly hyperinsulinemia [Welborn et al., 1969; Chlouverakis et al., 1967; Hayashi, 1980] and insulin resistance [Chen et al., 1985] do not depend on adiposity. When elderly individuals with normal glucose tolerance tests were given typical mixed meals, glucose levels were elevated 11% in the elderly vs. young controls, whereas insulin levels were elevated 40% [Fink et al., 1984b]. When nonobese, nondiabetic middle-aged individuals with slightly elevated plasma glucose and insulin responses to oral glucose were given meals similar in composition to the typical American diet, their glucose levels were normal but insulin levels were still elevated [Reaven et al., 1972]. Therefore hyperinsulinemia appears to precede hyperglycemia and be a more primary feature of senescence.

Pancreatic glucagon plasma levels after oral glucose increase with age in rats [Hayashi, 1982] and humans [Hayashi, 1980], even in older nondiabetic individuals who do not exhibit abnormal glucose tolerance [Hayashi, 1982, 1980]. This hypersecretion of glucagon is due to a loss of the suppressive effect of glucose on glucagon [Hayashi, 1980, 1982]; loss of glucagon suppression is accelerated in diabetics [Hayashi, 1980]. In portal vein blood from 12- and 24-month-old rats, glucagon actually *increased* two- to threefold in response to gastric glucose, in contrast to the suppressive effect of glucose on glucagon in young rats [Klug et al., 1979].

Pancreatic hypersecretion can cause glucose intolerance. Elevated physiological levels of insulin cause glucose intolerance [Ramirez and Friedman, 1982] and insulin resistance at both receptor and postreceptor levels [Martin et al., 1983], independent of increased adiposity [Martin et al., 1983], in both humans [Rizza et al., 1985; Nankervis et al., 1985; Soman and DeFronzo, 1980] and rats [Kobayashi and Olefsky, 1978; Martin et al., 1983]. Elevated insulin levels also cause desensitization in vitro at both receptor and postreceptor levels [Ercolani et al., 1985; Marshall and Olefsy, 1980; Davidson and Casanello-Ertl, 1979; Chang and Polakis, 1978]. Elevated pancreatic glucagon antagonizes the effect of insulin [Christ et al., 1986], is associated with increased hepatic glucose output and consequent hyperglycemia [Henry et al., 1986], and may be a cause of diabetes [Unger and Orci, 1795; Baron et al., 1987; Del Prato et al., 1987]. Increased glucagon in particular has been emphasized as contributing to age-correlated insulin resistance [Klug et al., 1979; Hayashi, 1980, 1982], whereas hypersecretion of insulin has been cited as a possible proximal cause of insulin resistance in humans [Haffner, 1987] and the Zucker rat [Jeanrenaud, 1985]. Finally, type II diabetes, pancreatic hypersecretion, and aging are associated with the deposition of a pancreatic amyloid that is similar to calcitonin gene-related peptide [Westermark et al., 1987; Leighton and Cooper, 1988]. This peptide can directly cause insulin resistance [Leighton and Cooper, 1988]. All these data taken together suggest that pancreatic hypersecretion may be a proximal cause of age-correlated insulin resistance and consequent hyperglycemia.

VMH Is Impaired With Age, and VMH Damage Causes Pancreatic Hypersecretion and Hyperglycemia

Evidence that the VMH loses neurons and exhibits other impairments during aging is presented above. Lesions of the VMH cause, in addition to reproductive impairments, pancreatic hypersecretion [Rohner-Jeanrenaud and Jeanrenaud, 1980; Tokunaga et al., 1986], insulin resistance [Kasuga et al., 1980; Penicaud et al., 1983], and hyperglycemia [Komeda et al., 1980; Matsuo and Shino, 1972; Cameron et al., 1976; Zarco de Coronada and Yepez Chamorro, 1987] similar in many respects to age-correlated impairments in glucose regulation. VMH lesions cause diabetes ten times more frequently in old than in young rats [Lazaris et al., 1985]. VMH lesions can also cause a high incidence of diabetes in monkeys [Hamilton and Brobek, 1963], and possibly in humans as well [Buzzi et al., 1987]. Lesions of the VMH cause increased pancreatic secretion of both insulin [Tokunaga et al., 1986; Berthoud and Jeanrenaud, 1979] and glucagon [Rohner-Jeanrenaud and Jeanrenaud, 1984]. The increased secretion of insulin and glucagon after VMH lesion can be detected even using the perfused pancreas in vitro [Rohner-Jeanrenaud and Jeanrenaud, 1980]. The pancreatic hypersecretion and glucose intolerance are not a result of hyperphagia and obesity since 1) increased insulin secretion occurs within minutes after an electrolytic lesion of the VMH in an anesthetized animal, in the absence of any food ingestion [Tokunaga et al., 1986]; 2) monosodium glutamate, which lesions both the ventromedial nucleus and arcuate nucleus in Chinese hamsters, causes diabetes without causing obesity [Komeda et al., 1980]; and 3) monosodium glutamate lesions induce hyperglycemia in KK mice at an age before body weight differs from nonlesioned controls [Cameron et al., 1976]. Furthermore, hyperphagia and obesity caused by VMH lesion can be prevented by cutting input of the vagus nerve to the pancreas [Inoue and Bray, 1977], or by destroying the pancreas and replacing with a pancreatic graft [Inoue et al., 1977], the direct effect of which is to prevent the onset of hyperinsulinemia. Conversely, stimulation of the vagus nerve directly causes immediate hyperinsulinemia and hyperglycemia [Ionesou et al., 1983]; the hyperglycemia is presumably due to glucagon secretion. Apparently, increased pancreatic secretion after VMH lesion is due to the loss of the inhibitory influence of the VMH neurons on the vagus nerve, which stimulates pancreatic secretion [Jeanrenaud, 1985]. The increased pancreatic secretion precedes and may cause the hyperglycemia.

Aging, VMH lesions, and certain diabetogenic mutations are all associated with pancreatic hypertrophy and hypersecretion of insulin and glucagon, both in vivo and in vitro [Jeanrenaud, 1985; Coleman, 1982]. In the diabetic mutants hyperinsulinemia and glucose intolerance develop early [Jeanrenaud, 1985; Coleman, 1982]. However, overstimulation appears to cause eventual exhaustion of the pancreas, leading to a transition from insulin-independent to insulin-dependent glucose intolerance [Coleman, 1982]. Furthermore, VMH lesions also cause hypertrophy and apparently eventual exhaustion, which is exaggerated in older individuals and leads to diabetes [Lazaris et al., 1985]. The hypertrophy/exhaustion phenomenon following VMH lesions may explain the decreased pancreatic responsiveness to glucose of individual beta cells that is associated with age-correlated hypertrophy of the pancreas [Reaven et al., 1979; Kithara and Adelman, 1979].

Elevated Glucose Causes Age-Correlated VMH Damage and Other Neuroendocrine Impairments

Hyperglycemia, due either to pancreatic damage (insulin insufficiency) [Bestetti and Rossi, 1980; Akmayev and Rabkina, 1976] or genetic defects (insulin resistance?) [Garris et al., 1982, 1985a; Smith-West and Garris, 1983], causes progressive neuronal damage in the VMH. Hyperglycemia resulting from alloxan-induced pancreatic damage specifically reduces the size of nuclei in neurons in the VMH, leaving nuclei in neurons in the dorsomedial nucleus and the periventricular nucleus unaffected [Akmayev and Rabkina, 1976]. This reduction is not due to normal responses to blood glucose, since the well-fed state is normally associated with *larger* ventromedial nuclei than the fasted state [Pfafff, 1972]. In another study, hyperglycemia resulting from streptozotocin-induced pancreatic damage caused many neuropathologies (fragmented endoplasmic reticulum, loss of organelles, irregular nuclei, etc.) in the arcuate nucleus [Bestetti and Rossi, 1980]. Diabetes may also be associated with hypothalamic impairments in humans [Morgan et al., 1937]. In the genetically diabetic db/db mouse, the number of degenerating neurons progressively increased, and neuronal density decreased, in the ventromedial and arcuate nuclei compared with controls [Garris et al., 1985a]. In the genetically diabetic Chinese hamster, the area, number of neurons, and neuronal density of the ventromedial nucleus were all significantly reduced compared with nondiabetic controls [Garris et al., 1982]. In another study with the Chinese hamster, diabetes was associated with an acceleration of the normal age-correlated loss of neurons in this nucleus [Smith-West and Garris, 1983]. Diabetes in mutant mice is associated with a progressive loss of estrogen receptors in the VMH [Garris and Coleman, 1984; Garris et al., 1985b]. Hyperglycemia in rats from streptozotocin also causes a progressive loss of estradiol receptors in the VMH, as well as a loss in estradiol-induced progesterone receptors, and a correlated loss of lordosis behavior [Ahdieh et al., 1983]. This study was notable in that lowering blood sugar with insulin after weeks of hyperglycemia did not restore receptors or behavior [Ahdeih et al., 1983]. Hyperglycemia causes impairments in LH regulation similar to impairments that occur after VMH lesions and during reproductive senescence [Katayama et al., 1984; Blades et al., 1985; Spindler-Vomachka and Johnson, 1985].

These data suggest a mutually destructive feedback cycle, in which normal transient (postprandial) hyperglycemia when young causes a slight decrease in VMH feedback inhibition (caused by damage or down-regulation of glucose transporters [Karnieli et al., 1981]), which in turn (probably via increased glucagon antagonism and hyperinsulinemia-induced desensitization) increases subsequent insulin resistance and hyperglycemia, in turn causing more damage to the VMH, etc. This cycle would lead to progressive insulin resistance exacerbated in some individuals by insulin insufficiency (after pancreatic exhaustion). In individuals with genetic predisposition, this insulin resistance could precipitate (perhaps by insulin insufficiency) overt development of clinical diabetes; in other individuals, other glucose- or insulin-sensitive impairments might develop. The hyperglycemia could also down-regulate glucose transporters in many cells, amplifying the destructive cycle [Unger and Grundy, 1985].

Therefore a primary lesion during senescence may be the glucose-induced damage to the VMH that results in progressive insulin resistance, in some cases precipitating type II diabetes. Many other age-correlated impairments could be precipitated by age-correlated hyperinsulinemia, insulin resistance, and hyperglycemia, since age-correlated impairments are remarkably similar to impairments associated with diabetes, especially in postmitotic tissues [Mooradian, 1988]. Glucose has been proposed to cause age-correlated impairments by nonenzymatic glycation of proteins and nucleic acids [Cerami, 1985, 1986]. Glucose intolerance is a major risk factor for cardiovascular complications, which are the major cause of human mortality in aging [West, 1978; Jarrett, 1984; Arbogast et al., 1984]; hyperinsulinemia is an independent risk factor for atherosclerosis [Stout, 1981] and hypertension [Modan et al., 1985; Zavaroni et al., 1989]. Fertility in women is the first physiological system to senesce and perhaps the most sensitive to hyperglycemia [Hamel et al., 1986; Tallarigo et al., 1986]. All of the following are also associated with senescence and with glucose intolerance: neuropathy [Vlassara, 1988; Brown et al., 1982]; impaired thermoregulation [Scott et al., 1987]; immunodeficiencies [Larkin et al., 1985; Nakano et al., 1984]; kidney damage [Steffes and Mauer, 1982]; cataracts [Leopold and Mosier, 1978; Gonzalez et al., 1981]; certain cancers [O'Mara et al., 1985]; collagen cross-linking [Schnider and Kohn, 1981]; impaired mitotic potential [Lorenzi et al., 1985]; and memory impairments [Gold et al., 1988]. Hyperglycemia can also damage the suprachiasmatic nucleus [Bestetti et al., 1987], which could lead to age-correlated changes in circadian rhythms [Ingram et al., 1982]. The almost universal age-correlated increase in adiposity has long been a puzzle, in view of the highly regulated manner in which the body defends its set point of adiposity, and is has been proposed that the adiposity set point increases with age [Woods et al., 1985]. Old animals will lose weight when deprived of food but will quickly grow back to their original weights when given renewed access to food [Quigley et al., 1987]. Thus age-related adiposity is characterized by a state similar to the "static" phase of the VMH-lesioned rat, which is not hyperphagic but will defend its new body fat level [Penicaud et al., 1983]. The mechanism of hyperglycemic hysteresis suggests that the set point of adiposity increases with age due to the gradual loss or suppression of neurons in the VMH; in turn this increased adiposity exacerbates insulin resistance.

Food Restriction Reduces Glucose and Delays Age-Correlated Impairments

Restricting food intake to 60% ad lib reduces the mean 24-hour levels of plasma glucose as well as plasma glucose at every time-point, even after feeding [Masoro et al., 1989]. Food restriction also reduces the levels of protein glycation [Masoro et al., 1989]. Since levels of glycation reflect integrated exposure to glucose [Cerami, 1985] food restriction, as expected, reduces average exposure to glucose. Food restriction also delays impairments in glucose regulation [Reaven et al., 1983]. Therefore some effects of caloric restriction on delaying many age-correlated impairments could plausibly be due to a reduction in glucose and a concomitant attenuation of hyperglycemia hysteresis.

GLUCOCORTICOID HYSTERESIS

Toxic effects of glucocorticoids on hippocampal neurons may mediate some age-correlated impairments in glucocorticoid secretion in male rats [Landfield, 1980; Landfield et al., 1981a; Sapolsky et al., 1986]. Since several recent reviews on this mechanism are available [Sapolsky et al., 1986, 1987; Sapolsky, 1987], only some general points and recent data will be mentioned here.

1. Regulation of corticosterone, especially suppression of corticosterone after stress, is impaired with age.
2. CA3 hippocampal neurons, which regulate the post-stress suppression of corticosterone, are damaged with age.
3. Damage to the hippocampus causes impaired ability to suppress corticosterone secretion after stress.
4. Elevated corticosterone levels damage CA3 hippocampal neurons.
5. Reduction of corticosterone by adrenalectomy when young decreases age-correlated hippocampal damage.

Although aging does not influence basal glucocorticoid secretion in most species [Riegle, 1983], Landfield reported that adrenal glucocorticoid secretion in male rats increases with age, and that this increase is correlated with hippocampal damage [Landfield et al., 1978]. Based on these and other studies Landfield was the first to formulate a theory of neurohumoral hysteresis, which he called the "adrenocortical hypothesis of brain and somatic aging" in which adrenal glucorticoid secretion causes hippocampal damage, impairing feedback inhibition, causing more secretion of adrenal steroids, causing more damage, leading to a "runaway positive feedback loop" [Landfield, 1980; Landfield et al., 1981a]. In a series of elegant studies in male rats Sapolsky et al. confirmed and greatly extended several elements of this hypothesis, focusing particularly on the age-correlated loss of CA3 hippocampal neurons and impaired glucocorticoid feedback inhibition after stress [reviewed in Sapolsky et al.,

TABLE 1. Forms of Neurohormonal Hysteresis and Selected References

Form	Species	Reference
Estrogen hysteresis		
1 Regulation of estrogen and/or gonadotropin secretion impaired during reproductive senescence		
	Humans	Mills and Mahesh, 1978
	Rats	Nass et al., 1984
		Depaolo and Chappel, 1986
		Gray et al., 1980
		Lu, 1983
		Wise, 1987
	Mice	Nelson et al., 1981
		Mobbs et al., 1984a
		Mobbs et al., 1985
2 VMH neurons lost or impaired during aging		
	Rats	Sabel and Stein, 1981
		Sartin and Lamperti, 1985
		Hsu and Peng, 1978
		Wise, 1987
3 Damage to VMH causes age-like reproductive impairments		
	Rats	Sridaran and Blake, 1978
		Blier and Byne, 1985
		Pfaff, 1980
	Guinea pigs	Joseph and Knigge, 1968
4 Elevated estrogen induces VMH damage and age-like reproductive impairments		
	Rats	Brown-Grant, 1975
		Brawer and Sonnenschein, 1975
		Brawer et al., 1980
		Brewer et al., 1983
		Morgan et al., 1985
		Gottschall and Meites, 1986
	Mice	Mobbs et al., 1984b
5 Reduction of estrogen by ovariectomy attenuates neuroendocrine reproductive impairments		
	Rats	Aschheim, 1983
		Blake et al., 1983
		Schipper et al., 1981
	Mice	Gee et al., 1983
		Felicio et al., 1983
		Mobbs et al., 1984b
		Mobbs et al., 1985
		Gordon et al., 1984
		Telford et al., 1986
6 Food restriction reduces estrogen and delays or reverses neuroendocrine reproductive impairments		
	Rats	Merry and Holehan, 1979
		Quigley et al., 1987
	Mice	Nelson et al., 1985

(continued)

TABLE 1. Forms of Neurohormonal Hysteresis and Selected References *(Continued)*

Form	Species	References

Hyperglycemic hysteresis
1 Elevated glucose and/or pancreatic hypersecretion during aging
    Humans — Harris et al., 1987
                  Chen et al., 1985
                  Chlouverakis et al., 1967
                  Welborn et al., 1969
                  Zimmet and Whitehouse, 1979
                  Hayashi, 1980
                  DeFronzo, 1982
                  Davidson, 1989
    Rats — Brancho-Romero and Reaven, 1977
                  Klug et al., 1979
                  Hayashi, 1982
    Mice — Leiter et al., 1988
2 VMH neurons lost or impaired during aging (See under *estrogen hysteresis*)
3 VMH damage causes pancreatic hypersecretion and elevated glucose
    Humans — Buzzi et al., 1987 (anecdotal)
    Monkeys — Hamilton and Brobek, 1963
    Hamsters — Komeda et al., 1980
    Rats — Lazaris et al., 1985
                  Tokunaga et al., 1986
                  Kasuga et al., 1980
                  Rohner-Jeanrenaud and Jeanrenaud, 1984
    Mice — Cameron et al., 1976
4 Elevated glucose damages VMH
    Humans — Morgan et al., 1937 (anecdotal)
    Rats — Bestetti and Rossi, 1980
                  Akmayev and Rabkina, 1976
                  Ahdieh et al., 1983
    Hamsters — Garris et al., 1982
                  Smith-West and Garris, 1983
    Mice — Garris et al., 1985b
                  Garris and Coleman, 1984
5 Food restriction reduces glucose and attenuates impairments in glucose regulation
    Rats — Masoro et al., 1989
                  Reaven et al., 1983

Glucocorticoid hysteresis
1 Glucocorticoid hypersecretion during aging
    Rats — Landfield et al., 1978
                  Sapolsky et al., 1983
2 Hippocampal neurons damaged during aging
    Rats — Landfield et al., 1978
                  Sapolsky et al., 1985
3 Damage to hippocampus causes glucocorticoid hypersecretion
    Rats — Sapolsky et al., 1984
4 Elevated glucocorticoids impair hippocampal neurons
    Rats — Sapolsky et al., 1985
5 Reduction of glucocorticoids by adrenalectomy attenuates hippocampal damage
    Rats — Landfield et al., 1981

1986, 1897; Sapolsky, 1987]. They showed that: 1) glucocorticoid feedback inhibition after stress was impaired during aging [Sapolsky et al., 1983]; 2) CA3 hippocampal neurons and glucocorticoid receptors decrease with age [Sapolsky et al., 1985]; 3) damage to hippocampal neurons impaired glucocorticoid feedback inhibition after stress [Sapolsky et al., 1984]; and 4) elevated corticosterone caused irreversible CA3 hippocampal damage [Sapolsky et al., 1985]. In an interesting study using a novel approach, glucocorticoid receptors in the hippocampus were increased by neonatal handling, and this protocol delayed impairments in glucocorticoid feedback inhibition and other age-correlated changes [Meaney et al., 1988]. Similarly, Landfield had shown that adrenalectomy when young attenuated age-correlated hippocampal damage and associated impairments [Landfield et al., 1981b].

Glucocorticoid hysteresis as presently formulated in terms of the hippocampal/stress axis may not apply to normal aging in humans [Sapolsky et al., 1986], but an alternate form of hysteresis, involving the VMH/cortisol rhythm axis, may apply to humans and other species. In an extensive review, Riegle concludes on the basis of his own work and the work of others that, as with cattle and goats, there is "no consistent change in basal corticosterone concentrations in aging rats of either sex" [Riegle, 1983]. In mice neither hippocampal receptors [Nelson et al., 1976] nor corticosterone levels [Finch et al., 1969] are altered during aging. In humans glucocorticoid feedback inhibition after stress is not impaired during aging [Blichert-Toft, 1978; Blichert-Toft and Hummer, 1976]. Furthermore, the CA3 region of the human hippocampus, which in the rat is specifically vulnerable to glucocorticoid damage [Sapolsky et al,. 1985], is specifically spared during human aging [Mani et al., 1986], even in senile dementia [Flood et al., 1987]. However, the circadian rhythm of cortisol in older humans is phase-advanced by 2 hours [Sherman et al., 1985]; the phase advance was clearly evident at every time-point because samples were taken every 20 minutes over a 24-hour period. The 24-hour mean cortisol levels are *lower* in older humans than in younger humans, but cortisol begins rising first in older humans [Sherman et al., 1985; Sharma et al., 1989]. Therefore when cortisol is at its lowest point (nadir) in young humans it is (temporarily) twice as high in older humans [Sherman et al., 1985]. Other studies, with less frequent samplings, have also suggested that older humans may have altered cortisol rhythms, especially what appear to be elevated nadir levels of cortisol [Friedman et al., 1969; Touitou et al., 1983; Stokes et al., 1984]. Aged male rats may also have elevated nadir but not zenith corticosterone levels [Klug and Adelman, 1979]. Interestingly, impairments in cortisol rhythms can result from VMH lesions [Dallman, 1984], and from diabetes in humans [Grad et al., 1971] and rats [Gibson et al., 1985]. VMH lesions in rats can also impair feedback inhibition of glucocorticoids [Filaretov and Filaretova, 1985]. Furthermore, high levels of glucocorticoids in neonates can cause impairments in cortisol rhythms [Krieger, 1972] and indications of VMH damage [Kiss, 1985; Palkovits and Mitro, 1968]. Thus, as with other forms of neurohumoral hysteresis, the VMH could play a role in glucocorticoid hysteresis as well.

Food restriction increases glucocorticoid levels; there is a highly significant inverse correlation between food intake and mean 24-hour glucocorticoid levels, so the more food is restricted, the higher the levels of glucocorticoids [Honma et al., 1983]. In a recent thorough study, food restriction caused an increase in basal corticosterone secretion in male rats that were 8, 16, and 23 months old, and restricted rats at every age had higher adrenal weights/gm body weight than nonrestricted rats [Stewart et al., 1988]. Corticosterone levels were also higher after stress in restricted young rats, but not in restricted older rats, compared with nonrestricted age-matched controls [Stewart et al., 1988]. In this study corticosterone levels after stress were actually suppressed *sooner* in normal old rats than in normal young rats [Stewart et al., 1988]. Since in this study no age-correlated impairments in glucocorticoid feedback inhibition after stress were observed, in contrast to previous studies [Sapolsky et al., 1986], the stress studies may require cautious interpretation. In any case, since food restriction increases glucorticoid levels, and a toxic effect of glucocorticoids is the essential feature of glucocorticoid hysteresis, it seems unlikely that glucocorticoids mediate effects of food restriction to delay age-correlated impairments.

CONCLUSIONS

These studies suggest that toxic effects of estrogen, glucose, and glucocorticoids on hypothalamic and hippocampal neurons may contribute to the development of age-correlated neuroendocrine impairments. In particular, hyperglycemic hysteresis, in which glucose damages neurons in the VMH, leading to pancreatic hypersecretion and hyperglycemia, leading to further damage, could contribute to the development of type II diabetes and other age-correlated impairments in humans. Since different forms of neurohumoral hysteresis may involve neurons in the VMH, it is possible that senescence of a given system may involve the interaction of different forms of hysteresis (i.e., some reproductive impairments could be due to glucose toxicity). It will be important to determine more definitively if normal, youthful levels of humoral substances can cause neuroendocrine damage or if this damage occurs only after these substances are increased by some other primary cause. The essence of neurohumoral hysteresis is specific neurotoxicity, so the mechanisms by which estrogen, glucose, and glucocorticoids damage VMH and hippocampal neurons merit much further study.

REFERENCES

Ahdieh HB, Hamilton J, Wade GN (1983): Copulatory behavior and hypothalamic estrogen and progestin receptors in chronically insulin-deficient female rats. Physiol Behav 31:219–223.
Akmayev IG, Rabkina AE (1976): CNS-pancreas system. The hypothalamic response to insulin deficiency. Endokrinologie 68:211–220.
Arbogast BW, Berry DL, Newell CL (1984): Injury of arterial endothelial cells in diabetic, sucrose-fed, and aged rats. Atherosclerosis 51:31–45.
Arendash GW, Leung PCK (1986): Alleviation of estrogen-induced hyperprolactinemia through intracerebral transplantation of hypothalamic tissue containing dopaminergic neurons. Neuroendocrinology 43:359–367.
Aschheim P (1983): Relation of neuroendocrine system to reproductive decline in female rats. In Meites J (ed): "Neuroendocrinology of Aging." New York: Plenum Press, pp 73–102.
Baron AD, Schaeffer L, Shragg P, Kolterman OS (1987): Role of hyperglucagonemia in maintenance of increased rates of hepatic glucose output in type II diabetes. Diabetes 36:274–284.
Bestetti G, Rossi GL (1980): Hypothalamic lesions in rats with long-term streptozotocin-induced diabetes mellitus. Acta Neuropathol 52:119–127.
Bestetti G, Hofer R, Rossi GL (1987): The preoptic-suprachiasmatic nuclei though morphologically heterogeneous are equally affected by streptozotocin diabetes. Exp Brain Res 66:74–82.
Berthoud HR, Jeanrenaud B (1979): Acute hyperinsulinemia and its reversal by vagotomy following lesions of the ventromedial hypothalamus in anesthetized rats. Endocrinology 105:146–151.
Blades RA, Bryant KR, Whitehead SA (1985): Feedback effects of steroids and gonadotropin control in adult rats with streptozotocin-induced diabetes mellitus. Diabetologia 28:348–354.
Blaha GC, Lamperti AA (1983): Estradiol target neurons in the hypothalamic arcuate nucleus and lateral ventromedial nucleus of young adult, reproductively senescent and monosodium glutamate-lesioned female golden hamsters. J Gerontol 35:335.
Blake CA, Elias KA, Huffmann LJ (1983): Ovariectomy of young adult rats has a sparing effect on the suppressed ability of aged rats to release luteinizing hormone. Biol Reprod 28:575.
Bleier R, Byne W (1985): Septum and hypothalamus. In Paxinos G (ed): "The Rat Brain Nervous System. New York: Academic Press.
Blichert-Toft M (1978): The adrenal glands in old age. Aging (Geriatr Endocrinol) 5:81–102.
Blichert-Toft M, Hummer L (1976): Immunoreactive cortitrophin reserve in old age in man during and after surgical stress. J Gerontol 31:539–545.
Brancho-Romero E, Reaven GM (1977): Effect of age and weight on plasma glucose and insulin responses in the rat. J Am Geriatr Soc 7:299–302.
Brawer JR, Sonnenschein C (1975): Cytopathological effects of estradiol on the arcuate nucleus of the female rat. Am J Anat 144:57–87.
Brawer JR, Naftolin F, Martin J, Sonnenschein C (1978): Effects of a single injection of estradiol valerate on the hypothalamic arcuate nucleus and on reproductive function in the female rat. Endocrinology 103:501–512.
Brawer JR, Schipper H, Naftolin F (1980): Ovary-dependent degeneration in the hypothalamic arcuate nucleus. Endocrinology 107:274–279.
Brawer JR, Schipper H, Robaire B (1983): Effects of long-term androgens and estradiol exposure on the hypothalamus. Endocrinology 112:510–515.
Brown MR, Dyck PJ, McClearn GE, Sima AF, Powell HC, Porte D (1982): Central and peripheral nervous system complications. Diabetes 31(Suppl):65–70.
Brown-Grant K (1975): On "critical periods" during the postnatal development of the rat. Int. Symposium on Sexual Endocrinology of the Perinatal Period. INSERM 32:357–376.
Buzzi S, Buzzi G, Guzzi A, Baccini C (1987): Hypothalamic syndrome in a woman with three sewing needles in the brain. Lancet 1:1313.

Cameron DP, Poon TK-Y, Smith GC (1976): Effects of monosodium glutamate administration in the neonatal period on the diabetic syndrome in KK mice. Diabetologia 12:621–626.

Cerami A (1985): Glucose as a mediator of aging. J Am Geriatr Soc 33:626–634.

Cerami A (1986): Aging of proteins and nucleic acids—what is the role of glucose? Trends Biochem Sci 11:311–314.

Chang T-H, Polakis SE (1978): Differentiation of 3T3-L1 fibroblasts to adipocytes. Effect of insulin and indomethacin on insulin receptors. J Biol Chem 253:4693–4696.

Christ B, Probst I, Jungermann K (1986): Antagonistic regulation of the glucose/glucose 6-phosphate cycle by insulin and glucagon in cultured hepatocytes. Biochem J 238:185–191.

Chen M, Bergman RN, Pacini G, Porte D (1985): Pathogenesis of age-related gluocses intolerance in man: Insulin resistance and decreased beta-cell function. J Clin Endo Metab 60:13–20.

Chlouverakis C, Jarrett RJ, Keen H (1967): Glucose tolerance, age, and circulating insulin. Lancet 1:806–809.

Clark JT, Simpkins JW, Kalra SP (1986): Long-term weekly gonadal steroid treatment: Effects on plasma prolactin, sexual behavior, and hypothalamic-preoptic area catecholamines. Neuroendocrinology 44:488–493.

Cocchi D, Castoldi C, Locatelli S, Novelli A, Colombo AM, Tammaro A, Muller EE (1986): Evaluation of hypothalamic dopaminergic function by neuropharmacologic means in aged women. J Neural Transm 64:199–210.

Coleman DL (1982)): Diabetes-obesity syndrome in mice. Diabetes 31:1–6.

Dallman M (1984): Viewing the ventromedial hypothalamus from the adrenal gland. Am J Physiol 246:R1–R12.

Davidon MD (1979): The effect of aging on carbohydrate metabolism. A review of the English literature and a practical approach to the diagnosis of diabetes mellitus in the elderly. Metabolism 28:688–705.

Davidson MB, Casanello-Ertl D (1979): Insulin antagonism on cultured rat myoblasts secondary to chronic exposure to insulin. Horm Metab Res 11:207–209.

DeFronzo RA (1979): Glucose tolerance and aging. Evidence for tissue insensitivity to insulin. Diabetes 28:1095–1101.

DeFronzo RA (1982): Glucose intolerance and aging. In Shimke R (ed): "Biological Markers of Aging." NIH Publication No. 81–2194, Washington, D.C. pp 98–119.

DeFronzo RA, Tobin JD, Andres R (1979): The glucose clamp technique. A method for the quantification of beta cell sensitivity to glucose and of tissue sensitivity to insulin. Am J Physiology 237:E214–E223.

Del Prato S, Castellino P, Simonson DC, DeFronzo RA (1987): Hyperglucogonemia and insulin-mediated glucose metabolism. J Clin Invest 79:547–556.

DePaolo LV, Chappel SC (1986): Alterations in the secretion and production of follicle-stimulating hormone precede age-related lengthening of estrous cycles in rats. Endocrinology 118:1127–1133.

DePaolo LV, Rowlands KL (1986): Deceleration of age-associated changes in the preovulatory but not secondary follicle-stimulating hormone surge by progesterone. Biol Reprod 35:320–326.

Dilman VM (1971): Age-associated elevation of hypothalamic threshold to feedback control, and its role in development, aging and disease. Lancet 1:1211–1218.

Ercolani L, Lin HL, Ginsberg BH (1985): Insulin-induced desensitization at the receptor and post-receptor level in mitogen-activated human T-lymphocytes. Diabetes 34:931–937.

Felicio LS, Nelson JF, Gosden RG, Finch CE (1983): Restoration of ovulatory cycles by young ovarian grafts in aging mice: Potentiation by long-term ovariectomy decreases with age. Proc Soc Natl Acad Sci USA 80:6076.

Filaretov AA, Filaretova LP (1985): Role of the paraventricular and ventromedial hypothalamic nuclear areas in the regulation of the pituitary-adrenocortical system. Brain Res 342:135–140.

Finch CE (1976): The regulation of physiological changes during mammalian aging. Qu Rev Biol 51:49–83.
Finch CE, Foster JR, Mirsky AE (1969): Aging and the regulation of cell activities during exposure to cold. J Gen Physiol 54:690.
Finch CE, Felicio LS, Mobbs CV, Nelson JF (1984): Ovarian and steroidal influences on neuroendocrine aging processes in female rodents. Endocr Rev 5:467–497.
Fink RI, Kolterman OG, Kao M, Olefsky JF (1984a): The role of the glucose transport system in the post-receptor defect in insulin action associated with human aging. J Clin Endocrinol Metab 58:721–725.
Fink R, Koleterman O, Olefsky J (1984b): The physiological significance of the glucose intolerance of aging. J Gerontol 39:273–278.
Flood DG, Guarnaccia M, Coleman PD (1987): Dendritic extent in human CA2-3 hippocampal pyramidal neurons in normal aging and senile dementia. Brain Res 409:88–96.
Friedman M, Green MF, Sharland DE (1969): Assessment of hypothalamic-pituitary-adrenal function in the geriatric age group. J Gerontol 24:292–297.
Garris DR, Coleman DL (1984): Diabetes-associated changes in estradiol accumulation in the aging C57BL/KsJ mouse brain. Neurosci Lett 49:285–290.
Garris DR, Diani AR, Smith C, Gerritsen GC (1982): Depopulation of the ventromedial hypothalamic nucleus in the diabetic chinese hamster. Acta Neuropathol 56:63–66.
Garris DR, West LR, Coleman DL (1985a): Morphometric analysis of medial basal hypothalamic neuronal degeneration in diabetes (db/db) mutant C57BL/KsJ mice: Relation to age and hyperglycemia. Dev Brain Res 20:161–168.
Garris DR, Coleman DL, Morgan CR (1985b): Age- and diabetes-related changes in tissue glucose uptake and estradiol accumulation in the C57B1/KsJ mouse. Diabetes 34:47–52.
Gee DM, Flurkey K, Finch CE (1983): Aging and the regulation of luteinizing hormone in C57BL/6J mice: Impaired elevations after ovariectomy and spontaneous elevations at advanced ages. Biol Reprod 28:598.
Gerli R, Rambotti P, Nicoletti I, Orlandi S, Migliorati G (1986): Reduced number of natural killer cells in patients with pathological hyperprolactinemia. Clin Exp Immunol 64:399–406.
Gibson MJ, DeNicola AF, Krieger DT (1985): Streptozotocin-induced diabetes is associated with reduced immunoreactive beta-endorphin concentrations in intermediate pituitary lobe and with disrupted circadian periodicity of plasma corticosterone levels. Neuroendocrinology 41:64–71.
Gold G, Karoly K, Freeman C, Adelman RC (1976): A possible role for insulin in the altered capability for hepatic enzyme adaptation during aging. Biochem Biophys Res Commun 73:1003–1010.
Gold PE, Stone WS (1988): Neuroendocrine effects on memory in aged rodents and humans. Neurobiol Aging 9:709–717.
Gonzales AM, Sochor M, Rowles PM, Wilson-Holt N, McLean P (1981): Sequential biochemical and structural changes occuring in rat lens during cataract formation in experimental diabetes. Diabetes 21:23.
Gordon MN, Mobbs CV, Morgan DG, Finch CE (1988): Pituitary and hypothalamic glucose-6-phosphate-dehydrogenase effects of estradiol and age in C57B116J mice. Endocrinol 122:726–733.
Gosden RG, Laing SC, Felicio LS, Nelson JF, Finch CE (1983): Imminent oocyte exhaustion and reduced follicular recruitment mark the transition to acyclicity in aging C57BL/6J mice. Biol Reprod 28:255–260.
Gottschall PE, Meites J (1986): Evidence for a permanent decline in tuberinfidibular dopaminergic neuronal function after chronic estrogen treatment is terminated in Fischer 344 rats. Neuroendocrinology 44:211–216.

Grad B, Rosenberg GM, Liberman H, Trachtengerg J, Kral VA (1971): Diurnal variation of the serum cortisol level of geriatric subjects. J Gerontol 26:351–357.

Gray GD, Wexler BC (1980): Estrogen and testosterone sensitivity of middle-aged female rats and the regulation of LH. Exp Gerontol 15:201.

Haffner SM (1987): Hyperinsulinemia as a possible etiology for the high prevalence of non-insulin-dependent diabetes in Mexican Americans. Diabete Metab 13:337–344.

Hamel EE, Santisteban GA, Ely JTA, Read DH (1986): Hyperglycemia and reproductive defects in non-diabetic gravidas: A mouse model test of a new theory. Life Sci 39:1425–1428.

Hamilton CL, Brobeck JR (1963): Diabetes mellitus in hyperphagic monkeys. Endocrinology 73:512–515.

Harris MI, Hadden WC, Knowler WC, Bennett PH (1987): Prevalence of diabetes and impaired glucose tolerance and plasma glucose levels in U.S. population aged 20–74. Diabetes 36:523–524.

Hayashi K (1980): Glucose tolerance in the elderly with special reference to insulin and glucagon responses. Wakayama Med Rep 23:29–39.

Hayashi K (1982): Insulin insensitivity and hyposuppressibility of glucagon by hyperglycemia in aged Wistar rats. Gerontology 28:10–18.

Henry RR, Wallace P, Olefsky JM (1986): Effects of weight loss on mechanisms of hyperglycemia in obese non-insulin-dependent diabetes mellitus. Diabetes 35:990–998.

Holehan AM, Merry BJ (1985): Modification of the oestrous cycle hormonal profile by dietary restriction. Mech Aging Dev 32:63–76.

Honma K-I, Honma S, Hiroshige T (1983): Critical role of food amount for prefeeding corticosterone peak in rats. Am J Physiol 245:R339–R344.

Hsu HK, Peng MT (1978): Hypothalamic neuron number in old female rats. Gerontology 24:434–440.

Ingram DK, London ED, Reynolds MA (1982): Circadian rhythmicity and sleep: Effects of aging in laboratory animals. Neurobiol Aging 3:287–292.

Inoue S, Bray GA (1977): The effect of subdiaphragmatic vagotomy in rats with ventromedial hypothalamic lesions. Endocrinology 100:108–114.

Inoue S, Bray GA, Mullen YS (1977): Effect of transplantation of pancreas on development of hypothalamic obesity. Nature 266:742–744.

Ionescu E, Rohner-Jeanrenaud F, Berthoud HR, Jeanrenaud B (1983): Increases in plasma insulin levels in response to electrical stimulation of the dorsal motor nucleus of the vagus nerve. Endocrinology 112:904–910.

Jarrett RJ (1984): Type 2 (non-insulin dependent) diabetes mellitus and coronary heart disease—chicken, egg, or neither? Diabetologia 26:99–102.

Jeanrenaud B (1985): An hypothesis on the aetiology of obesity: Dysfunction of the central nervous system as a primary cause. Diabetologia 28:502–513.

Joseph SA, Knigge KM (1968): Effects of VMH lesions in adult and newborn guinea pigs. Neuroendocrinology 3:309–331.

Karnieli E, Hissin PJ, Simpson IA, Salans LB (1981): A possible mechanism of insulin resistance in the rat adipose cell in streptozotocin-induced diabetes mellitus. Depletion of intracellular glucose transport systems. J Clin Invest 68:811–814.

Kasuga M, Inoue S, Akanuma Y, Kosaka K (1980): Insulin receptor function and insulin effects on glucose metabolism in adipocytes from ventromedial hypothalamus-lesioned rats. Endocrinology 107:1549–1555.

Katayama S, Brownscheidle CM, Wootten V, Lee JB, Shimaoka K (1984): Absent or delayed preovulatory luteinizing hormone surge in experimental diabetes mellitus. Diabetes 33:324–327.

Kawashima S (1960): Influence of continued injections of sex steroids on the estrous cycle in the female rat. Annot Zool Jpn 33:226.

Kiss A (1985): Ultrastructural changes in neurons of the anterior subdivision of the rat hypothalamic ventromedial nucleus following repeated immobilization stress. Anat Anz 158:125–134.

Kithara A, Adelman RC (1979): Altered regulation of insulin secretion in isolated islets of different size in aging rats. Biochem Biophys Rres Comm 87:1207–1213.

Klug TL, Adelman RC (1979): Altered hypothalamic-pituitary regulation of thyrotropin in male rats during aging. Endocrinology 104:1136–1142.

Klug TL, Freeman C, Karoly K, Adelman RC (1979): Altered regulation of pancreatic glucagon in male rats during aging. Biochem Biophys Res Commun 907–912.

Kobayashi M, Olefsky JM (1978): Effect of experimental hyperinsulinemia on insulin binding and glucose transport in isolated rat adipocytes. Am J Physiol 235: E53-E62.

Komeda K, Yokote M, Oki Y (1980): Diabetic syndrome in the Chinese hamster induced with monosodium glutamate. Experientia 36:232–234.

Krieger DT (1972): Circadian corticosteroid periodicity: Critical period for abolition by neonatal injection of corticosteroid. Science 178:1205–1207.

Landfield P (1980): Adrenocortical hypothesis of brain and somatic aging. In Shimke RT (ed): "Biological Mechanisms in Aging," publication No. 81-2194. Washington, DC: NIH, pp 658–672.

Landfield P, Waymire J, Lynch G (1978): Hippocampal aging and adrenocorticoids: A quantitative correlation. Science 202:1098–1102.

Landfield PW, Sundberg DK, Smith MS, Eldridge JC, Morris M (1981a): Mammalian aging: Theoretical implications of changes in brain and endocrine systems during mid- and late-life in rats. Peptides 1 (Suppl 1):185–196.

Landfield PW, Baskin RK, Pitler TA (1981b): Brain aging correlates: Retardation by hormonal-pharmacological treatments. Science 214:581–584.

Lapolt PS, Matt DW, Judd HL, Lu JKH (1986): The relation of ovarian steroid levels in young female rats to subsequent estrous cyclicity and reproductive function during aging. Biol Reprod 35:1131–1139.

Lapolt PS, Yu SM, Lu JKH (1988): Early treatment of young female rats with progesterone delays the aging-associated reproductive decline: A counteraction by estradiol. Biol Reprod 38:987–995.

Larkin JG, Frier BM, Ireland JT (1985): Diabetes mellitus and infection. Postgrad Med J 61:233–237.

Lazaris JA, Goldberg RS, Kozlov MP (1985): Studies on diabetes mellitus after ventromedial hypothalamic lesions in adult and aged rats. Endocrinol Exp 19:67–76.

Leighton B, Cooper GJS (1988): Pancreatic amylin and calcitonin gene-related peptide cause resistance to insulin in skeletal muscle in vitro. Nature 335:632–635.

Leiter EH, Premdas F, Harrison DE, Lipson LG (1988): Aging and glucose homeostasis in C57BL/6J male mice. FASEB J 2:2807–2811.

Leopold IH, Mosier MA (1978): Cataracts in diabetes mellitus. Geriatrics 33:33–41.

Lorenzi M, Cagliero E, Toledo S (1985): Glucose toxicity for human endothelial cells in culture. Delayed replication, disturbed cell cycle, and accelerated death. Diabetes 34:621–627.

Lu JKH (1983): Changes in ovarian function and gonadotropin and prolactin secretion in aging female rats. In Meites J (ed): "Neuroendocrinology of Aging." New York: Plenum Press, pp 103–122.

Lu JHK, Gilman DP, Meldrum DR, Judd HL, Sawyer CH (1981): Relationship between circulating estrogens and the central mechanisms by which ovarian steroids stimulate luteinizing hormone secretion in aged and young female rats. Endocrinology 108:836–841.

Mani RB, Lohr JB, Jeste DV (1986): Hippocampal pyramidal cells and aging in the human: A quantitative study of neuronal loss in sectors CA1 to CA4. Exp Neurol 94:20–40.

Marshall S, Olefsky JM (1980): Effects of insulin incubation on insulin binding, glucose transport, and insulin degradation by isolated rat adipocytes. J Clin Invest 66:763–772.

Martin C, Desai KS, Steiner G (1983): Receptor and post-receptor insulin resistance induced by in vivo hyperinsulinemia. Can J Pharmacol 61:802–807.

Masoro EJ (1988): Food restriction in rodents: An evaluation of its role in the study of aging. J Gerontol Biol Sci 43:B59–B64.

Masoro EJ, Katz MS, McMahan CA (1989): Evidence for the glycation hypothesis of aging from the food-restricted rodent model. J Gerontol 44:B20–B22.

Matsumoto A, Kobayshi S, Murakami S, Arai Y (1984): Recovery of declined ovarian function in aged female rats by transplantation of newborn hypothalamic tissue. Proc Jpn Acad [B]60:73–76.

Matsuo T, Shino A (1972): Induction of diabetic alterations by goldthioglucose—obesity in KK, ICR, and C57B1 mice. Diabetologia 8:391–397.

Meaney MJ, Aitken DH, van Berkel C, Bhatnagar S, Sapolsky RM (1988): Effect of neonatal handling on age-related impairments associated with the hippocampus. Science 239:766–768.

Merry BJ, Holehan AM (1979): Onset of puberty and duration of fertility in rats fed a restricted diet. J Reprod Fertil 57:253–259.

Meyer F, Brown JB, Morrison AS, MacMahon B (1986): Endogenous sex hormones, prolactin, and breast cancer. J Nat Cancer Inst 77:613–616.

Mills TM, Mahesh VB (1978): Pituitary function in the aged. Aging (Geriatr Endocrinol) 5:1–11.

Mobbs CV (1989): Neurohumoral hysteresis as a mechanism for senescence: Comparative aspects. In Scanes CG, Schriebman MP (eds): "Development, Maturation, and Senescence of the Neuroendocrine System." New York: Academic Press, pp 223–252.

Mobbs CV, Gee DM, Finch CE (1984a): Reproductive senescence in female C57BL/6J mice: Ovarian impairments and neuroendocrine impairments that are partially reversible and delayable by ovariectomy. Endocrinology 115:1653–1662.

Mobbs CV, Flurkey K, Gee DM, Yamamoto K, Sinha YN, Finch CE (1984b): Estradiol-induced adult anovulatory syndrome in female C57BL/6J mice: Age-like neuroendocrine, but not ovarian, impairments. Biol Reprod 30:556–563.

Mobbs CV, Cheyney D, Sinhan YN, Finch CE (1985): Age-correlated and ovary-dependent changes in relationships between plasma estradiol and luteinizing hormone, prolactin, and growth hormone in female C57BL/6J mice. Endocrinology 116:813–820.

Modan M, Helkin H, Almog S, Lusky A, Eshkol A, Shefi M, Shitrit A, Fuchs Z (1985): Hyperinsulinemia. A link between hypertension, obesity, and glucose intolerance. J Clin Invest 75:809–817.

Mooradian AD (1988): Tissue specificity of premature aging in diabetes mellitus. The role of cellular replicative capacity. J Am Ger Soc 36:831–839.

Mordes JP, Rossini AA (1981): Animal models of diabetes. Am J Med 70:353–360.

Morgan LO, Vonderahe AR, Malone EF (1937): Pathological changes in the hypothalamus in diabetes mellitus. J Nerv Ment Dis 85:125–138.

Morgan WW, Steger RW, Smith MS, Bartke A, Sweeney CA (1985): Time course of induction of prolactin-secreting pituitary tumors with diethylstilbestrol in male rats: Response of tuberinfundibular dopaminergic neurons. Endocrinology 116:17–24.

Nakano K, Yoshida T, Kondo M, Muramatsu S (1984): Immune responsiveness and phagocytic activity of macrophages in streptozotocin (SZ)-induced diabetic mice. Endocrinol Jpn 31:15–22.

Nankervis A, Proietto J, Aitken P, Alford F (1985): Hyperinsulinemia and insulin insensitivity: Studies in subjects with insulinoma. Diabetologia 28:427–431.

Nass TE, LaPolt PS, Judd HL, Lu JKH (1984): Alterations in ovarian steroid and gonadotrophin secretion preceding the cessation of regular estrous cycles in ageing female rats. J Endocrinol 100:43–50.

Nelson JF, Holinka CF, Latham KR, Allen JK, Finch CE (1976): Corticosterone binding in cytosol from brain regions of mature and senescent male C57B1/6J mice. Brain Res 115:345–351.

Nelson JF, Felicio LS, Osterburg HH, Finch CE (1981): Altered profiles of estradiol and progesterone associated with prolonged estrous cycles and persistent vaginal cornification in aging C57BL/6J mice. Biol Reprod 24:784.

Nelson J, Gosden R, Felicio L (1985): Effect of dietary restriction on estrous cyclicity and follicular reserves in aging C57BL/6J mice. Biol Reprod 32:515–522.

Nelson JF, Bergman MD, Karelus K, Felicio LS (1987): Aging of the hypothalamic-pituitary-ovarian axis: Hormonal influences and cellular mechanisms. J Steroid Biochem 27:699–705.

Odell WD, Swerdloff RS (1968): Progesterone-induced luteinizing and follicle-stimulating hormone surge in post-menopausal women: A simulated ovulatory peak. Proc Natl Acad Sci USA 61:529.

O'Mara BA, Byers T, Schoenfeld E (1985): Diabetes mellitus and cancer risk: A multisite case-control study. J Chron Dis 38:435–441.

Palkovits M, Mitro A (1968): Morphological changes in the rat hypothalamus aand adrenal cortex in the early postnatal period after ACTH and hydrocortisone administration, stress, and adrenalectomy. Neuroendocrinology 3:200–210.

Peng MT, Hsu HK (1982): No neuron loss from hypothalamic nuclei of male rats in old age. Gerontology 28:19–22.

Penicaud L, Larue-Achagiotis C, Le Magnen J (1983): Endocrine basis for weight gain after fasting or VMH lesion in rats. Am J Physiol 245:E246–E252.

Pfaff DW (1972): Histological differences between ventromedial hypothalamic neurons of well-fed and underfed rats. Nature 223:77–79.

Pfaff DW (1980): "Estrogens and brain function." New York: Springer-Verlag.

Quigley K, Goya R, Meites J (1987): Rejuvenating effects of 10-week underfeeding period on estrous cycles in young and old rats. Neurobiol Aging 8:225–232.

Ramirez I, Friedman MI (1982): Suppression of food intake by intragastric glucose in rats with impaired glucose tolerance. Physiol Behav 31:39–43.

Reaven GM, Reaven EP (1985): Age, glucose intolerance, and non-insulin-dependent diabetes mellitus. J Am Geriatr Soc 33:286–290.

Reaven GM, Olefsky J, Farquar JW (1972): Does hyperglycemia or hyperinsulinemia characterize the patient with chemical diabetes? Lancet 2:1247–1249.

Reaven EP, Gold G, Reaven GM (1979): Effect of age on glucose-stimulated insulin release by the beta cell of the rat. J Clin Invest 64:591–599.

Reaven E, Wright D, Mondon CE, Solomon R, Ho H, Reaven GM (1983): Effect of age and diet on insulin secretion and insulin action in the rat. Diabetes 32:175–180.

Richardson SJ, Senikas V, Nelson JF (1987): Follicular depletion during the menopausal transition: Evidence for accelerated loss and ultimate exhaustion.

Riegle GD (1983): Changes in hypothalamic control of ACTH and adrenal cortical functions during aging. In Meites J (ed): "Neuroendocrinology of Aging." New York: Plenum. pp 309–332.

Rizza RA, Mandarino LJ, Genest J, Baker BA, Gerich JE (1985): Production of insulin resistance by hyperinsulincmia in man. Diabetologia 28:70–75.

Rohner-Jeanrenaud F, Jeanrenaud B (1980): Consequences of ventromedial hypothalamic lesions upon insulin and glucagon secretion by subsequently isolated perfused pancreases in the rat. J Clin Invest 65:902–910.

Rohner-Jeanrenaud F, Jeanrenaud B (1984): Oversecretion of glucagon by pancreases of hypothalamic-lesioned rats: A re-evaluation of a controversial topic. Diabetologia 27:535–539.

Rose DP, Pruitt BT (1981): Plasma prolactin levels in patients with breast cancer. Cancer 48:2687–2691.

Rowe JW, Minaker KL, Pallota JA, Flier JS (1983): Characterization of the insulin resistance of aging. J Clin Invest 71:1581–1589.

Sabel BA, Stein DG (1981): Extensive loss of subcortical neurons in the aging rat brain. Exp Neurol 73:507–516.

Sandberg H, Yoshimine N, Maeda S, Symons D, Zavodnick J (1973): Effects of an oral glucose load on serum immunoreactive insulin, free fatty acid, growth hormone, and blood sugar levels in young and elderly subjects. J Am Geriatr Soc 10:433–438.

Sapolsky RM (1987): Glucocorticoids and hippocampal damage. TINS 10:346–349.

Sapolsky RM, Krey L, McEwen BS (1983): The adrenocortical stress-response in the aged male rat: Impairment of recovery from stress. Exp Gerontol 18:55.

Sapolsky RM, Krey LC, McEwen BS (1984): Glucocorticoid-sensitive hippocampal neurons are involved in terminating the adrenocortical stress response. Proc Natl Acad Sci USA 81:6274.

Sapolsky RM, Krey LC, McEwen BS (1985): Prolonged glucocorticoid exposure reduces hippocampal neuron number: Implications for aging. J Neurosci 5:1222–1227.

Sapolsky RM, Krey LC, McEwen BS (1986): The neuroendocrinology of stress and aging: The glucocorticoid cascade hypothesis. Endocr Rev 7:284–301.

Sapolsky RM, Armanini M, Packan D, Tombaugh G (1987): Stress and glucocorticoids in aging. Endocrinol Metab Clin N Am 16:965–980.

Sartin JL, Lamperti AA (1985): Neuron numbers in hypothalamic nuclei of young, middle-aged and aged male rats. Experientia 41:109–111.

Schipper H, Brawer JR, Nelson JF, Felicio LS, Finch CE (1981): The role of the gonads in the histologic aging of the hypothalamic arcuate nucleus. Biol Reprod 25:413–418.

Schnider SL, Kohn RR (1981): Effects of age and diabetes mellitus on the solubility and non-enzymatic glycosylation of human skin collagen. J Clin Invest 67:1630–1635.

Scott AR, Bennett T, MacDonald IA (1987): Diabetes and thermoregulation. Can J Physiol Pharmacol 65:1365–1376.

Sharma M, Palacios-Bois J, Schware G, Iskadar H, Thakur M, Quirion R, Nair NPV (1989): Circadian rhythms of melatonin and cortisol in aging. Biol Psychiatry 25:305–319.

Sherman B, Wysham C, Pfohl B (1985): Age-related changes in the circadian rhythm of plasma cortisol in man. J Clin Endocrinol Metab 61:439–443.

Simard M, Brawer JR, Farookhi R (1987): An intractable, ovary-independent impairment in hypothalamo-pituitary function in the estradiol-valerate-induced polycystic ovarian condition in the rat. Biol Reprod 36:1229–1237.

Smith-West C, Garris DR (1983): Diabetes-associated hypothalamic neuronal depopulation in the aging Chinese hamster. Dev Brain Res 9:385–389.

Soman VR, DeFronzo RA (1980): Direct evidence for downregulation of insulin receptors by physiologic hyperinsulinemia in man. Diabetes 29:159–163.

Spindler-Vomachka M, Johnson DC (1985): Altered hypothalamic-pituitary function in the adult female rat with streptozotocin-induced diabetes. Diabetologia 28:38–44.

Sridaran R, Blake CA (1978): Effects of neonatal monosodium glutamate (MSG) in pulsatile plasma LH and estrogen feedback on LH release in ovariectomized (OVX) rats. Physiologist 21:114.

Steffes MW, Mauer SM (1982): Diabetic glomerulopathy in man and experimental animal models. In Richter GW, Epstein MA (eds): "International Review of Experimental Pathology." New York: Academic Press, pp 147–159.

Stewart J, Meaney MJ, Aitken D, Jensen L, Kalant N (1988): The effects of acute and life-long food restriction on basal and stress-induced serum corticosterone levels in young and aged rats. Endocrinology 123:1934–1941.

Stokes PE, Stoll PM, Koslow SH, Maas JW, Davis JM, Swann AC, Robins E (1984): Pretreat-

ment DST and hypothalamic-pituitary-adrenocortical function in depressed patients and comparison groups. Arch Gen Psychiatry 41:257–267.
Stout RW (1981): The role of of insulin in atherosclerosis in diabetics and non-diabetics. A review. Diabetes 30(Suppl 2):54–57.
Tallarigo L, Giampietro O, Penno G, Miccoli R, Gregori G, Navalesi R (1986): Relation of glucose tolerance to complications of pregnancy in nondiabetic women. N Engl J Med 315:989–992.
Telford N, Mobbs CV, Sinha YN, Finch CE (1986): The increase of anterior pituitary dopamine in aging C57BL/6J mice is caused by ovarian steroids, not intrinsic pituitary aging. Neuroendocrinology 43:135–142.
Tokunaga K, Fukushima M, Kemnitz JW, Bray GA (1986): Effect of vagatomy on serum insulin in rats with paraventricular or ventromedial hypothalamic lesions. Endocrinology 119:1708–1711.
Touitou Y, Sulon J, Bogdan A, Reinberg A, Sodoyez J-C, Demey-Ponsart E (1983): Adrenocortical hormones, ageing, and mental condition: Seasonal and circadian rhythms of plasma 18-hydroxy-11-deoxycorticosterone, total and free cortisol, and urinary corticosteroids. J Endocrinol 93:53–64.
Unger RH, Orci L (1975): The essential role of glucagon in the pathogenesis of diabetes mellitus. Lancet 1:14–16.
Unger RH, Grundy S (1985): Hyperglycemia as an inducer as well as a consequence of impaired islet cell function and insulin resistance: Implications for the management of diabetes. Diabetologia 28:119–121.
Vlassara H (1988): Peripheral neuropathy and aging. Age 74–78.
Welborn TA, Stenhouse NS, Johnstone CG (1969): Factors determining serum insulin response in a population sample. Diabetologia 5:263–266.
West KM (1978): "Epidemiology of Diabetes and Its Vascular Lesions." New York: Elsevier North-Holland.
Westermark P, Wernstedt C, Wilander DW, O'Brien TD, Johnson KH (1987): Amyloid fibrils in human insulinoma and islet of Langerhans of the diabetic cat are derived from a neuropeptide-like protein also present in normal islet cells. Proc Natl Acad Sci USA 84:3881–3885.
Wise PM (1987): The role of the hypothalamus in aging of the female reproductive system. J Steroid Biochem 27:713–719.
Wise PM, Parsons B (1984): Nuclear estradiol and cytosol progestin receptor concentrations in the brain and pituitary gland and sexual behavior in ovariectomized estradiol-treated middle-aged rats. Endocrinology 115:810–816.
Wise PM, McEwen BS, Parsons B, Rainbow TC (1984): Age-related changes in cytoplasmic estradiol receptor concentrations in microdissected brain nuclei: Correlations with changes in steroid-induced sexual behavior. Brain Res 321:119–126.
Woods SC, Porte D, Bobbioni E, Ionescu E, Sauter J-F, Rohner-Jeanrenaud F, Jeanrenaud B (1985): Insulin: Its relationship to the central nervous system and to the control of food intake and body weight. Am J Clin Nutr 42:1063–1071.
Zarco de Coronado I, Yepez Chamorro MC (1987): Increase of circulating levels of glucose after the electrolytic lesion of the arcuate hypothalamic nucleus. Soc Neurosci Abstr 13:403.
Zavaroni I, Bonora E, Pagleria M, Dell'Aglio E, Luchetti L, Buonanno G, Bonah A, Bergonzeni M, Gnudi L, Passen M, Reaven G (1989): Risk factors for coronary artery disease in healthy persons with hyperinsolinemia and normal glucose tolerance. New Engl J Med 320:702–706.
Zimmet P, Whitehouse S (1979): The effect of age on glucose tolerance. Diabetes 28:617–628.

SECTION IV
ENDOCRINOLOGY OF AGING

Problems in Design and Interpretation of Aging Studies: Illustration by Reports on Hormone Secretion and Actions

J.R. Florini and F.J. Mangiacapra

INTRODUCTION

This review represents the third time that the senior author has reviewed this area during the current decade, and it is written from the perspectives gained in those repeated surveys of the literature. Those perspectives include observations of some major problems in experimental design that appear with distressing regularity in aging research. In particular, in spite of many admonitions on these points in review articles and meeting discussions, much of the current gerontological literature remains flawed in two significant ways: 1) use of animals housed in conventional animal colonies; and 2) comparisons of only two age groups (young/old comparisons).

Use of Appropriate Animal Models

Although the National Institute on Aging has made available and advocated the use of specific pathogen-free (SPF) barrier-protected animals for gerontological research, a large part of the literature on effects of aging still describes studies on animals housed in conventional animal colonies, which frequently include number of diseases not detected by the experimenters or animal attendants. Of course, it can be argued that humans do not spend their lives protected from disease, but it is clearly impossible to know if observed changes are inherent aspects of aging or result from unknown diseases unless the latter are eliminated from the animal colony to the greatest extent possible. The availability of microisolator cages and laminar flow hoods at modest prices has now made it possible to maintain the SPF status of animal subjects for treatment periods of modest length, and we [Florini and Ewton, 1989] have recently completed a series of experiments in which treatment of the animals with hormones for 6 months was done successfully at the vendor's barrier

Biology Department, Syracuse University, Syracuse, New York 13244

facility without the expense of duplicating such an expensive structure in our laboratories.

Comparisons of Only Two Age Groups

If, as is too frequently the case, measurements are restricted to young (2–4 months) versus old (24–30 months) animals, it is impossible to know what is happening during the interval when no measurements are made. It is generally the investigators' assumption that any changes occur progressively with age (or are localized late in the life span), but it is equally possible that all of the change occurs during the early (developmental) portion of the life span. There are a number of instances in which the major changes occur early in the life span, so this possibility cannot be dismissed as too unlikely to consider seriously. In other cases (such as that described by Geary and Florini [1972]), there is an early increase in a process, so the subsequent decrease with age is even greater than that found in comparisons restricted to young and old animals. It is easy to understand why many investigators limit their experiments to binary young/old comparisons. Middle-aged animals are quite expensive for the amount of gerontological information they provide, and many kinds of experiments are much more readily done with two rather than three or more experimental groups. In practice, it is sometimes possible to show in initial experiments that the changes of interest occur progressively throughout the life span (as was done in the experiments summarized by Sartin et al. [1986]), and then do more intensive mechanistic studies in comparisons of two ages. Even here, it is best to return to experiments including a series of ages to test the most important conclusions.

Limitations and Organization of This Review

This review covers only those aspects of the endocrine literature not involving pituitary hormones, as that aspect of the subject is covered by Meites in another chapter in this volume. We have emphasized publications that have appeared since the last review of this literature was published from this laboratory [Florini and Regan, 1985]. To avoid presenting yet another catalog of reports of age-related changes in secretion and circulation of hormones, we have chosen to emphasize papers that illustrate an important aspect of gerontological literature. Specifically, we will illustrate the two problems mentioned above, and consider two other important but not obvious points: 1) age-related decreases in hormone secretion might not be attributable to a single cause, but might be the cumulative result of several changes; and 2) changes in hormone levels are not necessarily reflected by changes in composition or function of target tissues for those hormones. Some additional clinical reports of interest are also mentioned at the end of the review.

DIFFERENCES IN OBSERVATIONS WITH CONVENTIONAL AND BARRIER-PROTECTED ANIMALS
Vasopressin

There is direct disagreement about whether levels of vasopressin increase or decrease with aging; SPF barrier-maintained animals gave results opposite to those obtained with animals from conventional colonies. Zbuzek's group has extensively investigated changes in vasopressin secretion with aging. They have shown decreases in both plasma and hypothalmic levels of vasopressin, but could not detect a change in the neurophyophysis of Sprague–Dawley [Zbuzek and Wu, 1982] and Fischer 344 [Zbuzek et al., 1983] rats, all SPF animals in barrier-protected colonies. Later studies revealed reduced secretion of vasopressin by superfused neurohypopheses from old rats [Zbuzek et al., 1986]. Zbuzek et al. [1987] subsequently reported a decreased rate of uptake of [^{3}H]-arginine into vasopressin and its subsequent release from the neurophyophysis.

In contrast, Miller [1987] reported that in vitro vasopressin release by isolated hypothalmic–neurohypophyseal units from conventionally housed Long–Evans rats *increased* with age, with the increase being detectable as early as age 7 months. He suggested that the elevated level of hormone could induce a decreased sensitivity in the renal tubule, affecting urine-concentrating capacity. However, changes in this action appears to be tissue-specific. Frolkis et al. [1982] found greater sensitivity to vasopressin in aged coronary vessels, and Beck and Yu [1982] reported an opposite effect in kidney, stating that vasopressin-dependent cAMP production was reduced in aged rats. The net effect could help to explain some age-related changes in water balance and urinary output.

Glucocorticoids

Early reports in this area indicated little or no change in circulating levels of glucocorticoids, including cortisol. However, more recent animal studies have led to some disagreement about changes in levels of ACTH and cortisol. In an extensive study with Long–Evan rats (not SPF barrier-maintained) at several ages throughout the life span, Malamed and Carsia [1983] reported a decrease with age in adrenal response to ACTH or cAMP by suspended adrenocortical cells. This report is a clear warning of the hazards of young/old comparisons; maximum corticosterone secretion decreased primarily between 6 and 12 months of age, but sensitivity to ACTH showed the largest decrease between 12 and 18 months. Popplewell et al. [1987] attempted to isolate changes in the steroidogenic enzymes with age, but found no differences between young and aged animals to account for the decrease in corticosterone secretion.

In contrast, Sonntag et al. [1987] found no changes in the diurnal secretion for corticosterone in SPF barrier-maintained Fischer 344 rats aged 2–3, 10–12, and 22–24 months, but did observe reduction of ACTH levels for older rats of about 64% of that secreted by young animals. The authors, in direct disagreement with the report of Malamed and Carsia [1983] cited above, concluded that there was a greater adrenal sensitivity for the maintenance of in vivo corticosterone levels. Here again, it appears that an age-related decrement may be a result of housing in a conventional colony rather than an inherent property of aging.

Such effects of animal housing are not restricted to changes in hormone secretion. The literature cited under Changes in Hormone Levels Do Not Necessarily Predict Changes in Target Tissues below illustrates instances in which changes in skeletal muscle composition and function have been reported frequently for animals housed in conventional colonies but have not been confirmed in more recent research with SPF barrier-protected animals.

CHANGES RESTRICTED TO THE EARLY PORTION OF THE LIFE SPAN
Vasopressin

The observations described above also provide a striking example of the situation in which "young/old" comparisons would be misleading. Zbuzek et al. [1986] showed that secretion of vasopressin (also designated antidiuretic hormone) by superfused neurohyphyses from 2-, 12-, and 30-month-old Fischer 344 rats resulted in an about 50% reduction in vasopressin secretion by between 2 and 12 months of age, but no subsequent change to 24 months. Thus a simple comparison of young and old animals might suggest a gerontologically significant change, but in fact the entire change occurs during the first half of the normal life span. The observations of Malamed and Carsia [1983] on corticosterone secretion mentioned above also illustrate an example of a situation in which changes occur primarily during the early part of the life span.

Glucocorticoids

The report of Malamed and Carsia [1983] cited above is a clear example of a case in which the major loss in cortisol secretion occurred between 6 and 12 months of age, not later in the life span.

Other examples of developmental rather than gerontological changes in endocrine processes are cited in previous reviews in this series [Florini and Regan, 1985; Sartin, 1983].

MULTIPLE CONTRIBUTIONS TO AN OBSERVED IMPAIRMENT

It is regrettable but understandable that most investigators are satisfied upon finding a single change that appears to explain an observed age-related change.

However, as illustrated here, it is quite possible that more than one decrement might make a significant contribution to the change, and it is difficult to be assess the relative contributions made by a number of demonstrated age-related changes.

Insulin

In contrast to the problems cited in the section above, a firm foundation for binary studies on effects of age on carbohydrate utilization was established by Adelman's group [reviewed by Sartin et al., 1980], who showed a progressive increase with age in the lag period prior to response to glucose administration. Subsequently, the same group [Chaudhuri et al., 1983] cited increased levels of somatostatin in the pancreas as a reason for the impairment in adaptive secretion of insulin. They demonstrated a very large increase with age in somatostatin secretion by islets from 24-month-old compared with 2-month-old rats. Administration of antibodies to somatostatin restored the age-related decrease of insulin secretion to glucose challenge from 27% to 61% of that produced in young animals. The observation of *partial* recovery was explained by the presence of two isoforms of rat pancreatic somatostatin (one blocking glucagon biosynthesis, the other insulin biosynthesis) [Mandarino et al., 1986]. However, lack of full recovery to young levels is compatible with the possibility that other mechanisms might contribute to the observed age-related decrease in insulin secretion.

A change in sensitivity to glucose may be one of those mechanisms. Sartin et al. [1986] reported decreased sensitivity to glucose with age by isolated perifused islets. Comparing Sprague–Dawley rats of 2 and 24 months, they found that aged rats required higher threshold glucose concentrations. In addition, maximal secretory response decreased by 33–50% compared with that in young animals. Higher concentrations of glucose in the perfusion medium compensated for the lower sensitivity to glucose in aged rats at submaximal insulin response.

Elahi et al. [1985] compared the sensitivity of isolated perfused β-cells from 12- and 23-month-old Wistar rats to varying glucose concentrations and got results similar to those cited in the preceding paragraph. It was found that first phase insulin secretion was lower in old rats in the presence of lower levels (150 and 220 mg/dl) of glucose, but higher levels (360 mg/dl) of glucose eliminated the difference. A third possibility was raised by Goodman et al. [1986], who examined the secretion of insulin in 2- and 18-month-old Fischer 344 rats using a sequential gating islet perfusion technique. They found that lower "glucose-stimulated secretion vesicle margination appears to be the abnormality accounting for diminished insulin release in aged animals." They reported little or no impairment in biosynthesis of insulin, or in packaging or lysis of secretory vesicles at the cell membrane; indeed, islet insulin content was *increased* with age. However, vesicles did not aggregate as well at the plasma membrane in older rats. The authors proposed that this phenomenon might be the result of changes in calcium–calmodulin actions or other factors that could affect secretion vesicle margination, such as changes in glucose metabolism.

A fourth possibility was raised by Lipson's group [Morina et al., 1985; Premdas et al., 1983], based on the obligatory role of glucose metabolism in the glucose-stimulated secretion of insulin suggested by Malaisse et al. [1976]. Lipson and coworkers found that glucose-stimulated secretion of insulin by isolated islets is impaired in 13-month-old rats compared with 2.5-month-old animals, but addition of 5.0, 10.0, or 14.0 mM D-glyceraldehyde with nonstimulatory glucose concentrations eliminated the age-related impairment. These findings indicate that "the major rate-limiting step in stimulus-secretion coupling in aging is before the metabolism of the trioses." To be sure, this comparison is limited to a relatively early part of the life span, but it has been demonstrated (see above) that the impairment of insulin secretion is progressive throughout the life span.

CHANGES IN HORMONE LEVELS DO NOT NECESSARILY PREDICT CHANGES IN TARGET TISSUES

The thyroid hormone–cardiac/skeletal muscle system presents a striking example of lack of parallel changes in target tissues in response to age-related decreases in circulating levels of hormones. (This topic is considered in greater detail in a current minireview [Florini, 1989].) Over the years, a number of reports [reviewed by Florini and Regan, 1985] have suggested that there is a small but significant decrease in circulating levels of both thyroxine (T_4) and triiodothyronine (T_3) with age. Two recent reports [Carter et al., 1987; Effron et al., 1987] have confirmed these observations, using modern radioimmunoassay procedures and studying barrier-protected animals. The dramatic effect of thyroid status on expression of the genes for the principal protein component of cardiac and skeletal muscle, the myosin heavy chain (MHC), was strikingly demonstrated in a series of mRNA measurements published by Izumo et al. [1986]. In general, thyroid deficiency leads to expression of the isoform associated with slower contraction rates and lower ATPase activity (the β-MHC [cardiac] or type I [skeletal] isozyme), while increasing thyroid hormone levels give increased expression of the gene for the faster-contracting, higher ATPase (the α-MHC [cardiac] or type II [skeletal] isozyme). Similar effects of thyroid hormones on protein composition of cardiac and skeletal muscles indicate that most of the regulation of MHC gene expression is at the transcriptional level.

On this basis, it would be reasonable to expect a decrease in muscle content of the faster isoforms in cardiac and skeletal muscles from older animals, and the literature provides evidence for such changes. A recent group of papers has rather thoroughly characterized changes in the heart; there is a decrease with age in cardiac myosin ATPase activity and a corresponding decrease in relative content of the α-MHC isozyme [Carter et al., 1987; Effron et al., 1987; Florini and Ewton, 1989] as well as in α-MHC/β-MHC mRNA ratio. This change was correlated with a change in physiological function; time to peak tension also decreased with age [Effron et al., 1987]. Treatment of the rats with large doses of T_4 [Effron et al., 1987] or smaller doses of T_3 [Carter et al., 1987] substantially reversed the age-related changes.

The older literature on skeletal muscle [reviewed by Florini, 1987a,b] indicates that similar changes toward expression of the slower MHC isoforms (at least in muscles containing predominantly fast-twitch type II fibers) occur with age, as would be expected from the thryoid hormone changes described above. However, more recent experiments done with SPF barrier-maintained animals do not confirm these earlier studies (all done on animals from conventional colonies). The first such report was by Eddinger et al. [1985], who found only very small, barely significant changes in relative numbers of fast fibers of extensor digitorum longus (EDL), soleus, and diaphragm muscles from Fischer 344 rats. We [Florini and Ewton, 1989] have recently found no significant change in either myofibrillar Ca^{2+}-activated ATPase activity or percent of type II (fast-twitch) fibers of Fischer rats in an extensive series of comparisons involving at least 25 animals at each of three ages throughout the life span. This lack of change is not an unusual property of the Fischer 344 rat; we also found a similar lack of change with age in muscles of C57Bl/NNia mice, as well.

Determinations of physiological functions have given results in agreement with the histological and biochemical indications that there is no change in myosin composition or fiber types as a function of age in skeletal muscle. Using a variety of different measurements, Eddinger et al. [1986], McCarter and McGee [1987], and Brooks and Faulkner [1988] have all reported little or no change in contraction rate-related physiological parameters. All of these investigators used SPF barrier-protected animals.

On the basis of all of these results it seems reasonable to conclude that the previously reported age-related changes in skeletal muscle MHC composition can be attributed to a lifetime of undetected and unregulated minor disease states, however difficult it may be to devise a mechanism to explain such changes. Thus the thyroid hormone–cardiac/skeletal muscle system illustrates two of the major points being made in this review.

CLINICAL OBSERVATIONS
Aldosterone

Although there have been few reports of significant changes in the level of circulating aldosterone, some other interesting observations have been made. Heystad et al. [1983] reported that "while plasma aldosterone levels remained unchanged with respect to age if the subject was recumbant, it decreased with age if the subject was upright, and aging humans exhibited a lower urinary excretion rate for aldosterone." In contrast, a recent review [Gregerman, 1986] stated that levels of aldosterone decrease with age as much as 70% under a sodium-restricted condition, leading to a diminished ability to retain sodium [Epstein and Hollenberg, 1976].

Adrenal Medullary Hormones

The hormones of the adrenal medulla and of the sympathetic nervous system, epinephrine and norepinephrine, form one of the few cases in which an age-related *increase* in secretion occurs. The decline in sleep quality with age has been attributed to elevated levels of norepinephrine secretion by Prinz et al. [1984], who found a close correlation between plasma norepinephrine levels and sleep. Higher levels of plasma norepinephrine were always found in the aged group, but levels were unaffected by changes in the sleep/wake pattern. The authors' suggest that "heightened sympathetic activity may fragment sleep with wakefulness in the aged."

Testosterone

It is generally found that the androgen levels in men fall in response during aging. Tsitouras et al. [1982] attributed decreases in serum testosterone, luteinizing hormone (LH), and follicle-stimulating hormone (FSH) to lack of availability of chorionic gonadotropin to Leydig cells. Upon long-term treatment with chorionic gonadotropin, they found a reversal of diminished testosterone levels, increased cAMP levels both in vitro and in vivo, and chorionic gonadotropin binding capacity seen in old animals. They concluded "prolonged in vivo exposure to hCG appeared to reverse both the *in vivo* and *in vitro* age related Leydig cell secretory defect, despite gonadotropin receptor down regulation," and that "the aging defects are caused by chorionic gonadotropin deprivation."

Others have demonstrated not only a decrease in available testosterone (i.e., unbound,), but a decrease in protein bound levels of the hormone as well. For example, Nankin and Calkins [1986] found levels of testosterone bound to sex hormone binding globulin significantly decreased between young (22–39-year-old) and old (65–83-year-old) men. Healthy impotent men at both ages also showed a significant decrease in bound testosterone.

Winters et al. [1984] measured feedback inhibition on secretion of LH and FSH by infusion of testosterone, dihydrotestosterone (DHT), or estradiol. DHT and testosterone did not reduce serum LH levels significantly in old (65–80-year-old) compared with young (18–32-year-old) men. Infusion of estradiol reduced LH and FSH levels in both groups. Infusion of DHT also decreased the level of testosterone binding globulin. The authors concluded that "elderly men are more responsive than are young men to the gonadotropin-suppressive effects of androgens, but not to DHT effects on circulating testosterone bound levels. The more pronounced declaration of spontaneous LH secretion episodes during DHT infusion in aged men provides evidence for an alteration in hypothalmic function in male senescence."

CONCLUSIONS

We have used this review to illustrate some important generalizations on the design and interpretation of gerontological studies, as well as to list some of the changes with age in hormone secretion and actions. Two major points of experimental design emerge repeatedly in consideration of this literature: 1) it is necessary to compare more than two ages to obtain a clear picture of the gerontological significance of observed differences between "young" and "old" subjects; and 2) the long-held suspicion that animals from conventional colonies might show different age-related changes than those protected by uncharacterized infections is well confirmed in some of the work cited here. The multiplicity of factors that apparently contribute to the age-related impairment of insulin secretion mandate caution that the first explanation is not necessarily the complete one—a point applicable to many endeavors beyond biological gerontology. Finally, variability in target tissue sensitivity can be a substantial modifier of the physiological consequences of age-related changes in hormone secretion and circulation; clearly (at least in the case of the thyroid hormones) not all target tissues respond in parallel to the same decrease. In spite of the emphasis placed here on problems in interpretation of some studies, it is heartening to see a general improvement in the kinds of studies now being published. In recent years, there has been a significant increase in the reports of research incorporating several ages throughout the life span, using well-characterized SPF barrier-protected animals, employing up-to-date and appropriate measurement techniques. We hope that this trend will continue.

REFERENCES

Beck N, and Yu BP (1982): Effect of aging on urinary concentrating mechanism and vasopressin-dependent cyclic AMP in rats. Am J Physiol 243:F121–F125.

Brooks SV, Faulkner, JA (1988): Contractile properties of skeletal muscles from young, adult, and aged mice. J Physiol (Lond) 404:71–82.

Carter WJ, Kelly WF, Faas FH, Lynch ME, Perry CA (1987): Effect of graded doses of triiodothyronine on ventricular myosin ATPase activity and isomyosin profile in young and old rats. Biochem J 247:329–334.

Chaudhuri M, Sartin JL, Adelman RC (1983): A role for somatostatin in the impaired insulin secretory response to glucose by islets from aging rats. J Geron 38:431–437.

Eddinger TJ, Moss RL, Cassens RL (1985): Fiber number and type composition in extensor digitorum longus, soleus, and diaphragm muscles with aging in Fisher 344 rats. J Histochem Cytochem 33:1033–1041.

Eddinger TJ, Cassens RG, Moss RL (1986): Mechanical and histochemical characterization of skeletal muscles from senescent rats. Am J Physiol (Cell Physiol) 251, in press-C430.

Effron MB, Bhatnagar GM, Spurgeon HA, Ruano-Arroyo G, Lakatta EG (1987): Changes in myosin isoenzymes, ATPase activity and contraction duration in rat cardiac muscle with aging can be modulated by thyroxine. Circ Res 60:238–245.

Elahi D, Muller DC, Andersen DK, Tobin JD, Andres R (1985): The effect of age and glucose concentration on insulin secretion by the isolated perfused rat pancreas. Endocrinology 116:11–16.

Epstein M, Hollenberg NK (1976): Age as a determinant of renal sodium conservation in normal man. J Lab Clin Med 87:411–417.

Florini JR (1987a): Hormonal control of muscle growth. Muscle Nerve 7:577–598.

Florini JR (1987b): Effects of aging on skeletal muscle composition and function. In Rothstein M (ed): "Review of Biological Research in Aging," Vol. 3. New York: Alan R. Liss, Inc., pp 337–358.

Florini JR (1989): Limitations in interpretation of age-related changes in hormone levels. Illustration by effects of thyroid hormones on cardiac and skeletal muscle. J Gerontol 44:B107–109.

Florini, JR, Regan JF (1985): Age-related changes in hormone secretion and action. In Rothstein M (ed): "Review of Biological Research in Aging," Vol. 2. New York: Alan R. Liss, Inc., pp 227–250.

Florini JR, Ewton DZ (1989): Age and growth hormone administration do not change skeletal muscle fiber composition or myosin ATPase activity. J Gerontol, 44:B110–117.

Frolkis VV, Golovchenko SF, Medved VJ, Froklis RA (1982): Vasopressin and cardiovascular system in aging. Gerontology 28:290–302.

Geary S, Florini JR (1972): Change in rate of protein synthesis in isolated perfused hearts as a function of age in the mouse. J Gerontol 27:325–332.

Goodman M, Leitner JW, Sussman KE, Draznin, B (1986): Insulin secretion and aging: Studies with sequential gating of secretion vesicle margination and lysis. Endocrinology 119:827–832.

Gregerman RI (1986): Mechanisms of age-related alterations of hormone secretion and action. An overview of 30 years of progress. Exp Gerontol 21:345–365.

Heystad BA, Brown RD, Jiang NS, et al. (1983): Aging and aldosterone. Am J Med 74:442–448.

Izumo S, Nadal-Ginard B, and Mahdavi V (1986): All members of the MHC multigene family respond to thyroid hormone in a highly tissue-specific manner. Science 231:597–600.

Malaisse WJ, Sener A, Levy J (1976): The stimulus-secretion coupling of glucose-induced insulin release. J Biol Chem 251:1731–1737.

Malamed S, Carsia RV (1983): Aging of the rat adrenocortical cell: response to ACTH and cyclic AMP in vitro. J Gerontol 38:130–136.

Mandarino L, Stenner D, Blanchard W, et al. (1981): Selective effects of somatostatin 14 and 28 on in vitro insulin and glucagon secretion. Nature 291:76–77.

McCarter R, McGee J (1987): Influence of nutrition and aging on the composition and function of rat skeletal muscle. J Gerontol 42:432–441.

Miller M (1987): Increased vasopressin secretion: An early manifestation of aging in the rat. J Gerontol 42:3–7.

Morina JM, Premdas FH, Lipson LG (1985): Insulin release in aging: Dynamic response of isolated islets of Langerhans of the rat to D-glucose and D-glyceraldehyde. Endocrinology: 116:821–826.

Nankin HR, Calkins JH (1986): Decreased bioavailable testosterone in aging normal and impotent men. Endocrinology 63:1418–1420.

Popplewell PY, Butte J, Azhar S (1987): The influence of age on steroidogenic enzyme activities of the rat adrenal gland: Enhanced expression of cholesterol side-chain cleavage activity. Endocrinology 120:2521–2528.

Premdas FH, Molina JM, Lipson LG (1983): Insulin release in aging: the role of glyceraldehyde. Acta Endocrinol 103:539.

Prinz PN, Vitiello MV, Smallwood RG (1984): Plasma norepinephrine in normal young and aged men: Relationship with sleep. J Gerontol 39:561–567.

Sargin J (1983): Endocrine physiology. In Rothstein M (ed): "Review of Biological Research in Aging," Vol. 1. New York: Alan R. Liss, Inc., pp 181–193.

Sartin JL, Chadhuri M, Obenrader M, Adelman RC (1980): The role of hormones in changing adaptive mechanisms during aging. Fed Proc 39:3163–3167.

Sartin JL, Chaudhuri M, Farina S, Adelman, RC (1986): Regulation of insulin secretion by glucose during aging. J Gerontol 41:30–35.

Sonntag WE, Goliszek AG, Brodish A (1987): Diminished diurnal secretion of adrenocorticotropin (ACTH), but not corticosterone, in old male rats: Possible relation to increased adrenal sensitivity to ACTH in vivo. Endocrinology 120:2308–2315.

Tsitouras PD, Martin CE, Harman SM (1982): Relationship of serum testosterone to sexual activity in healthy elderly men. J Gerontol 37:288–293.

Winters SJ, Sherins RJ, Troen P (1984): The gonadotropin-suppressive activity of androgen is increased in elderly men. Metabolism 33:1052–1059.

Zbuzek VK, Wu WH (1982): Age-related vasopressin changes in rat plasma and the hypothalo–hypophyseal system. Exp Gerontol 17:133–138.

Zbuzek VK, Zbuzek V, Wu WH (1983): The effect of aging on the vasopression system in Fischer 344 rats. Exp Gerontol 18:305–331.

Zbuzek VK, Zbuzek V, Wu WH (1986): Decremental vasopressin release after repeated stimulation of superfused neurohypophyses of Fisher 344 rats of different ages. J Gerontol 41:140–146.

Zbuzek VK, Zbuzek V, Wu WT (1987): Age related differences in the incorporation of ^{3}H-arginine into vasopressin in Fischer 344 rats. Exp Gerontol 22:113–125.

Hormone/Neurotransmitter Action During Aging: The Calcium Hypothesis of Impaired Signal Transduction

George S. Roth

INTRODUCTION

For many years, there has been general agreement that normal aging is accompanied by an impaired ability for adaptation to environmental influences [Shock, 1962; Roth, 1985]. Numerous investigators have sought to elucidate the mechanism(s) by which such alteration occurs. As physiological regulation became better understood by cellular and molecular biologists, gerontologists quickly applied the newest theoretical and technical advances to the problem of age-related adaptive changes. Thus, it can now be said that senescence is associated with alterations in cellular and molecular components that were not even known to exist a few years ago. For example, who could have predicted that loss of certain classes of neuronal dopamine receptors during aging might result in deterioration of motor control, [for a review, see Roth et al., 1986] when such receptors were not discovered until the middle of the last decade [Creese et al., 1977]. Moreover, the exact signal transduction sequence by which interaction of dopamine with these receptors ultimately leads to proper regulation of movement is unknown to this day.

Clearly, biogerontological research is now in a position to ride the crest of the wave between discovery and application. We can advance most efficiently only by a proper mixture of basic and applied studies. Adaptive regulation of physiological functions, mediated by hormones and neurotransmitters, provides an extremely useful phenomenon for which to employ this strategy. Our own laboratory, as well as a number of others around the world, began in the early 1970s to attempt to understand how the mechanisms of hormone/neurotransmitter action became altered with age [for reviews, see Roth and Hess, 1982; Burchinsky, 1984; Kalimi, 1984; Dax, 1987]. Not surprisingly, initial

Molecular Physiology and Genetics Section, Gerontology Research Center, National Institute on Aging, Francis Scott Key Medical Center, Baltimore, Maryland 21224

studies of age changes in such signal transduction focused heavily at the receptor level [Roth and Hess, 1982]. A great number of alterations in hormone and neurotransmitter receptors were reported during aging [Roth and Hess, 1982; Burchinsky, 1984; Kalimi, 1984; Dax, 1987]. However, as it became clear that responsiveness to these agents could be modulated by altered physiological and pathological states at many levels, gerontologists shifted some of their attention to "postreceptor" components and events, [Roth and Hess, 1982; Dax, 1987; Roth, 1988].

CALCIUM FLUX

One of the most exciting and best studied of these events over the past few years has been the stimulated movement of calicum ions into, out of, and within various cell types. At least nine different mechanisms by which calcium flux occurs have been described [Meldolesi and Pozzan, 1987]. Proper calcium movement is required for a multitude of biological processes, ranging from secretion to neurotransmission, muscle contraction, or cell division [Carafoli and Penniston, 1987; Meldolesi and Pozzan, 1987]. All of these processes have been shown to change during aging under certain conditions [Roth, 1989]. Particular emphasis has been placed on the regulation of release of calcium ions from intracellular storage sites by "second messenger" molecules such as inositol triphosphate (IP$_3$) [Berridge, 1984]. The exact mechanism(s) by which this molecule interacts with its receptor and postulated effector systems, associated with the endoplasmic reticulum, is currently the subject of intense interest.

Another recent focal area is the regulation of plasma membrane channels by which calicum enters the cell. At least two major classes (in addition to the second messenger-operated channels associated with the endoplasmic reticulum) have been characterized: voltage-operated channels and receptor-operated channels [Meldolesi and Pozzan, 1987]. The former, in turn, have been divided into T (transient) L (long-lasting), and N (negative holding potential activiated depolarization) channels. Clearly then, the entire field of calcium movement has shifted from the exclusive domain of the electrophysiologists to that of the molecular biologists and biochemists.

THE CALCIUM HYPOTHESIS OF AGE CHANGES IN HORMONE/NEUROTRANSMITTER ACTION

Until the early 1980s, biogerontological interest in calcium metabolism was centered almost exclusively at the level of the blood–bone axis. Control of muscle contraction really offered the first opportunities to examine age changes in soft-tissue calcium movement with age [Guarnieri et al., 1980; Cohen and

Berkowitz, 1976]. Collaborative studies between our laboratory and NIA cardiovascular scientists suggested that the age-related decline in beta-adrenergic-stimulated cardiac muscle contraction was quite specific to a "postreceptor" impairment in calcium movement, since elevation or extracellular calcium levels enabled unstimulated muscle from aged rats to contract equally as well as isoproterenol-stimulated muscle from young animals [Guarnieri et al., 1980]. Although the precise molecular nature of this age-associated defect is still uncertain, a surprising number of similar phenomena has now been reported for other calcium-dependent responses to hormones and neurotransmitters [Roth, 1989].

The truly remarkable aspect of these studies is that even though cellular calcium movement occurs through a wide variety of mechanisms, almost every age-related impairment in hormone/neurotransmitter-stimulated, calcium-dependent processes can be at least partially reversed if sufficient calcium is moved to/at the appropriate site [Roth, 1989]. Table I shows the most up to date list of systems exhibiting impaired calcium mobilization during aging, most of which exhibit this phenomenon.

Despite the apparently generalized nature of impaired stimulation of calcium movement during aging and the simplicity of the hypothesis, some conceptual problems exist. For example, calcium concentrations in some cell types actually increase, rather than decrease with age. Table II lists some examples of systems in which calcium levels have been examined during aging. The apparent discrepancy between calcium flux [Meyer et al., 1986; Peterson and Gibson, 1983a, Peterson et al., 1985; Peterson and Gibson, 1983b; Davis et al., 1983] and calcium concentration [Landfield and Pitler, 1984; Landfield and Morgan, 1984; Battaini et al., 1985] presented a particular problem for neuronal systems. More recently, however, Reynolds and Carlen [1989] have provided evidence that seems to reconcile these divergent findings. They have reported that neuronal calcium currents, especially a high threshold, slowly inactivating L-type current, are depressed in hippocampi of aged animals concomitant with elevated calcium concentrations. They further suggest that the influx of calcium through L-type channels may be quite sensitive to chronic changes in free intracellular calcium concentration [Reynolds and Carlen, 1989].

Another area of possible disagreement involves the role of stimulated calcium movement in age changes in the immune response. Decline in immune function with increasing age constitutes one of the best characterized and most studied physiological manifestations of aging [Adler and Nordin 1981; Kay and Makinodan 1981]. Consequently, studies of Miller [1986, Miller et al., 1987] suggesting that the age-related decrement in calcium-dependent mouse thymic lymphocyte mitogenesis could be reversed by administration of phorbol esters and ionophores have inspired a multitude of further investigations [Proust et al., 1987; Chopra et al., 1987; Grossman et al., 1989; Lustijik and O'Leary,

TABLE I. Systems Exhibiting Impaired Stimulation of Calcium Mobilization During Aging

Stimulus	Species	Tissue	Response	References
Alpha-adrenergic	Rat	Parotid	Electrolyte secretion	Ito et al., 1982
				Bodner et al., 1983
		Parotid	Glucose oxidation	Ito et al., 1981
				Gee et al., 1986
		Aorta	Contraction	Cohen and Berkowitz, 1976
Beta-adrenergic	Rat	Heart	Contraction	Guarnieri et al., 1980
Cholinergic	Rat	Brain (striatum)	Dopamine release	Joseph et al., 1988
Depolarization	Rat	Heart	Contraction	Elfellah et al., 1986
		Brain (forebrain (and cortex)	Acetylcholine release	Meyer et al., 1986
				Peterson and Gibson, 1983a
		Brain (hippocampus)	Calcium current	Reynolds and Carlen, 1989
	Mouse	Brain (forebrain)	Acetylcholine release	Peterson et al., 1985
		Whole animal	Motor function	Peterson and Gibson, 1983a
	Rat	Whole animal	Maze learning	Davis et al., 1983
Serotonin	Rat	Aorta	Contraction	Cohen and Berkowitz, 1976
Gonadotropin releasing hormone	Rat	Pituitary	Gonadotropin secretion	Chuknyiska et al., 1987
Lectin	Rat	Lymphocyte	Mitogenesis	Wu et al., 1985
				Segal, 1986
	Mouse	Lymphocyte	Mitogenesis	Miller et al., 1987
				Miller, 1986
				Proust et al., 1987
	Human	Lymphocyte	Mitogenesis	Chopra et al., 1987
				Grossman et al., 1989

Compound	Species	Cell type	Effect	Reference
Compound 48-80	Rat	Mast cell	Histamine release	Orida and Feldman, 1982
Formyl-methionyl-leucyl-phenylalanine	Human	Neutrophil	Superoxide generation	Lipschitz et al., 1987; Lipschitz et al., 1988
Thyroid hormones	Human	Erythrocyte	Activation of calcium ATPase	David et al., 1987
Low-density lipoprotein	Human	Polymorphonuclear leukocytes	Release of β-glucuronidase	Fulop et al., 1985
Cytochalasin B	Human	Polymorphonuclear leukocytes	Release of β-glucuronidase	Fulop et al., 1985
Immune complexes	Human	Polymorphonuclear leukocytes	Release of β-glucuronidase	Fulop et al., 1985
Elastin peptides	Human	Polymorphonuclear leukocytes	Cytosolic free calcium	Varga et al., 1988
Formyl-methionyl-leucyl-phenylalanine	Human	Polymorphonuclear leukocytes	Cytosolic free calcium	Varga et al., 1988
Opsonized zymojan	Human	Polymorphonuclear leukocytes	Respiratory burst	Fulop et al., 1988
Formyl-methionyl-leucyl-phenylolomine	Human	Polymorphonuclear leukocytes	Respiratory burst	Fulop et al., 1988
Carbachol	Human	Polymorphonuclear leukocytes	Respiratory burst	Fulop et al., 1988
Phosphatidylserine	Rat	Brain	Protein kinase activation	Calderini et al., 1984

TABLE II. Systems Examined for Changes in Nonstimulated Calcium Movement, Binding, or Content During Aging

Change	Species	Tissue	Parameter	References
Decrease	Rat	Brain	Sodium-calcium exchange	Michaelis et al., 1984
			Calcium ATPase	Michaelis et al., 1984
			Mitochondrial sodium and H^+-calcium exchange	Victoria and Satrustegui, 1986
		Brain (cortex)	Mitochondrial and synaptosomal calcium uptake	Farrar et al., 1984
		Brain (cortex) (hippocampus) (cerebellum, forebrain, midbrain, brainstem)	Calcium uptake	Gibson et al., 1986
		Brain (midbrain brainstem)	Calcium content	Gibson et al., 1986
		Brain (cortex)	Nitrendipine binding affinity to calcium channels'	Govoni et al., 1985
	Mouse	Brain (forebrain)	Calcium uptake	Peterson et al., 1985
	Human	Skin fibroblast	Bound calcium content	Govoni et al., 1985
			Calcium uptake	Govoni et al., 1985
	Rat	Heart	Sarcoplasmic reticulum calcium ATPase	Peterson and Goldman 1986 Gafni and Yuh, 1985
No change	Rat	Lymphocyte	Calcium uptake	Segal, 1986
		Neuromuscular junction	Free calcium content	Blumberg et al., 1986
		Brain (cortex, striatum, hippocampus, cerebellum forebrain)	Calcium content	Gibson et al., 1986
Increase	Rat	Brain (hippocampus)	Calcium content (as reflected in after hyperpolarization)	Landfield and Pitler 1984 Landfield and Morgan 1984
		Brain (cortex)	Berapamil binding to calcium channels	Battaini et al., 1985
	Rat	Aorta	Calcium uptake and binding	Williams, 1984

1989]. Although there appears to be general agreement that some subtypes of thymic lymphocytes exhibit impaired mitogen-stimulated calcium mobilization during aging, it is unclear whether the same phenomena occur in the same cell populations in rodents and humans and whether the alterations in calcium movement are casually related to the alterations in mitogenesis. In fact, the postulated causal relationship between calcium flux and mitogenesis, independent of aging, has itself recently been questioned [Sussman et al., 1988]. These issues need to be resolved in order to determine whether or not impairments in stimulated calcium mobilization are truly responsible for immune dysfunction.

Perhaps the most perplexing problem with the calcium hypothesis is the fact that the aging process seems to impact negatively on so many highly diverse calcium-transporting systems. As mentioned above, calcium is moved by cells in at least nine different ways [Meldolesi and Pozzan, 1987]. Why then, should essentially all of these mechanisms be negatively affected in much the same way? Perhaps in this case, one manifestation of the aging process can be used to gain basic information concerning the regulation of calcium movement. Rather than biogerontologists relying solely on the process of nongerontological counterparts for technological and theoretical advances, it may be possible to utilize aging as a naturally occurring perturbation of an essential physiological process, through which a better mechanistic understanding might be achieved.

CONCLUSIONS

Impaired regulation of cellular calcium mobilization constitutes a rather generalized manifestation of senescence. In many cases, apparent inability of aged cells to respond to certain hormones, neurotransmitters, and related stimuli can be at least partially reversed by selective manipulation of calicum fluxes. Although some inconsistencies may exist, it seems clear that this natural perturbation of a critical signal transduction event, essential for many biological processes, may provide an ideal model system with which to gain fundamental knowledge regarding basic regulation of calcium movement. Moreover, through a better understanding of such mechanisms and their alterations during aging, it may be possible to derive potentially useful therapeutic strategies to halt, prevent, or reverse age-related dysfunctions in calcium-dependent processes.

REFERENCES

Adler WH, Nordin AA (1981): "Immunological Techniques Applied to Aging Research." Boca Raton, FL: CRC Press.

Battaini F, Govoni S, Rius RA, Trabucchi M (1985): Age-dependent increase in [^{3}H]verapamil binding to rat cortical membranes. Neurosci Lett 61:67–71.

Berridge MJ (1984): Inositol trisphosphate and diacylglycerol as second messengers. Biochem J 220:345–360.

Blumberg D, Rosenheimer JL, Smith DO (1986): Calcium entry, utilization and clearance at the neuromuscular junction of aged rats. Soc Neurosci Abstr 12:733.

Bodner L, Hoopes MT, Gee M, et al. (1983): Multiple transduction mechanisms are likely involved in calcium mediated exocrine secretory events in rat parotid cells. J Biol Chem 258:2774–2777.

Burchinsky SG (1984): Neurotransmitter receptors in the central nervous system and aging: Pharmacological aspect (review). Exp Gerontol 19:227–239.

Calderini G, Bellini F, Bonetti AC, Galbiati E,Teolato S, Toffano G (1984): Effect of aging on phospholipid sensitive Ca^{++} dependent protein kinase in the rat brain. Soc Neurosci Abstr 10:275.

Carafoli E, Penniston JT (1987): The calcium signal. Sci Am 257:270–278.

Chopra R, Nagel J, Adler W (1987): Decreased response of T cells from elderly individuals of phytohemogglutinin (PHA) stimulation can be augmented by phorbol myristate acetate (PMA) in conjunction with CA-ionophore A23187. Gerontologist 27:204a.

Chuknyiska RS, Blackman MR, Roth GS (1987): Ionophore A23187 partially reverses LH secretory defect of pituitary cells from old rats. Am J Physiol 258:E233–E237.

Cohen ML, Berkowitz BA (1976): Vascular contraction: Effect of age and extracellular calcium. Blood Vessels 67:139–149.

Creese I, Burt DR, Snyder SH (1977): Dopamine receptor binding enhancement accompanies lesion-induced behavioral supersensitivity. Science 197:596–598.

Davis HP, Idowu A, Gibson GE (1983): Improvement of 8-arm maze performance in aged Fischer 344 rats with 3, 4-diaminopyridine. Exp Aging Res 9:211–214.

Davis PJ, Davis FB, Blas SD (1987): Donor age-dependent decline in response of human red cell Ca^{++}-ATPase activity to thyroid hormone in vitro. J Clin Endocrinol Metab 64:921–925.

Dax EM (1987): Age-related changes in membrane receptor interactions. Endocrinol Metab Clin 16:947–963.

Elfellah MS, Johns A, Shepherd AMM (1986): Effect on age of responsiveness of isolated rat atria to carbachol and on binding characteristics of atria muscarinic receptors. J Cardiovasc Pharmacol 8:873–877.

Farrar R, Chandler LJ, Barr EM, Spirduso WW, Leslie SW (1984): Reduced calcium uptake by rat brain mitochondria and synaptosomes in response to aging. Soc Neurosci Abstr 10:449.

Fulop T, Foris G, Wocum I, Paragh G, Leovey A (1985): Age related variations of some polymorphonuclear leukocyte functions. Mech Ageing Dev 29:1–8.

Fulop T, Varga Z, Nagy JT, Foris G (1988): Studies on opsonized zymogan, FMLP, carbachol, PMA and A23187 stimulated respiratory burst on human PMLs. Biochem Int 17:419–426.

Gafni A, Yuh K (1985): Age-related deterioration in the sarcoplasmic reticulum Ca^+ pump. Gerontologist 25:215–216.

Gee MV, Ishikawa Y, Baum BJ, Roth GS (1986): Impaired adrenergic stimulation of rat parotid cell glucose oxidation during aging: The role of calcium. J Gerontol 41:331–335.

Gibson G, Perrino P, Dienel GA (1986): In vivo brain calcium homeostasis during aging. Mech Ageing Dev 37:1–12.

Govoni S, Rius RA, Battaini F, Bianchi A, Trabucchi M (1985): Age-related reduced affinity of [^{3}H]nitrendipine labeling of brain voltage dependent calcium channels. Brain Res 333:374–377.

Grossman A, Ledbetter JA, Rabinovitch PS (1989): Reduced proliferation in T-lymphocytes in aged humans is predominant in the CD8$^+$ subset, and is unrelated to defects in transmembrane signalling which are predominantly in the CD4$^+$ subset. Exp Cell Res 180:367–382.

Guarnieri T, Filburn CR, Zitnik G, Roth GS, Lakatta EG (1980): Mechanisms of altered cardiac inotropic responsiveness during aging in the rat. Am J Physiol 239:H501–H508.

Ito H, Hoopes MT, Roth GS, Baum BJ (1981): Adrenergic and cholinergic mediated glucose oxidation by rat parotid gland acinar cells during aging. Biochem Biophys Res Commun 98: 275–282.

Ito H, Baum BJ, Uchida T, Hoopes MT, Bodner L, Roth GS (1982): Modulation of rat parotid cell α-adrenergic responsiveness at a step subsequent to receptor activation. J Biol Chem 257:9532–9538.

Joseph JA, Dalton TK, Roth GS, Hunt WA (1988): Alterations in muscarinic control of striatal dopamine autoreceptors in senescence: A deficit at the ligand-muscarinic receptor interface? Brain Res 454:149–155.

Kalimi M (1984): Glucocorticoid receptors from development to aging. Mech Ageing Dev 24:129–138.

Kay MMB, Makinodan T (1981): "Handbook of Immunology and Aging." Boca Raton, Florida: CRC Press.

Landfield PW, Pitler TA (1984): Prolonged CA^{2+}-dependent after hyperpolarizations in hippocampal neurons of aged rats. Science 226:1089–1092.

Landfield PW, Morgan GA (1984): Chronically elevating plasma Mg^{2+} improves hippocampal frequency potentiation and reversal learning in aged and young rats. Brain Res 322:167–171.

Lipschitz DA, Udupa KB, Boxer LA (1987): Evidence that microenvironmental factors account for the age-related decline in neutrophil function. Blood 70:1131–1135.

Lipschitz DA, Udupa KB, Boxer LA (1988): The role of calcium in the age related decline of neutrophil function. Blood (in press).

Lustyik O, O'Leary JJ (1989): Aging and the mobilization of intracellular calcium by phytohemaglutinin in human T cells. J Gerontol 44:B30–36.

Meldolesi J, Pozzan T (1987): Pathways of Ca^{++} influx at the plasma membrane: Voltage-, receptor-, and second messenger-operated channels. Exp Cell Res 171:271–283.

Meyer EM, Crews FT, Otero DH, Larson K (1986): Aging decreases the sensitivity of rat cortical synaptosomes to calcium ionophore-induced acetylcholine release. J Neurochem 47:1244–1246.

Michaelis ML, Johe K, Kitos TE (1984): Age-dependent alterations in synaptic membrane systems for calcium regulation. Mech Aging Dev 25:215–225.

Miller RA (1986): Immunodeficiency of aging: Restorative effects of phorbol ester combined with calcium ionophore. J Immunol 137:805–808.

Miller RA, Jacobson B, Weil G, Simons ER (1987): Diminished calcium influx in lectin-stimulated T cells from old mice. J Cell Physiol 132:337–342.

Orida N, Feldman JD (1982): Age related deficiency in calcium uptake by mast cells. Fed Proc 41:822.

Peterson C, Gibson GE (1983a): Aging and 3, 4-diaminopyridine alter synaptosomal calcium uptake. J Biol Chem 258:11482–11486.

Peterson C, Gibson GE (1983b): Amelioration of age-related neurochemical and behavioral deficits by 3, 4-diamiopyridine. Neurobiol Aging 4:25–30.

Peterson C, Goldman JE (1986): Alterations in calcium content and biochemical processes in cultured skin fibroblasts from aged and Alzheimer donors. Proc Natl Acad Sci USA 83:2758–2762.

Peterson C, Nicholls DG, Gibson GE (1985): Subsynaptosomal distribution of calcium during aging and 3, 4-diaminopyridine treatment. Neurobiol Aging 6:297–304.

Proust JJ, Filburn CR, Harrison SA, Buchholz MA, Nordin AA (1987): Age-related defect in signal transduction during lecting activation of murine T-lymphocytes. J Immunol 139:1472–1478.

Reynolds JN, Carlen PL (1989): Diminished calcium currents in aged hippocampus dentate gyrus granule neurons. Brain Res 479:384–390.

Roth GS (1985): Changes in hormone/neurotransmitter action during aging. In Davis BB, Wood WG (eds): "Homeostatic Function and Aging." New York: Raven Press, pp 41–58.

Roth GS (1989): Changes in hormone action with age; altered calcium mobilization and/or responsiveness impairs signal transduction. In Armbrecht HJ (ed): "Endocrine Function and Aging." New York: Springer-Verlag (in press).

Roth GS, Hess GD (1982): Changes in the mechanisms of hormone and neurotransmitter action during aging: Current status of the role of receptor and post receptor alterations. Mech Ageing Dev 20:175–194.

Roth GS, Henry JM, Joseph JA (1986): The striatal dopaminergic system as a model for modulation of altered neurotransmitter action during aging: Effects of dietary and neuroendocrine manipulation. Prog Brain Res 70:473–484.

Segal J (1986): Studies on the age-related decline in the response of lymphoid cells to mitogens: measurements of concanavalin A binding and stimulation of calcium and sugar uptake in thymocytes from rats of varying ages. Mech Ageing Dev 33:295–303.

Shock NW (1962): The physiology of aging. Sci Am 206:100–110.

Sussman JJ, Mercep M, Saito T, Germain RN, Bonvini E, Ashwell JD. (1988): Dissociation of phosphoinositide hydrolysis and Ca^{2+} fluxes from the biological responses of a T-cell hybridoma. Nature 334:625–628.

Varga Z, Kovacs EM, Paragh G, Jacob M-P, Robert L, Fulop T (1988): Effect of elastin peptides and N-formyl-methionyl leucyl phenylalanine on cytosolic free calcium in polymorphonuclear leukocytes of healthy middle-aged and elderly subject. Clin Biochem 21:127–130.

Vitorica J, Satrustegui J (1986): The influence of age on the calcium-efflux pathway and matrix calcium buffering power in brain mitochondria. Biochem Biophys Acta 851:209–216.

Williams PB (1984): Effect of age upon the uptake and binding of calcium in rat aorta. Biochem Pharmacol 33:3097–3099.

Wu W, Pahlavani M, Richardson A, Cheung HT (1985): Effect of maturation and age on lymphocyte proliferation induced by A23187 through an interleukin independent pathway. J Leukocyte Biol 38:531–540.

Effects of Aging on the Hypothalamic–Pituitary Axis

Joseph Meites

INTRODUCTION

In this chapter I shall present the evidence for the view that the neuroendocrine system, particularly its hypothalamic component, has a critical role in regulating the onset and progression of some specific declines in body functions with age. Research on this subject began in the mid-1960s as a result of the growing knowledge of the importance of the hypothalamus in regulating hormone secretion by the pituitary gland, and via the pituitary, secretion of hormones by the target glands (thyroid, gonads, adrenals, pancreas, etc.) The hypothalalmus secretes hypophysiotropic hormonal peptides that are released into the portal vessels to act directly on the pituitary, and also neurotransmitters that modulate the release of the hormonal peptides. It will be seen that dysfunctions develop with age in the capacity of the hypothalamic neurons to secrete hormones and neurotransmitters, resulting in the onset and progression of several important declines in body functions. The pituitary and its target glands, as well as other tissues of the body, often exhibit a reduced responsiveness to hormones with age due to loss of receptors or to postreceptor changes in cells, but these are secondary to the faults that develop in the hypothalamus. We define aging in terms of a decline in body functions associated with reduced ability to maintain homeostasis.

Most of the knowledge of the role of the hypothalamus in aging events has come from studies in the rat. Relatively little is as yet known of possible changes with age in functions of the hypothalamus in human subjects. Obviously it is difficult to measure hormones or neurotransmitters in the hypothalamus of living individuals, and the few such reported assays have been made after autopsy and removal of the brain. Some indirect information has also come from investigations that utilized challenging stimuli or drugs that act via the hypothalamus to alter pituitary hormone secretion.

I shall briefly review the role of the hypothalamus on the decline in repro-

Department of Physiology, Michigan State University, East Lansing, Michigan 48824

ductive functions, development of numerous mammary and pituitary tumors, decrease in growth hormone (GH) and somatomedin secretion and their relation to the decline in protein synthesis, and reduction in size and function of the thymus gland, the key component of the immune system. It will be seen that correction of the hypothalamic faults that develop with age can delay or reverse these declines in body functions. I shall also point out some similarities and differences in the role of the neuroendocrine system in regulating aging events in the rat and human.

EVIDENCE THAT HYPOTHALAMIC DYSFUNCTIONS ARE MAINLY RESPONSIBLE FOR THE DECLINE IN SEVERAL BODY FUNCTIONS WITH AGE

Reproduction

Female rats have a life span of about 3 years under laboratory conditions. They begin to exhibit estrous cycles between 35 and 40 days of age. Beginning about 7–8 months of age their cycles tend to become irregular, usually lengthened, and by 10–15 months of age they enter into a persistent estrous state characterized by ovaries with well-developed follicles but no ovulation, as indicated by the absence of corpora lutea. Some rats then enter into a state of prolonged and irregular pseudopregnancies with ovaries that contain active corpora lutea that secrete progesterone. The oldest rats, 2–3 years of age, become anestrous and have shrunken ovaries with only small follicles and little evidence of estrogen secretion as observed by the atrophic uterus. In aging male rats, there is a decrease in testosterone secretion and perhaps a small reduction in spermatogenesis [Harman and Talbert, 1985; Meites, et al., 1987; Meites, 1988].

Neither the gonads nor the pituitary of the rat are primarily responsible for the decline in reproduction in the rat. Thus, when the ovaries of old rats were transplanted to young ovariectomized rats, many of the young rats resumed cycling. When ovaries from young rats were transplanted to old ovariectomized rats, cycling did not resume. These observations demonstrate that the ovaries of old rats are capable of functioning under appropriate stimulation [Aschheim, 1976; Peng, 1983]. The pituitary also is not responsible for loss of estrous cycles in old rats. It has been shown that when the pituitary of old noncycling rats was transplanted into young hypophysectomized rats, many of the young rats resumed cycling [Peng, 1983]. However, there is evidence that both the gonads and pituitary become less responsive to hormonal stimulation with age [Harman and Talbert, 1985; Meites, 1988].

Several kinds of evidence demonstrate that the hypothalmus is primarily responsible for the decline in reproduction in old rats. Thus, when old female or male rats were castrated, the rise in circulating LH and FSH was much lower than in castrated young rats [Shaar et al., 1975; Meites, 1988]. Also, when the positive feedback action of ovarian hormones was tested in old as compared with young ovariectomized rats, the rise in LH levels in the old rats was significantly lower than in young rats [Meites, 1988]. The inability of the pituitary to release amounts of LH sufficient to induce ovulation is due to failure of the hypothalamus to release adequate amounts of gonadotropin releasing hormone (GnRH). This is demonstrated by the failure of ovariectomy to induce release of GnRH in old rats in contrast to the large release of GnRH in young rats following ovariectomy [Wise and Ratner, 1980]. Some workers have also reported lower GnRH content in the hypothalamus of old as compared with young rats [Simpkins, 1983].

The decrease in GnRH release from the hypothalamus of old rats is due primarily to the decline in hypothalamic norepinephrine (NE), which normally promotes GnRH release in the rat [Simpkins et al., 1977; Wise, 1983]. NE and dopamine (DA) concentrations decline in many areas of the hypothalamus of old rats [Simpkins et al., 1977; Simpkins, 1983]. Daily administration of L-dopa, the precursor of DA and NE, delayed loss of estrous cycles in aging rats and reinitiated cycling in old persistent estrous rats [Quadri et al., 1973]. Other drugs or hormones that increase NE in the hypothalamus were also effective. Such treatments have also elevated testosterone levels in old rats [Meites, 1988]. The possible causes for the decrease in hypothalamic catecholamines (CAs) with age will be considered elsewhere.

The decline in reproductive functions with age in women and men appears to originate in faults that develop in the gonads. Beginning a few years prior to the menopause, there is a decline in secretion of ovarian hormones, leading to a rise in secretion of gonadotropic hormones that are less effective in stimulating the ovaries. This results in irregular menstrual cycles. In the postmenopausal period, the ovaries gradually lose their follicles and ova, become atrophic and fibrotic, and result in a further rise in secretion of gonadotrophic hormones [Gosden, 1985]. In men there is a gradual decline in testosterone secretion with age and a resultant rise in secretion of gonadotropins. Sperm production also appears to decrease. However, healthy elderly men, like healthy old male rats, may continue to reproduce even into advanced old age [Harman and Talbert, 1985].

Mammary and Pituitary Tumors

Aging female rats and mice develop numerous mammary and pituitary tumors [Russfield, 1966]. These tumors are believed to result mainly from the decrease in hypothalamic DA activity, which leads to elevation of prolactin (PRL) secretion [Meites et al., 1987; Meites, 1988]. It is well established that hypothalamic DA is the major inhibitor of PRL secretion [MacLeod and Lehmeyer, 1974]. In addition to the decline in hypothalamic DA with age, the prolonged and unopposed action of estrogen during the constant estrous state in female rats probably also contributes to development of mammary and pituitary tumors. Estrogen not only reduces hypothalamic DA activity but also acts directly on the pituitary to promote PRL secretion [Meites, 1988].

Administration of L-dopa, dopaminergic ergot drugs, or other agents that inhibit PRL secretion, produces regression of mammary and pituitary tumors in old rats [Meites et al., 1987; Meites, 1988]. C_3H female mice normally exhibit a 80–90% incidence of spontaneous mammary adenocarcinomas with age, but daily administration of bromocryptine, an ergot drug, prevented development of these tumors [Welsch and Nagasawa, 1977].

In humans, PRL does not appear to have a major role in the development of breast cancer, and there is no significant change with age in PRL secretion [Welsch and Nagasawa, 1977]. Although PRL-secreting microadenomas of the pituitary appear frequently in women and men, there is no evidence at present for a role of DA [Post et al., 1980].

GROWTH HORMONE SECRETION, PROTEIN SYNTHESIS, AND IMMUNE FUNCTION

Growth hormone (GH) is essential for protein synthesis throughout life and has important effects on fat, carbohydrate, mineral, and vitamin metabolism. It promotes growth and function of muscle, bone, heart, kidneys, liver, gastrointestinal tract, pancreas, thymus, spleen, etc. With age, there is a significant decline in GH secretion in rats of both sexes [Sonntag and Meites, 1988; Takahashi et al., 1987]. In female rats, there is about a 50% reduction in GH secretion by 11 months of age, which may account for the body growth stasis in mature female rats. Somatomedin secretin is normally stimulated by GH and is also decreased in old rats [Florini et al., 1985; Takahashi and Meites, 1987]. Somatomedin is believed to act directly on tissues to induce protein synthesis.

In aging men and women, GH and somatomedin secretion both decline, as in rats [Florini et al., 1985]. The surge of GH secretion that normally occurs during deep sleep completely disappears or is markedly reduced [Florini et al., 1985]. In animals and humans, there is a significant decline in protein synthesis in many tissues [Richardson, 1981], and there is some evidence that this is related to the decrease in GH and somatomedin secretion.

GH secretion is regulated by hypothalamic GH releasing hormone (GHRH), which stimulates GH sercretion, and somatostatin, which inhibits GH secretion. There is some evidence that GHRH release is decreased in old rats [Morimoto et al., 1988] and that somatostatin release in increased [Sonntag and Meites, 1988]. DA and particularly NE both promote GH release in rats and man [Martin, 1976], perhaps by elevating GHRH release and reducing somatostatin release. Both of these CAs are decreased in the hypothalamus of old rats [Simpkins et al., Meites, 1988], and this is believed to account for the decline in GH secretion in aging rats. There is also evidence for a decline of DA and NE in the hypothalamus of elderly human subjects [Hornykiewicz, 1986], but whether this is related to the reduction in GH secretion is unknown at present.

Proof that the decrease in the two CAs largely accounts for the decline in GH secretion in rats comes from the observation that twice-daily injections of L-dopa for 8 days to old male rats restored GH levels in the circulation to the same values as in young male rats [Sonntag and Meites, 1988]. The pituitary of old rats also shows a smaller response to injections of GHRH than the pituitary of young rats. This is believed to be due to the elevated release of somatostatin, which diminishes the response to GHRH [Sonntag and Meites, 1988].

Daily injections of GH to old male rats restored protein synthesis in diaphragm muscle to the same level as in young male rats, and L-dopa injections partially elevated protein synthesis in this tissue [Sonntag and Meites, 1988]. Injections of bovine GH for 10 days significantly increased the weights of the liver, kidneys, heart, and spleen in old male rats, and also elevated the weight of the thymus in old mice [Sonntag and Meites, 1988]. Kelly et al. [1986] demonstrated that GH administration not only restored thymus weight in old rats to the same weight as in young rats, but also reinitiated full *functional capacity* of the thymic cells. This is particularly significant since the thymus is considered to be the key component of the immune system [Fabris and Piantanelli, 1982].

EFFECTS OF REDUCING HYPOTHALAMIC CAs IN YOUNG ANIMALS

If the reduction in hypothalamic CAs with age is largely responsible for development of specific declines in body functions in old rats, can a decrease in hypothalamic CAs in young rats produce the same decreases in body functions? The answer appears to be in the positive. Chronic administration of neuroleptic drugs (reserpine, phenathiazines, haloperidol, etc.) decreases hypothalamic CAs in young rats and mice and results in reduced gonadotropin secretion and cessation of estrous cycles, increased PRL secretion, and development of mammary tumors [Welsch and Nagasawa, 1977] and probably lowers GH secretion. Chronic administration of estrogen to young rats or mice also decreases hypothalamic CAs and similarly results in reduced gonadotropin secretion and cessation of estrous cycles, and increased PRL secretion and development of both mammary and pituitary tumors [Welsch and Nagasawa, 1977]. Estrogen also acts directly on the pituitary to promote PRL secretion.

Since the reduction in hypothalamic CAs in young rats or mice can produce some of the same aging events observed in older animals, this finding suggests that the decline in hypothalamic function rather than age per se is primarily responsible for the onset of these developments. It also suggests that individual differences in hypothalamic function may partially account for differences in the onset of aging changes among individuals. This notion appears to receive support from a study we [Steger et al., 1980] reported in female white-footed mice (*Peromyscus leucopus*), which have an average life span of more than 48 months, in contrast to the much shorter-lived laboratory mouse (*Mus musculus*). The white-footed mice between 12 and 48 months of age continued to undergo estrous cycles and exhibited unchanged serum levels of LH, estradiol, and progesterone, pituitary content of LH and PRL, and hypothalamic content of GnRH, NE, and DA. This finding is in marked contrast to the declines with age observed in some of these parameters in the laboratory mouse.

POSSIBLE CAUSES FOR THE DECLINE IN HYPOTHALAMIC CAs WITH AGE

The role of the genome in the fall in hypothalamic CAs with age is unknown at present, but there is evidence that environmental and metabolic factors can adversely affect neurons in the arcuate nucleus (a major source of dopaminergic neurons), medial preoptic area, and ventromedial and lateral hypothalamus of old rats [Peng, 1983]. A significant loss of neurons has also been reported in the locus coeruleus of elderly human subjects [Hornykiewicz, 1986], a major source of NE in the hypothalamus. Several investigators have observed a decrease with age in hypothalamic tyrosine hydroxylase, the rate-limiting enzyme for CA synthesis, and an increase in monoamine oxidase, the major enzyme responsible for catabolism of CAs [Hornykiewicz, 1986; O'Neill et al., 1986].

Hormones, particularly estrogen, have been found to damage neurons in the hypothalamus. Brawer et al. [1978] reported that administration of a long-acting estrogen to young rats induced degeneration of dendrites and axons, increased gliosis, and enhanced accumulation of lipofuscin in neurons of the arcuate nucleus and medial basal hypothalamus. This was similar to the neuronal damage observed in the hypothalamus of old rats. Sarkar et al. [1982] reported that chronic treatment of young rats with estrogen specifically damaged dopaminergic neurons in the arcuate nucleus. The arcuate nucleus is the major site of dopaminergic neurons that inhibit PRL secretion, and also serves as an intermediate pathway between the preoptic area, which regulates ovulation in the rat, and the median eminence from which GnRH is released into the portal vessels. The ability of estrogen to damage hypothalamic neurons explains the results observed when young rats [Aschheim, 1976] or mice [Finch et al., 1984] were ovariectomized and many months later received fresh grafts of ovaries from young rats, enabling these rats to resume cycling that continued for many months beyond the period when intact control rats had ceased to exhibit estrous cycles. The absence of estrogens as a result of ovariectomy had a sparing effect on the hypothalamic neurons regulating estrous cycles.

Chronic elevation of PRL also was found to damage dopaminergic neurons in the arcuate nucleus [Sarkar et al., 1984]. Since estrogen stimulates PRL secretion, it is possible that estrogen damages hypothalamic neurons partly via enhanced PRL secretion. Glucocorticoid hormones also have been observed to damage neurons in the hippocampus [Landfield et al., 1978], an area that may influence hypothalamic functions.

"Free radicals" have been reported to damage cells in many body tissues. Davison (1987) has suggested that damage to neurons may be the result of the metabolism of CAs, with production of hydrogen peroxide, superoxide anions, and hydroxyl radicals. Toxins from the external or internal environment also may damage neurons. Toxins have been synthesized that specifically damage catecholaminergic and serotonergic neurons. It has been shown that there is an increase with age in cellular debris (lipofuscin) in cells, including neurons and cells of the endocrine glands. This may limit their functional capacity [Strehler, 1977]. Also, it is possible that "wear and tear" through chronic usage over much of the life span may damage hypothalamic neurons. This could explain why prolonged caloric restriction in young or mature rats or mice delays aging and pathology and extends the normal life span. Food restriction reduces CAs in the hypothalamus [Wurtman and Wurtman, 1983] and decreases release of hormones from the hypothalamus, pituitary, and target glands [Campbell et al., 1977], thereby preserving the morphological and functional integrity of these tissues.

CONCLUSIONS

The studies reviewed here demonstrate that the neuroendocrine system has a primary role in regulating the onset and progression of a number of important aging developments in the rat. These include loss of estrous cycles in females and decline in testosterone secretion in males, appearance of numerous mammary and pituitary tumors in females, fall in GH and somatomedin secretion and the associated decline in protein synthesis, and probably the decrease in thymic and immune competency. The reduction in hypothalamic CA activity appears to be particularly important, since CAs control release of several important hypophysiotropic hormones. Administration of L-dopa or other agents that elevate CAs in the hypothalamus of old rats can delay or reverse several specific aging developments in the rat. It is probable that other neurotransmitters in the hypothalamus are involved in the decline in hormone secretion with age. Serotonin, acetylcholine, brain opiates, and other neurotransmitters have been shown to influence hormone secretion, but little is yet known of their role in aging processes.

The role of the hypothalamus in the decline of body functions with age in human subjects remains to be elucidated. Although cessation of menstrual cycles in women and decrease in testosterone secretion in men have been attributed mainly to faults that develop in the gonads, this does not mean that hypothalamic function is unchanged in elderly individuals. Hypothalamic involvement appears most likely in relation to the decline in GH and somatomedin secretion, since CAs are decreased in the hypothalamus and other brain areas of elderly individuals [Hornykiewicz, 1986]. Both NE and DA are known to be potent stimulators of GH secretion in humans [Martin, 1976]. The decrease in GH and somatomedin secretion with age in the rat and human may be largely responsible for the decline in protein synthesis and immune competence, and perhaps contribute to the reduction in kidney, liver, and cardiac function.

REFERENCES

Aschheim P (1976): Aging in the hypothalamic-hypophyseal ovarian axis in the rat. In Everitt AV, Burgess JA (eds): "Hypothalamus, Pituitary and Aging." Springfield, IL: Charles C. Thomas, pp 376–418.

Brawer JR, Naftolin F, Martin J, Sonnenschein C (1978): Effects of a single injection of estradiol valerate on the hypothalamic arcuate nucleus and on reproductive function in the female rat. Endocrinology 103:501–512.

Campbell GA, Kurcz M, Marshall S, Meites J (1977): Effects of starvation in rats on serum levels of follicle stimulating hormone, luteinizing hormone, thyrotropin, growth hormone and prolactin: Response to LH-releasing hormone and thyrotropin-releasing hormone. Endocrinology 100:580–587.

Davison AN (1987): Functional morphology of neurons during normal and pathological aging. In Govoni S, Battaini F (eds): "Modification of Cell to Cell Signals during Normal and Pathological Aging." Berlin: Springer-Verlag, pp 1–16.

Fabris N, Piantanelli L (1982): Thymus-neuroendocrine interactions during development and aging. In Adelman RC, Roth GS (eds): "Endocrine and Neuroendocrine Mechanisms of Ageing." Boca Raton, FL: CRC Press, pp 167–184.

Finch CE, Felicio LS, Mobbs CV, Nelson JF (1984): Ovarian and steroidal influences on neuroendocrine aging processes in female rodents. Endocr Rev 5:467–497.

Florini J, Prinz P, Vitiello M, Hinz R (1985): Somatomedin-C levels in healthy young and old men: Relationship to peak and 24-hour integrated levels of GH. J Gerontol 40:2–7.

Florini J, Roberts S (1980): Effects of rat age on blood levels of somatomedin-like growth factors. J Gerontol 35:23–30.

Gosden RG (1985): "Biology of the Menopause." New York: Academic Press.

Harman SM, Talbert GB (1985): Reproductive aging. In Finch CE, Schneider EL (eds): "Handbook of the Biology of Aging." New York: Van Nostrand Reinhold, pp 457–510.

Hornykiewicz O (1986): Neurotransmitter changes in human brain during aging. In Govoni S, Battaini F (eds): "Modification of Cell to Cell Signals during Normal and Pathological Aging." Berlin: Springer-Verlag, pp 169–182.

Kelley KW, Brief S, Westley HG, Novakofski J, Bechtel PJ, Simon J, Walker EB (1986): GH_3 pituitary adenoma cells can reverse thymic aging in rats. Proc Natl Acad Sci 83:5663–5667.

Landfield PW, Waymire JL, Lynch G (1978): Hippocampal aging and adrenocorticoids: Quantitative correlations. Science 202:1098–1102.

MacLeod RM, Leymeyer JE (1974): Studies on the mechanism of the dopamine-mediated inhibition of prolactin secretion. Endocrinology 94:1077–1085.

Martin EJ (1976): Brain regulation of growth hormone secretion. In Martini L, Ganong WF (eds): "Frontiers in Neuroendocrinology," Vol. 4. New York: Raven Press, pp 129–168.

Meites J (1988): Neuroendocrine basis of aging in the rat. In Everitt AV, Walton JR (eds): "Regulation of neuroendocrine Aging." Basel: Karger, pp 37–50.
Meites J, Goya R, Takahashi S (1987): Why the neuroendocrine system is important in aging processes. Exp Gerontol 22:1–15.
Morimoto N, Kawakami F, Makino S, Chihara K, Hasegawa M, Ibata Y (1988): Age-related changes in growth hormone releasing factor and somatostatin in the rat hypothalamus. Neuroendocrinology 47:459–464.
O'Neill C, Marcusson J, Nordberg A, Winblad B (1986): The influence of aging on neurotransmitters in the human brain. In Govoni S, Battaini F (eds): "Modification of Cell to Cell Signals during Normal and Pathological Aging." Berlin: Springer-Verlag, pp 183–198.
Peng MT (1983): Changes in hormone uptake and receptors in the hypothalamus during aging. In Meites J (ed): "Neuroendocrinology of Aging." New York: Plenum Press, pp 61–72.
Post KD, Jackson JMD, Reichlin S (1980): "The Pituitary Adenoma." New York: Plenum Press.
Quadri SK, Kledzik GS, Meites J (1973): Reinitiation of estrous cycles in old constant estrous rats by central-acting drugs. Neuroendocrinology 11:248–255.
Richardson A (1981): Comprehensive review of the scientific literature on the effects of aging on protein synthesis. In Schinke RT (ed): "Biological Mechanisms of Aging." Washington DC: US Department of Health and Human Services, pp 339–358.
Russfield AB (1966): "Tumors of Endocrine Glands and Secondary Sex Organs." Washington, DC: US Government, Pub. No. 1332.
Sarkar DK, Gottschall PE, Meites J (1982): Damage to hypothalamic dopaminergic neurons is associated with development of prolactin-secreting pituitary tumors. Science 218:684–686.
Sarkar DK, Gottschall PE, Meites J (1984): Decline of tuberoinfundibular dopaminergic function resulting from chronic hyperprolactinemia in rats. Endocrinology 115:1269–1274.
Shaar CJ, Euker JS, Riegle GD, Meites J (1975): Effects of castration and gonadal steroids on serum luteinizing hormone and prolactin in old and young rats. J Endocrinol 66:45–51.
Simpkins JW (1983): Changes in hypothalamic hypophysiotropic hormones and neurotransmitters during aging. In Meites J (ed): "Neuroendocrinology of Aging." New York: Plenum Press, pp 41–60.
Simpkins JW, Mueller GP, Huang HH, Meites J (1977): Evidence for depressed catecholamine and enhanced serotonin metabolism in aging male rats: Possible relation to gonadotropin secretion. Endocrinology 100:1672–1678.
Sonntag WE, Meites J (1988): Decline in growth hormone secretion in aging animals and man. In Everitt AV, Walton JR (eds): "Regulation of Neuroendocrine Aging." Basel: Karger, pp 111–124.
Steger RW, Peluso JJ, Huang HH, Hudson CA, Leung FC, Meites J, Sacher J (1980): Effects of advancing age on the hypothalamic, pituitary, ovarian axis of the female white footed mouse (*Peromyscus leucopus*). Exp Aging Res 6:329–339.
Strehler BL (1977): "Time, Cells, and Aging." New York: Academic Press, pp 5–30.
Takahashi S, Meites J (1987): GH binding to liver in young and old female rats: Relation to somatomedin-C secretion. Proc Soc Exp Biol Med 186:229–233.
Takahashi S, Gottschall P, Quigley K, Goya R, Meites J (1987): Growth hormone secretory patterns in young, middle-aged and old female rats. Neuroendocrinology 46:137–142.
Welsch CW, Nagasawa H (1977): Prolactin and murine tumorigenesis: A review. Cancer Res 37:951–963.
Wise PM (1983): Aging of the female reproductive system. In Rothstein M (ed): "Review of Biological Research in Aging," Vol. 1. New York: Alan R. Liss, Inc., pp195–222.
Wise PM, Ratner A (1980): Effect of ovariectomy on plasma LH, FSH, estradiol, and progesterone and medial basal hypothalamic LHRH concentrations in old and young rats. Neuroendocrinology 30:15–19.
Wurtman RJ, Wurtman JJ (1983): "Nutrition and the Brain," Vol. 2. New York: Raven Press.

SECTION V
CELL BIOLOGY OF AGING

Recent Advances in Cellular Aging Research: Understanding the Limited Life Span of Normal Human Fibroblasts

Paul D. Phillips and Vincent J. Cristofalo

HISTORICAL BACKGROUND

Up until about 1961, it was generally thought that vertebrate cells, once established in culture, could be propagated indefinitely so long as contaminations and laboratory accidents could be avoided. Aging was not viewed as a cellular phenomena but rather as an organismic one. The classic study of Hayflick and Moorhead [1961] was the first serious challenge to this view. They demonstrated that cells from a variety of normal human tissues could be propagated in culture for various periods of time, but eventually the cultures stopped proliferating and the cultures eventually died. The cultures always followed a similar sequence of events. After outgrowth from the explant, a period of rapid proliferation followed during which frequent subcultivations were possible. In a predictable way, the proliferative capacity of the cultures then declined, intracellular debris accumulated, and nuclear abnormalities appeared. The cultures were slowly reduced in cell number as cells died but were not replaced by new mitotic events. This fundamental observation has subsequently been confirmed in numerous laboratories for normal cells from a variety of tissues and organisms. Hayflick also proposed that the limited replicative potential of normal human cells in culture was an expression of senescence at the cellular level [Hayflick, 1965].

Fibroblast cultures have been used to test several of the major theories of the mechanism of aging. Two of the most easily testable ones have been DNA repair and free radicals.

The Wistar Institute, Philadelphia, Pennsylvania 19104

INTRODUCTION

There have been literally thousands of studies aimed at gaining some understanding of the limited proliferative life span of normal human fibroblasts in culture ever since the phenomenon was first rigorously described [Hayflick and Moorhead, 1961], and later proposed as a model of cellular aging [Hayflick, 1965]. Most of this work has been descriptive, as it has been in all areas of biological aging research. In more recent years, however, more and more studies have been done aimed at experimentally manipulating cell life span to better understand the processes that control cell growth and replication. In addition, in the last few years an ever increasing number of laboratories have utilized the techniques of molecular biology to try and understand the mechanism(s) that regulates this process.

In this review we will focus primarily on the emerging body of knowledge dealing with understanding the mechanisms of cell aging. This is a particularly exciting time in this field. As we begin to get some rudimentary understanding of the phenomenon it becomes increasingly clear that normal cell aging is a fundamental process that plays a role in the pathobiology of diseases such as cancer.

DNA Repair

It has been attractive to speculate that the limited life span of normal human cells might result from accumulated damage to DNA that old cells in culture could not repair. Using human fibroblasts, several early studies established a correlation between increasing in vitro age and the ability to repair UV-induced damage [Goldstein, 1971; Painter et al., 1973; Hart and Setlow, 1976; Bowman et al., 1976; Milo and Hart, 1976]. Despite these reports, the relationships between accumulated DNA damage, DNA repair, and in vitro aging have not been definitely established [for a brief review see Phillips and Cristofalo, 1987]. In many cases, most of the characteristics of senescent cells develop before a decline in repair can be detected.

Mayer et al. [1986, 1987] have recently examined DNA single-strand breaks (spontaneous and induced) and DNA double-strand breaks (spontaneous and induced) by the filter elution technique. They report no change with in vitro age in the accumulation of spontaneous breaks of either type, nor in the kinetics or completeness of DNA strand rejoining after gamma-irradiation. This was true at 34°C, 37°C, and 39°C. They did observe an increase in excision repair with age at 34°C, but his had no effect on cell survival and in fact the proliferative life span of these cultures was significantly less (cumulative population doubling level [CPDL] 55 ± 6 at 34°C, compared with CPDL 67 ± 6 at 37°C).

In another study, Tamamoto and Fujiwara [1987] measured the activity of uracil-DNA glycosylase in cell-free extracts from young and old cells. They found that although the enzyme activity declined with age it also showed an altered regulation in that the G_0 and G_1 levels were elevated in senescent as compared with young cells. These are interesting observations, but taken together with other reports, they probably reflect some general regulatory changes rather than specific alterations in actual DNA repair. In our view, there is no clear evidence that generalized differences exist in DNA repair as a function of donor age or in vitro age.

Free Radicals

Several studies have recently appeared addressing the role that free radicals play in the process of cellular aging. Honda and Matsuo [1988] report that N-acetylcysteine (NAC), which increases intracellular glutathione (GSH) levels by acting as a cysteine carrier, increases in vitro life span, while L-buthionine-(R,S)-sulfoximine (BOS), an inhibitor of GSH synthesis, decreases life span. However, this is not the general finding [Mehlhorn and Cole, 1985]. Gutman et al. [1987] found no significant correlation between donor age and the activities of superoxide dismutase or aryl hydrocarbon hydroxylase or susceptibility to damage by oxygen metabolites as measured by cell viability or lactate dehydrogenase leakage.

Poot et al. [1987] used cumene hydroperoxide and 4-hydroxynonenal as peroxidative challenges upon the GSH system in skin fibroblasts. They found that the amount of NADPH needed to maintain cellular GSH levels increased with age, but the capacity to respond to the challenge was not affected. This group also found [Poot et al., 1988a,b] that lipid peroxidation was not the cause of loss of proliferative capacity during in vitro aging.

All of this is not to say that the action of free radicals has nothing to do with the aging of the whole organism. Schroder et al. [1987] reported that superoxide radicals induced the release of immature mRNA from its intranuclear binding sites, resulting in the appearance of immature mRNA in the cytoplasm. This could result in both qualitative and quantitative changes in protein synthesis and provide an interesting mechanism by which gene expression would be altered.

Growth Factors and Receptors

The growth factor requirements for cells such as WI-38 change very little throughout the culture's proliferative life span; this work has been recently reviewed [Phillips and Cristofalo, 1987]. Since that time several reports have appeared dealing with cell aging. The functional equivalency of some mitogens has recently been demonstrated [Phillips and Cristofalo, 1988]. Analysis of the proliferative response of WI-38 cells to nine mitogens, which in various specific combinations stimulate DNA synthesis in these cultures, delineated three classes of mitogens. Class I includes epidermal growth factor (EGF), fibroblast growth factor (FGF), platelet-derived growth factor (PDGF), and thrombin; class II includes insulin-like growth factor-I (IGF-I), multiplication stimulating activity, and insulin (INS); class III includes hydrocortisone or the synthetic analog dexamethasone (DEX). In cultures arrested at low density, members of each of the three classes act synergistically in stimulating DNA synthesis. Any class I mitogen in combination with any class II and either class III mitogen stimulates DNA synthesis to levels observed in 10% serum supplemented medium. The class I mitogens all act through separate receptor systems, while the class II mitogens all stimulate DNA synthesis by their varying ability to bind to IGF-I receptors. The class III mitogens act through the glucocorticoid receptor system.

Hosokawa et al. [1986] showed a synergistic effect between DEX and EGF and INS in stimulating young but not old cells into DNA synthesis. This was also the case when IGF-I was substituted for INS [Phillips et al., 1987; Conover et al., 1987]. Clemmons et al. [1986] showed that increasing donor age at low culture density was associated with an increase in IGF-I receptor number and a decrease in receptor affinity. At high culture densities, these differences were not found. An effect of culture density on IGF-I binding was also observed by Phillips et al. [1987] with WI-38 cells. Binding can be traced to two separate cell proteins. Binding to the α subunit of the IGF-I transmembrane receptor may increase slightly with age, while the 50% displacement remains unchanged. The remainder of the IGF-I-specific binding (5- to 30-fold more) is to a low molecular weight cell-associated binding protein whose 50% displacement is 10 times higher, but also remains unchanged with age. Specific binding to the lower affinity sites decreases slightly with age at equal cell densities. IGF-I binding to the α subunit of the transmembrane receptor is independent of cell density, while binding to the low molecular weight binding protein is inversely proportional to cell density and may vary by as much as ten-fold.

At this time, there is no evidence to indicate that some failure in the IGF-I receptor system is involved in the loss of proliferative responsiveness of normal cells in culture. However, this possibility has not really been examined in depth.

Several years ago Carlin et al. [1983] showed that EGF receptors isolated from senescent WI-38 cells, by NP-40 solubilization and specific antibody immunoprecipitation, lacked their normally inherent tyrosine-specific autophosphorylating activity. Subsequently, Chua et al. [1986] reported that the EGF receptor autokinase activity was intact when it was measured in plasma membranes prepared from cultures of both young and senescent cells. This apparent contradiction has been at least partially resolved by Brooks et al. [1987], who showed that in fact both observations are correct. The mechanism by which this occurs is unclear, but several possibilities exist that could account for the phosphorylation of receptors in intact plasma membranes from senescent cells, but no autophosphorylation following detergent solubilization of the membranes or whole senescent cells: 1) the catalytic site may be differentially altered in the senescent EGF receptor during solubilization; 2) an inhibitory factor peculiar to old cells may be activated on solubilization of the senescent receptor; 3) the kinetic parameters may be altered in the senescent receptor; 4) a membrane cofactor necessary for phosphorylation may be lost preferentially in senescent cells during solubilization; and 5) there may be proteolytic cleavage of a small portion of the receptor that alters autocatalytic activity (possibly by conformational changes) but that is too small to be resolved on SDS–PAGE gels. All of these possibilities have been examined to some extent, but the exact mechanism is still obscure.

Overall processing of the EGF-receptor complex was examined several years ago [Phillips et al., 1983] by comparing degradation of the ligand and was found to be essentially the same in young and old cells. However, Matrisian et al. [1987] have confirmed this early work in much more detail by following the acidification and breakdown of the complex as it is processed. They found no evidence for alterations that could be associated with the loss of proliferative responsiveness.

Proliferation Inhibitors

Following in the wake of numerous studies dealing with a variety of oncogenes and normal cellular proto-oncogenes has been a small but growing number of investigations aimed at identifying normal cell genes and gene products that serve as inhibitors of cell proliferation. They have been variously referred to as anti-oncogenes and gerontogenes. We prefer to refer to them simply as replication inhibitory genes (RIGs) or proteins (RIPs).

The existence of inhibitors of DNA synthesis in senescent cells has been indicated since the work of Norwood et al. [1974, 1975; for a more complete review see Phillips and Cristofalo, 1987]. Proliferation inhibitory activity has been identified in plasma membranes isolated from young quiescent cells and senescent cells. Stein and Atkins [1986] found that this activity was abolished in young quiescent cells 20 hours after the addition of fresh serum containing medium. The inhibitory activity appeared to be constitutively expressed in senescent cells, and not reduced by refeeding with fresh serum containing medium. Spiering et al. [1988] have shown that the immortal cell line SUSM-1 (derived from normal liver fibroblasts following exposure to a carcinogen) produces an inhibitory activity that cannot be distinguished from that produced by senescent cells; however, this transformed cell line does not respond to the inhibitory activity. This activity is active at levels approximately ten-fold lower than similar inhibitors prepared from senescent cells. They hypothesize that SUSM-1 cells may have escaped senescence through the loss of a receptor or cofactor for the inhibitor.

Lumpkin et al. [1986a], by a series of microinjection experiments, have now shown that an inhibitory activity can be traced to the poly A^+ RNA isolated from senescent human fibroblasts, and to a lesser extent from young quiescent cells. The abundance level of inhibitor mRNA from senescent cells was estimated at 0.8% and that of quiescent young cells at 0.005%. Although the accuracy of these abundance levels may be questioned, it is quite clear that the inhibitory activity is real.

Strategies for identifying and isolating RIGs are now being developed. Padmanabhan et al. [1987] transfected DNA from quiescent human fibroblasts into HeLa cells. They found that this was more growth inhibitory than *E. coli* DNA or HeLa cell DNA. Most surprisingly, they found that the inhibitory activity depended on the growth state of the normal fibroblasts from which it was taken. It was strongest in DNA prepared from serum-deprived quiescent cells. This implies a role for DNA modification(s) in regulating the activity of the inhibitory sequences that were detected.

Wynford et al. [1989] formed hybrids between normal human foreskin fibroblasts and the transformed Chinese hamster fibroblast line V79-8. They found that approximately 30% of the hybrids showed stable reversion to normal morphology and growth control as demonstrated by serum dependence and anchorage dependence. In one-third of these clones the proliferative life span was limited to the extent of the normal human parental cell line. This demonstrates the separation of morphological transformation from immortalization.

Transformation and Immortalization

The transformation of normal diploid fibroblasts, which possess a limited and predictable life span, will be a useful approach for studies of both the biology of aging and tumorigenesis. Traditionally, the only successful approach was to infect cultures with DNA tumor viruses such as SV-40. However, several recent reports demonstrate other means by which these cells can be transformed. Also emerging from this work is the apparent multistep character of the process.

Carcinogen treatment was used by Zimmerman and Little [1983] to treat normal human fibroblasts in an attempt to isolate cells with one or more transformed phenotypes. This produced cultures with extended but not indefinite life spans. These cultures routinely produced colonies that were anchorage-independent. One such colony produced a large tumor in a nude mouse. The cells recovered from the tumor were human, apparently diploid, and of fibroblast morphology, but the cells had a finite life span when returned to culture. Although this tumor appeared invasive, similar tumors regressed and disappeared in animals that were left untouched for an additional 2–4 weeks. In a recent review of such studies, McCormick and Maher [1988] argue that there are no true examples of malignant transformation of human fibroblasts by carcinogen treatment. At best, the acquisition of all of the phenotypic characteristics required for the complete malignant transformation of human diploid cells is a rare event when initiated by chemical carcinogens.

Another approach to transformation and tumorigenesis comes by transfection of DNA into normal cells. This has been done by introducing whole cell DNA as well as only specific sequences such as those coding for certain oncogenes or proto-oncogenes. The results of these experiments suggest at least two stages, or requirements, for the expression of complete transformation: 1) immortalization; and 2) the ability to form invasive tumors in vivo.

Doniger et al. [1983] transfected human fibroblasts from a patient with Bloom's syndrome with DNA from a mouse cell line carrying a single copy of the Harvey murine sarcoma virus. The parental cells have a finite life span and do not express any known markers of transformation, but they are characterized by a high rate of sister chromatid exchange. The transformed cells derived from the transfections had an extended life span, but they were not immortalized. They did, however, grow in nude mice and contain the Harvey murine sarcoma virus sequences as well as the human ras sequence. Sutherland et al. [1985] then reported the transfection of human foreskin fibroblast with DNA from several human tumor-derived cell lines that are inefficient in transforming standard NIH 3T3 cells. The isolated clones had extended but not indefinite life spans and exhibited anchorage-independent growth but did not grow in nude mice. The transforming sequences appeared to be new and could represent early stages in the process, which presumably has spontaneously occurred in 3T3 cells.

Using the technique of microinjection, Lumpkin et al. [1986b] were able to induce DNA synthesis in quiescent young normal human fibroblasts by injecting c-H-ras DNA. However, c-H-ras DNA alone or in combination with adenovirus E1A DNA did not cause terminally nondividing senescent cells to enter DNA synthesis. Tubo and Rheinwald [1987] transfected normal human fibroblasts with the cancer-derived c-H-ras mutation EJ-ras. The isolated clones expressed the p21EJ-ras protein, became independent of EGF for growth, and secreted an EGF-like activity into the medium in sufficient quantities to stimulate normal cell proliferation. However, these cells were normal with respect to all other growth requirements. Eleven clones were tested, and all exhibited a finite life span similar to the parent cell line. Three clones were tested and would not grow in nude mice. Even though p21EJ-ras can activate an important mitogenic pathway in normal human cells, it would not otherwise transform them. In a similar study, Stevens et al. [1988] transfected normal human fibroblasts with a human c-sis cDNA. Transfectants were isolated that exhibited densely packed colony morphology and were found to overexpress a sis mRNA as well as PDGF proteins. Beyond a certain threshold, a linear positive correlation was found between sis overexpression and acquired anchorage independence. These cells would form transient, regressing tumors when injected into nude mice. All of the isolated transfectants tested retained a finite life span.

These studies demonstrate the acquisition of a partially tranformed phenotype by the overexpression of an autocrine growth factor. However, it is clear that this represents only one step towards the development of a truly malignant cell type.

Some structure has been put to this problem by a well conceived series of studies by Pereira-Smith and Smith [1988]. When normal human fibroblasts were fused with a variety of immortal human cell lines, the hybrids obtained exhibited limited life spans. From this they concluded that the senescent phenotype is dominant, and that immortal cells must arise as a result of recessive changes in the proliferation regulatory mechanisms in normal cells. They then fused various immortal human cell lines with each other and were able to assign 21 cell lines to at least four complementation groups based on indefinite proliferation. Their results showed that the cell type, embryonal layer of origin, or type of tumor did not affect group assignment. There was also no apparent correlation between the expression of a particular activated oncogene and group assignment. All of the SV40 transformed cell lines (except one xeroderma pigmentosum derived line), however, assigned to the same group. This is consistent with the virus immortalizing various human cell types by the same mechanism. This approach may provide the framework for sorting out the various functional requirements for immortalization.

The Cell Cycle and Gene Expression

As described earlier, there are similarities in DNA synthesis inhibitory activities isolated from both quiescent young cells and senescent cells. This poses the question of whether senescent cells and quiescent cells share a common G_0/G_1 arrest point. Stein's data [Stein et al., 1985; Stein, 1985] are consistent with an aging mechanism that is manifested as a progressive decline in the ability of cells to recognize or respond to mitogens. This would result in cells that are physiologically mitogen-deprived at the end of their life span. Consequently, they would arrest in a senescent state by the same mechanism that causes young cells to arrest in the quiescent state when they are mitogen-deprived. This would appear to be a reasonable hypothesis, which gains supportive data from fusion studies of senescent cells with various types of transformed cells [Stein et al., 1982, 1985; Stein, 1985]. However, there are several complicating factors suggesting both similarities and differences between quiescent and senescent cells.

Evidence from several laboratories indicates that senescent cells are not refractory to growth factor signals and are not permanently arrested in G_0. Work from our own laboratory [Cristofalo, 1973] found that the level of thymidine kinase activity in senescent cultures of WI-38 cells was similar to that of young rapidly growing cultures. This finding appeared anomalous since thymidine kinase is known to be cell cycle-regulated and induced just before the S phase; it first raised the possibility that senescent cells may be arrested in late G_1. Olashaw et al. [1983] found that thymidine triphosphate (TTP)synthesis was not impaired in senescent cells. The addition of serum to density-arrested populations of young and old cells induced TTP synthesis to a similar extent after 12 hours in both aged cultures. However, a far greater percentage of young cells subsequently entered DNA synthesis as compared with old cells. TTP synthesis is also known to be a late G_1 event. Additional support for a late G_1 block came from examining nuclear fluorescence following staining with quinacrine dihydrochloride [Gorman and Cristofalo, 1986]. As serum-stimulated cells traverse G_1 and reach S, the staining pattern changes from bright and homogenous to dimmer and segregated. The fluorescence pattern of senescent cells is typical of late G_1. Work by Ohno et al. [1986] showed that G_0 arrested cells and senescent cells could be distinguished by the pattern of staining of microtubule-associated proteins via monoclonal antibodies. In addition to these studies, there is evidence that the serum-deprived or contact-inhibited G_0 states of young and old cells are similar. Wang and coworkers [Wang and Lin, 1986; Wang, 1987; Muggleton-Harris and Wang, 1989] have shown that the nonproliferation-associated protein statin is expressed in both young and old cells that enter such arrested states. Statin's function is not known; however, it shows high sequence similarity with elongation factor 1α (EF1α) [Moutsatsos et al., 1988]. However, it does not appear to be EF1α since EF1α has been shown to decrease in senescent cells [Cavallius et al., 1986; Rattan et al., 1988]. Thus, one picture that emerges is that of normal G_0 and G_1 phases in senescent cells with a proliferation block late in G_1 possible just prior to the S phase.

Experiments designed to examine the G_0/G_1 progression of mitogen-stimulated young and senescent human fibroblasts have been reported by several groups. Rittling et al. [1986] examined the expression of 11 cell cycle-dependent genes in young and senescent WI-38 cells. Following serum stimulation of quiescent cultures, the mRNA levels of every gene studied was expressed in senescent cells at similar times and at similar levels as they were in young cells. This included the mRNAs for c-myc, c-ras, and ornithine decarboxylase (ODC), which are expressed in mid-G_1, and thymidine kinase and histone H3, both known to be maximally expressed near the G_1/S boundary. These data strongly suggest that young and old cells may become quiescent by responding physiologically in the same way to mitogen deprivation. Furthermore, both young and old cells appear to respond to fresh mitogens by carrying out many of the same processes in the same time frame with the salient exception of their abilities to synthesize DNA. However, subsequent work suggests a more complicated picture.

Paulsson et al. [1986], using AG1523 human foreskin fibroblast cultures, also found comparable levels of c-myc expression in young and old cells as well as comparable levels of c-fos following PDGF stimulation. However, they did not detect the expression of the nuclear antigen K67 in stimulated senescent cells. This antigen is expressed in the S phase of young cells. Using IMR-90 cells, Delgado et al. [1986] reported reduced levels of c-ras mRNA expression. Chang and Chen [1988] confirmed the similar expression of ODC and c-myc mRNAs using young and old IMR-90 cells but did not see equal expression of thymidine kinase mRNA. Interestingly, although ODC mRNA levels were similar in young and old cells, there was fivefold decrease in ODC enzyme activity in senescent cultures. Kihara et al. [1986] did not see increased thymidine kinase enzyme activity in senescent TIG-1 embryonic lung fibroblasts, and Zambetti et al. [1987] did not detect an increase in histone H3 mRNA levels in stimulated senescent CF3 foreskin fibroblasts. Seshadri and Campisi [1988] were unable to detect significant levels of c-fos mRNA in stimulated senescent WI-38 cells, although c-myc expression was equivalent in both young and old cells. We have also found minimal c-fos expression in senescent WI-38 cells [Phillips and Cristofalo, unpublished].

At this point several observations do seem consistent and independent of the cell line studied. The levels of at least c-myc and ODC mRNA seem to be mitogen-regulated to the same extent and with the same time course in both young and senescent cells. There are varying reports on the regulation of other cell cycle genes in senescent cells, such as c-fos, c-ras, thymidine kinase, and histone H3. These apparent inconsistencies could be due to cell line-dependent differences, or differences in cell handling, or differences in the specificity of probes. What does appear clear is that senescent cells do respond to growth factors to some extent, at the very least by the expression of c-myc and ODC approximately 6 hours after stimulation. Thus, at least some of the growth factor-initiated signals are operating in senescent cells hours after the initial growth factor–receptor interactions.

ACKNOWLEDGMENTS

This work was supported by NIH grant AG-00378.

REFERENCES

Bowman PD, Meek RL, Daniel CW (1976): Decreased unscheduled DNA synthesis in nondividing aged WI-38 cells. Mech Ageing Dev 5:251–257.

Brooks KM, Phillips PD, Carlin CR, Knowles BB, Cristofalo VJ (1987): EGF-dependent phosphorylation of the EGF receptor in plasma membranes isolated from young and senescent WI-38 cells. J Cell Physiol 133:523–531.

Carlin CR, Phillips PD, Knowles BB, Cristofalo VJ (1983): Diminished in vitro tyrosine kinase activity of the EGF receptor of senescent human fibroblasts. Nature 306:617–620.

Cavallius J, Rattan SI, Clark BF (1986): Changes in activity and amount of active elongation factor 1 alpha in aging and immortal human fibroblast cultures. Exp Gerontol 21:149–157.
Chang ZF, Chen KY (1988): Regulation of ornithine decarboxylase and other cell cycle-dependent genes during senescence of IMR-90 human diploid fibroblasts. J Biol Chem 263:11431–11435.
Chua CC, Geiman DE, Ladda RL (1986): Receptor for epidermal growth factor retains normal structure and function in aging cells. Mech Ageing Dev 34:35–55.
Clemmons DR, Elgin RG, James PE (1986): Somatomedin-c binding to cultured human fibroblasts is dependent on donor age and culture density. J Clin Endocrinol Metab 63:996–1001.
Conover CA, Rosenfeld RG, Hintz RL (1987): Somatomedin-c/insulin like growth factor I binding and action in human fibroblasts aged in culture: Impaired synergism with dexamethasone. J Gerontol 42:308–314.
Cristofalo VJ (1973): Cellular senescence: Factors modulating cell proliferation in vitro. INSERUM 27:65–92.
Delgado D, Raymond L, Dean R (1986): C-ras expression decreases during in vitro senescence in human fibroblasts. Biochem Biophys Res Commun 137:917–921.
Doniger J, DiPaolo JA, Popescu NC (1983). Transformation of Bloom's syndrome fibroblasts by DNA transfection. Science 222:1144–1146.
Goldstein S (1971): The role of DNA repair of aging of cultured fibroblasts from xeroderma pigmentosum and normals. Proc Soc Exp Biol Med 137:730–734.
Gorman SD, Cristofalo VJ (1986): Analysis of the G_1 arrest position of senescent WI-38 cells by quinacrine dihydrochloride nuclear fluorescence. Evidence for a late G_1 arrest. Exp Cell Res 167:87–94.
Gutman RL, Cohen MR, McAmis W, Ramchand CN, Sailer V (1987): Free radical scavenging systems and the effect of peroxide damage in aged human skin fibroblasts. Exp Gerontol 22:373–378.
Hart RW, Setlow RB (1976): DNA repair in late-passage human cells. Mech Ageing Dev 5:67–77.
Hayflick L (1965): The limited in vitro lifetime of human diploid cell strains. Exp Cell Res 37:614–636.
Hayflick L, Moorhead PS (1961): The serial cultivation of human diploid cell strains. Exp Cell Res 25:585–621.
Honda S, Matsuo M (1988): Relationships between the cellular glutathione level and in vitro life span of human diploid fibroblasts. Exp Gerontol 23:81–86.
Hosokawa M, Phillips PD, Cristofalo VJ (1986): The effect of dexamethasone on epidermal growth factor binding and stimulation of proliferation in young and senescent WI-38 cells. J Exp Cell Res 164:408–414.
Kihara F, Tsuji Y, Miura M, Ishibashi S, Ide T (1986): Events blocked in prereplicative phase in senescent human diploid cells TIG-1, following serum stimulation. Mech Ageing Dev 37:103–117.
Lumpkin CK, McClung JK, Pereira-Smith OM, Smith JR (1986a): Existence of high abundance antiproliferative mRNAs in senescent human diploid fibroblasts. Science 232:393–395.
Lumpkin CK, Knepper JE, Butel JS, Smith JR, Pereira-Smith OM (1986b): Mitogenic effects of the proto-oncogene and oncogene forms of c-H-ras DNA in human diploid fibroblasts. Mol Cell Biol 6:2990–2993.
Matrisian LM, David D, Magun BE (1987): Internalization and processing of epidermal growth factor in aging human fibroblasts in culture. Exp Geront 22:81–90.
Mayer PJ, Bradley MO, Nichols WW (1986): No change in DNA damage or repair of single and double-strand breaks as human diploid fibroblasts age in vitro. Exp Cell Res 166:497–509.
Mayer PJ, Bradley MO, Nichols WW (1987): The effect of mild hypothermia (34°C) and mild hyperthermia (39°C) on DNA damage, repair and aging of human diploid fibroblasts. Mech Ageing Dev 39:203–222.

McCormick JJ, Maher VM (1988): Towards an understanding of the malignant transformation of diploid human fibroblasts. Mutat Res 199:273–291.

Mehlhorn RJ, Cole G (1985): The free radical theory of aging: A critical review. Adv Free Rad Biol Med 1:165–223.

Milo GW, Hart RW (1976): Age-related alterations in plasma membrane glycoprotein content and scheduled or unscheduled DNA synthesis. Arch Biochem Biophys 176:324–333.

Moutsatsos IK, Nakumura T, Wang E (1988): Abstracts. Fourth Int. Congress of Cell Biology, Montreal, p 78.

Muggleton-Harris AL, Wang E (1989): Statin expression associated with terminally differentiating and post-replicative lens epithelial cells. Exp Cell Res 189:152–159.

Norwood TH, Pendergrass WR, Sprague CA, Martin GM (1974): Dominance of the senescent phenotype in heterokaryons between replicative and post-replicative human fibroblast-like cells. Proc Natl Acad Sci USA 71:2231–2235.

Norwood TH, Pendergrass WR, Martin GM (1975): Retention of DNA synthesis in senescent human fibroblasts upon fusion with cells of unlimited growth potential. J Cell Biol 64:551–556.

Ohno T, Kako R, Sato C, Ohkawa A (1986): Distinction of G_0 cells from senescent cells in cultures of noncycling human fetal lung fibroblasts by anti-MAP-1 monoclonal antibody staining. Exp Cell Res 163:309–316.

Olashaw NE, Dress ED, Cristofalo VJ (1983): Thymidine triphosphate synthesis in senescent WI-38 cells. Exp Cell Res 149:547–554.

Padmanabhan R, Howard TH, Howard BH (1987): Specific growth inhibitory sequences in genomic DNA from quiescent human embryo fibroblasts. Mol Cell Biol 7:1894–1899.

Painter RB, Clarkson JM, Young BR (1973): Ultraviolet-induced repair replication in aging diploid human cells (WI-38). Radiat Res 56:560–564.

Paulsson Y, Bywater M, Pfeifer-Ohlsson S, Ohlsson R, Nilsson S, Heldin CH, Westermark B, Betsholtz C (1986): Growth factors induce early prereplicative changes in senescent human fibroblasts. EMBO J 5:157–162.

Pereira-Smith OM, Smith JR (1988): Genetic analysis of indefinite division in human cells: Identification of four complementation groups. Proc Natl Acad Sci USA 85:6042–6046.

Phillips PD, Cristofalo VJ (1987): A review of recent research on cellular aging in culture. In Rothstein M (ed): ''Review of Biological Research in Aging,'' Vol. 3 New York: Alan R. Liss, Inc., pp 385–415.

Phillips PD, Cristofalo VJ (1988): Classification system based on the functional equivalency of mitogens that regulate WI-38 cell proliferation. Exp Cell Res 175:396–403.

Phillips PD, Kuhnle E, Cristofalo VJ (1983): ^{125}I-EGF binding ability is stable throughout the replicative life span of WI-38 cells. J Cell Physiol 114:312–316.

Phlips PD, Pignolo RJ, Cristofalo VJ (1987): Insulin-like growth factor-I: Specific binding to high and low affinity sites and mitogenic action throughout the life span of WI-38 cells. J Cell Physiol 133:135–143.

Poot M, Verkerk A, Koster JF, Esterbauer H, Jongkind JF (1987): Influence of cumene hydroperoxide and 4-hydroxynonenal on the glutathione metabolism during in vitro aging of human skin fibroblasts. Eur J Biochem 162:287–291.

Poot M, Verkerk A, Koster JF, Esterbauer H, Jongkind JF (1988a): Reversible inhibition of DNA and protein synthesis by cumene hydroperoxide and 4-hydroxynonenal. Mech Ageing Dev 43:1–9.

Poot M, Esterbauer H, Rabinovitch PS, Hoehn H (1988b): Disturbance of cell proliferation by two model compounds of lipid peroxidation contradicts causative role in proliferative senescence. J Cell Physiol 137:421–429.

Rattan SI, Cavallius J, Clark BF (1988): Heat shock-related decline in activity and amounts of active elongation factor 1 alpha in aging and immortal human cells. Biochem Biophys Res Commun 152:169–176.

Rittling SR, Brooks KM, Cristofalo VJ, Baserga R (1986): Expression of cell cycle-dependent genes in young and senescent WI-38 fibroblasts. Proc Natl Acad Sci USA 83:3316–3320.

Schroder HC, Messer R, Bachmann M, Bernd A, Muller WE (1987): Superoxide radical-induced loss of nuclear restriction of immature mRNA: A possible cause for aging. Mech Ageing Dev 41:251–266.

Seshadri T, Campisi J (1988): Serum-inducible gene expression in senescent human fibroblasts: Possible selective deficiencies in transcriptional and translational control. Abstract presented at the 1988 meeting of the Tissue Culture Association. In Vitro 24 (Part II): 47A.

Spiering AL, Smith JR, Pereira-Smith OM (1988): A potent DNA synthesis inhibitor expressed by the immortal cell line SUSM-1. Exp Cell Res 179:159–167.

Stein GH (1985): SV40-transformed human fibroblasts: Evidence for cellular aging in pre-crisis cells. J Cell Physiol 125:36–44.

Stein GH, Atkins L (1986): Membrane-associated inhibitor of DNA synthesis in senescent human diploid fibroblasts: Characterization and comparison to quiescent cell inhibitor. Proc Natl Acad Sci USA 83:9030–9034.

Stein GH, Yanishevsky RM, Gordon L, Beeson M (1982): Carcinogen-transformed human cells are inhibited from entry into S phase by fusion to senescent cells but cells transformed by DNA tumor viruses overcome the inhibition. Proc Natl Acad Sci USA 79:5287–5291.

Stein GH, Namba M, Corsaro CM (1985): Relationship of finite proliferative life span, senescence, and quiescence in human cells. J Cell Physiol 122:343–349.

Stevens CW, Brondyk WH, Burgess JA, Manoharan TH, Hane BG, Fahl WE (1988): Partially transformed, anchorage-independent human diploid fibroblasts result from overexpression of the c-sis oncogene: Mitogenic activity of an apparent monomeric platelet-derived growth factor 2 species. Mol Cell Biol 8:2089–2096.

Sutherland BM, Bennett PV, Freeman AG, Moore SP, Strickland PT (1985): Transformation of human cells by DNAs that are ineffective in transformation of NIH 3T3 cells. Proc Natl Acad Sci USA 82:2399–2403.

Tubo RA, Rheinwald JG (1987): Normal human mesothelial cells and fibroblasts transfected with the EJ-ras oncogene become EGF-independent, but are not malignantly transformed. Oncogene Res 1:407–421.

Wang E (1987): Contact-inhibition-induced quiescent state is marked by intense nuclear expression of statin. J Cell Physiol 133:151–157.

Wang E, Lin SL (1986): Disappearance of statin, a protein marker for non-proliferating and senescent cells, following serum-stimulated cell cycle entry. Exp Cell Res 167:135–143.

Wynford TD, Bond JA, Paterson H (1989): Suppression of transformation and immortality in human/Chinese hamster fibroblast hybrids: A model for suppressor gene isolation. Int J Cancer 43:293–299.

Yamamoto Y, Fujiwara Y (1987): Culture-age effect on uracil-DNA glycosylase activity in normal human skin fibroblasts. J Gerontol 42:470–500.

Zambetti G, Dell Orco R, Stein G, Stein J (1987): Histone gene expression remains coupled to DNA synthesis during in vitro cellular senescence. Exp Cell Res 172:397–403.

Zimmerman RJ, Little JB (1983): Characteristics of human diploid fibroblasts transformed in vitro by chemical carcinogens. Cancer Res 43:2183–2189.

SUGGESTED BIBLIOGRAPHY

Adam G, Simm A, Braun F (1987): Level of ribosomal RNA required for stimulation from quiescence increases during cellular aging in vitro of mammalian fibroblasts. Exp Cell Res 169:345–356.

Angello JC, Pendergrass WR, Norwood TH, Prothero J (1987): Proliferative potential of human fibroblasts: An inverse dependence on cell size. J Cell Physiol 132:125–130.

Autilio-Gambetti L, Morandi A, Tabaton M, Schaetzle B, Kovacs D, Perry G, Greenberg B, Gambetti P (1988): The amyloid precursor protein of Alzheimer disease is expressed as a 130 kDa polypeptide in various cultured cell types. FEBS Lett 241:94–98.

Azzarone B, Chaponnier C, Krief P, Mareel M, Suarez H, Macieira-Coelho A (1988): Human fibroblasts from cancer patients: Life span and transformed phenotype in vitro and the role of mesenchyme in vivo. Mutat Res 199:313–325.

Balin AK, Baker AC, Leong IC, Blass JP (1988): Normal replicative life span of Alzheimer skin fibroblasts. Neurobiol Aging 9:195–198.

Busbee D, Sylvia V, Stec J, Cernosek Z, Norman J (1987): Lability of DNA polymerase alpha correlated with decreased DNA synthesis and increased age in human cells. J Natl Cancer Inst 79:1231–1239.

Chang PL, Gunby JL, Tomkins DJ, Mak I, Rosa NE, Mak S (1986): Transformation of human cultured fibroblasts with plasmids carrying dominant selection markers and immortalizing potential. Exp Cell Res 167:407–416.

Chen KY, Chang ZF (1986): Age dependency of the metabolic conversion of polyamines into amino acids in IMR-90 human embryonic lung diploid fibroblasts. J Cell Physiol 128:27–32.

Chen KY, Chang ZF (1987): A marked increase of fucosylation of glycoproteins in IMR-90 human diploid fibroblasts during senescence in vitro. Biochem Biophys Res Commun 142:767–774.

Ching G, Wang E (1988): Absence of three secreted proteins and presence of a 57-kDa protein related to irreversible arrest of cell growth. Proc Natl Acad Sci USA 85:151–155.

Conover CA, Rosenfeld RG, Hintz RL (1986): Hormonal control of the replication of human fetal fibroblasts: Role of somatomedin c/insulin-like growth factor I. J Cell Physiol 128:47–54.

Dell Orco RT, Whittle WL, Macieira-Coelho A (1986): Changes in the higher order organization of DNA during aging of human fibroblast-like cells. Mech Ageing Dev 35:199–208.

Donahue LM, Stein GH (1988): High efficiency DNA-mediated transformation of human diploid fibroblasts. Oncogene 3:221–224.

Edelstein SB, Breakefield XO (1986): Monoamine oxidases A and B are differentially regulated by glucocorticoids and aging in human skin fibroblasts. Cell Mol Neurobiol 6:121–150.

Fairweather AU, Fox M, Margison GP (1987): The in vitro life span of MRC-5 cells is shortened by 5-azacytidine-induced demethylation. Exp Cell Res 168:153–159.

Fazio MJ, Olsen DR, Kuivaniemi H, Chu ML, Davidson JM, Rosenbloom J, Uitto J (1988): Isolation and characterization of human elastin cDNAs, and age-associated variation in elastin gene expression in cultured skin fibroblasts. Lab Invest 58:270–277.

Fong TC, Makinodan T (1988): Decreased in vitro polymerase II activity of whole-cell extracts from late passage human diploid fibroblasts. Mech Ageing Dev 46:219–223.

Green L, Whittle W, Dell Orco RT, Stein G, Stein J (1986): Histone gene stability during cellular senescence. Mech Ageing Dev 36:211–215.

Gurley R, Dice JF (1988): Degradation of endocytosed proteins is unaltered in senescent human fibroblasts. Cell Biol Int Rep 12:885–894.

Hall MD, Flickinger KS, Cutolo M, Zardi L, Culp LA (1988): Adhesion of human dermal reticular fibroblasts on complementary fragments of fibronectin: Aging in vivo or in vitro. Exp Cell Res 179:115–136.

Holliday R (1986): Strong effects of 5-azacytidine on the in vitro life span of human diploid fibroblasts. Exp Cell Res 166:543–552.

Icard-Liepkalns C, Doly J, Macieira-Coelho A (1986): Gene reorganization during serial divisions of normal human cells. Biochem Biophys Res Commun 141:112–123.

Ishimi Y, Kojima M, Takeuchi F, Miyamoto T, Yamada M, Hanaoka F (1987): Changes in chromatin structure during aging of human skin fibroblasts. Exp Cell Res 169:458–467.

Kondo H, Yonezawa Y, Monaguchi TA (1988): Effects of serum collected from rats of different ages on in vitro cell proliferation. Mech Ageing Dev 42:159–172.

Kondo H, Nomaguchi TA, Sakurai Y, Yonezawa Y, Kaji K, Matsuo M, Okabe H (1988): Effects of serum from human subjects of various ages on proliferation of human lung and skin fibroblasts. Exp Cell Res 178:287–295.

Lee MG, Norbury CJ, Spurr NK, Nurse P (1988): Regulated expression and phosphorylation of a possible mammalian cell-cycle control protein. Nature 333:676–679.

Liu AY, Chang ZF, Chen KY (1986): Increased level of cAMP-dependent protein kinase in aging human lung fibroblasts. J Cell Physiol 128:149–154.

Mann PL, Swartz CM, Holmes DT (1988): Cell surface oligosaccharide modulation during differentiation: IV. Normal and transformed cell growth control. Mech Ageing Dev 44:17–33.

Monticone RE, Nick RJ, Eichhorn GL (1987): The effect of cellular age on zinc levels in untreated and zinc-treated human diploid fibroblasts. J Inorg Biochem 30:291–298.

Mukherjee AB, Weinstein ME (1986): Culture media variation as related to in vitro of aging of human fibroblasts: I. Effects on population doubling, nuclear volume and nuclear morphology. Mech Ageing Dev 37:55–67.

Oliver CN, Ahn BW, Moerman EJ, Goldstein S, Stadtman ER (1987). Age-related changes in oxidized proteins. J Biol Chem 262:5488–5491.

Pereira-Smith OM, Smith JR (1987): Functional simian virus 40 T antigen is expressed in hybrid cells having finite proliferative potential. Mol Cell Biol 7:1541–1544.

Peterson C, Goldman JE (1986): Alterations in calcium content and biochemical processes in cultured skin fibroblasts from aged and Alzheimer donors. Proc Natl Acad Sci USA 83:2758–2762.

Peterson C, Ratan RR, Shelanski ML, Goldman JE (1986): Cytosolic free calcium and cell spreading decrease in fibroblasts from aged and Alzheimer donors. Proc Natl Acad Sci USA 83:7999–8001.

Pigoelet E, Raes M, Houbion A, Remacle J (1988): Effect of procaine on cultivated human WI-38 fibroblasts. Exp Gerontol 23:87–96.

Poot M, Verkerk A, Jongkind JF (1986): Accumulation of a high molecular weight glycoprotein during in vitro ageing and contact inhibition of growth. Mech Ageing Dev 34:219–232.

Puvion-Dutilleul F, Sarasin A (1989): Chromatin and nucleolar changes in xeroderma pigmentosum cells resemble aging-related nuclear events. Mutat Res 219:57–70.

Raes M, Remacle J (1987): Alteration of the microtubule organization in aging WI-38 fibroblasts. A comparative study with embryonic hamster lung fibroblasts. Exp Gerontol 22:47–58.

Scarpace PJ (1987): Characterization of beta-adrenergic receptors throughout the replicative life span of IMR-90 cells. J Cell Physiol 130:163–167.

Sherwood SW, Rush D, Ellsworth JL, Schimke RT (1988): Defining cellular senescence in IMR-90 cells: A flow cytometric analysis. Proc Natl Acad Sci USA 85:9086–9090.

Shevitz J, Jenkins CS, Hatcher VB (1986): Fibronectin synthesis and degradation in human fibroblasts with aging. Mech Ageing Dev 35:221–232.

Sottile J, Mann DM, Diemer V, Millis AJ (1989): Regulation of collagenase and collagenase mRNA production in early and late passage human diploid fibroblasts. J Cell Physiol 138:281–290.

Stanulis-Praeger BM (1987): Cellular senescence revisited: A review. Mech Ageing Dev 38:1–48.

Yamamoto K, Yamamoto M, Ooka H (1988): Changes in negative surface charge of human diploid fibroblasts, TIG-1, during in vitro aging. Mech Ageing Dev 42:183–195.

In Vitro Studies of Aging Human Epidermis: 1975–1990

Barbara A. Gilchrest

INTRODUCTION

The vast majority of in vitro studies on aging human epidermis to date have utilized fibroblast-like cells, principally derived from fetal lung tissue (e.g., WI-38, IMR-90) or skin biopsies, as extensively reviewed elsewhere [Stanulis-Praeger, 1987; Phillips and Cristofalo, 1987]. While these studies have yielded much valuable information over the past 25 years, they are limited in two important regards. First, studies performed with a single cell type, the fibroblast, cannot establish the general applicability of their findings for other somatic cell types. Second, the fibroblast has a limited number of well-studied differentiated functions that are amenable to in vitro investigation, with the consequence that the literature focuses on proliferative senescence to the virtual exclusion of other possibly more relevant age-associated changes in cell function.

Among the available nonfibroblastic cell aging models are epidermal keratinocytes and melanocytes. They offer the advantage of easy accessibility through skin biopsy and have well-studied differentiated behaviors of clear clinical relevance. For the melanocyte these include pigment production and transfer; for the keratinocyte, production of immune modulators as well as production of high molecular weight keratins, formation and cross-linking of a cornified envelope, and other discrete steps in formation of the stratum corneum, the epidermal barrier layer. In addition, although experimentally adequate cell culture techniques have been available only 15 years for the keratinocyte [Rheinwald and Green, 1975] and less than 10 years for the melanocyte [Eisinger and Marko, 1982; Wilkins et al., 1982; Gilchrest et al., 1984], in contrast to nearly 70 years for the fibroblast, the methodology has evolved rapidly, so that a variety of completely defined systems already exists, at least for the keratinocyte [Boyce and Ham, 1983; Gordon et al., 1988a and b, 1989].

USDA Human Nutrition Research Center on Aging at Tufts University, Boston, Massachusetts 02111

This article summarizes work to date using human keratinocytes and melanocytes to study cellular aging processes.

THE KERATINOCYTE MODEL

Rheinwald and Green [1975], in their seminal paper on human keratinocyte cultivation, first presented data suggesting the appropriateness of this cell type for aging studies. They reported that keratinocyte lines derived from seven newborn donors underwent a calculated 25–52 population doublings (PD), while lines derived from three donors aged 3, 12, and 34 years underwent 20–27 PD. Plating efficiency (visible colonies per 100 cells innoculated) ranged up to 15.7% for the newborn keratinocytes and usually exceeded 2%, while the highest plating efficiency observed for the older donor keratinocytes was 0.7%. Total keratinocyte population expansion in culture averaged 5,737-fold for the newborn keratinocytes and exceeded 10,000-fold for two of these donors, while population expansion averaged 33-fold for the three older donors and was 2.6-fold for the only adult donor. These data strongly suggested that donor age influences keratinocyte life span and proliferative capacity in vitro and that quantitative effects are comparable to those observed for dermal fibroblasts.

The same culture system was subsequently employed to assess photoaging, that aging-like process that occurs in chronically sun-exposed skin. Keratinocytes derived from paired skin biopsies of the lateral (sun-exposed) and medial (nonexposed) aspects of the upper arm in adult donors were maintained under identical conditions and serially passaged until senescence [Gilchrest, 1979]. In all instances, total PD were fewer for the cultures derived from sun-exposed skin, and the discrepancy in culture life span between cultures from the two sites increased linearly with donor age and clinical evidence of sun-induced "premature aging." Similar reductions in the culture life span of dermal fibroblasts attributable to chronic sun exposure were measured in the same skin specimens [Gilchrest, 1980], further emphasizing the comparability of keratinocytes and fibroblasts for in vitro studies of aging. Absolute values for keratinocyte generations were comparable to those reported by Rheinwald and Green [1975], although no consistent relationship between donor age and culture life span was apparent in the nine subjects. Confirmatory keratinocyte data were obtained in a parallel study utilizing paired biopsies of preauricular (sun-exposed) and postauricular (non-exposed) facial skin from eight middle-aged donors [Gilchrest et al., 1983].

More recently, Barrandon and Green [1987] have reported that donor age strongly influences the average proliferative capacity of colony-forming keratinocytes. They defined three patterns of clonal growth for single keratinocytes seeded at second passage on lethally irradiated 3T3 cells and allowed to form a colony over 12 days that was subsequently disaggregated, repassaged, and allowed to proliferate for an additional 12 days. Holoclones gave rise to fewer than 5% abortive or terminal colonies (by definition, less than 5mm^2 in area and composed only of terminally differentiated cells); paraclones gave rise only to terminal colonies; and meroclones gave rise to more than 5% but fewer than 100% terminal colonies. Two newborn cultures contained 28% and 31% holoclones, 49% and 66% meroclones, and 23% and 3% paraclones, while in contrast, cultures from two donors aged 54 and 78 years contained 3% and 0% holoclones, 38% and 9% meroclones, and 59% and 91% paraclones.

Serum-free keratinocyte culture systems independent of 3T3 feeder layer support have also been employed for gerontologic studies. Early studies compared the proliferative abilities of newborn and adult keratinocytes [Gilchrest, 1983]. Despite identical plating efficiencies of approximately 35% and similar survival rates in a serum-free hormone-supplemented medium, newborn keratinocytes responded strikingly better to mitogenic stimulation, attaining confluence over 1 week, while identically treated adult-derived cells remain quite sparse. Separate experiments examined the individual impacts of epidermal growth factor (EGF) and a highly mitogenic bovine hypothalamic extract, subsequently resolved into inositol, a phospholipid precursor [Gordon et al., 1988b], and a protein growth factor of approximately 29 kd [Gordon et al., 1989; Finch et al., 1989]. In this system, newborn keratinocyte cultures responded to 10–50 ng/ml EGF with a 10–40% increase in cell yield after 1 week, while adult cultures were completely nonresponsive. The hypothalamic extract increased newborn keratinocyte yield nearly 20-fold, while adult keratinocyte yields increased on average less than 10-fold.

While these data are interesting in the context of cellular aging, interpretation is complicated by the fact that foreskin, the only tissue readily available from healthy newborns, was compared with skin biopsies obtained from the medial upper arm, introducing a site variable. Furthermore, it is widely agreed that aging studies should properly examine early versus late stages of maturation, not a developmental stage versus maturity. Hence, these studies were repeated using skin biopsies derived from the medial upper arm of healthy young adult versus old adult volunteers [Stanulis-Praeger and Gilchrest, 1986]. Again, an age effect was observed. Neither young nor old adult-derived keratinocytes responded to EGF, but optimal concentration of the hypothalamic extract in young adult cultures increased cell yield approximately ninefold relative to unsupplemented controls, while in old adult cultures, it increased cell yield only twofold, a statistically significant difference in proliferative response.

A crude extract of newborn calf thymus also elicited an age-associated decrease in keratinocyte proliferative response, particularly between early and late adulthood [Stanulis-Praeger et al., 1988]; the responsible mitogens were not identified.

Even more striking was the effect of donor age on the ability of keratinocytes to generate and to respond to autocrine growth factors, as measured by conditioned media responses [Stanulis-Praeger and Gilchrest, 1986]. Confluent cultures derived from newborn, young adult, or old adult donors were incubated for 48 hours with serum-free medium that was then aspirated, lyophilized, and reconstituted in 5 times the original volume of fresh medium to yield newborn, young adult, and old adult conditioned media (N-CM, Y-CM, and O-CM). Control-CM was prepared in an identical manner following incubation of medium for 48 hours in empty dishes. The media were then used to maintain paired newborn keratinocyte cultures for 1 week following inoculation at low density. Compared with control cultures in the baseline hormone-supplemented medium, N-CM increased cell yield approximately 30%, while Y-CM, O-CM and control-CM had no effect. Furthermore, when the same CM preparations were assayed on paired young adult and old adult keratinocyte cultures, none increased cell yield; indeed, adult-derived CM had a slight but statistically insignificant inhibitory effect. In order to compare the overall growth response of cultured keratinocytes as a function of donor age, an "epidermal proliferative index" was arbitrarily defined as the product of cell yields in three assays: response to defined mitogens, elaboration of autocrine growth factors into the media, and response to such factors. In this comparison there were major age-associated differences, particularly between the newborn and adult cohorts, but also between the young and old adults (Fig. 1).

Very similar results were obtained in an independent study [Sauder et al., 1988] performed to examine the possible autocrine stimulatory role of epidermal cell-derived thymocyte activating factor (ETAF), subsequently shown to be identical to interleukin-1. Partially purified column fractions prepared from keratinocyte-CM and shown to contain ETAF activity in the conventional thymocyte bioassay were assayed at varying dilutions on newborn and adult-derived keratinocyte cultures. At optimal dilution, ETAF increased average cell yield for newborn cultures by 47% versus 27% for adult cultures, compared with paired controls supplemented with heat-inactivated ETAF preparations. As well, bioassay of CM prepared from confluent newborn and adult keratinocyte cultures revealed an approximately sevenfold greater ETAF activity per cell for newborn compared with adult donors. These data again suggest a marked decline between the newborn and adult periods in the ability of keratinocytes to generate and to respond to autocrine growth factors. Because interleukin-1 plays a major role in immunomodulation, this age-associated decrease in ETAF production by keratinocytes may also be relevant to immune dysfunction in old skin.

Most recently, the keratinocyte model has been used to investigate whether aging is associated with increased sensitivity to growth inhibitors as well as decreased responsiveness to growth stimulators at the cellular level. It had previously been observed that interferon (IFN) profoundly and reversibly inhib-

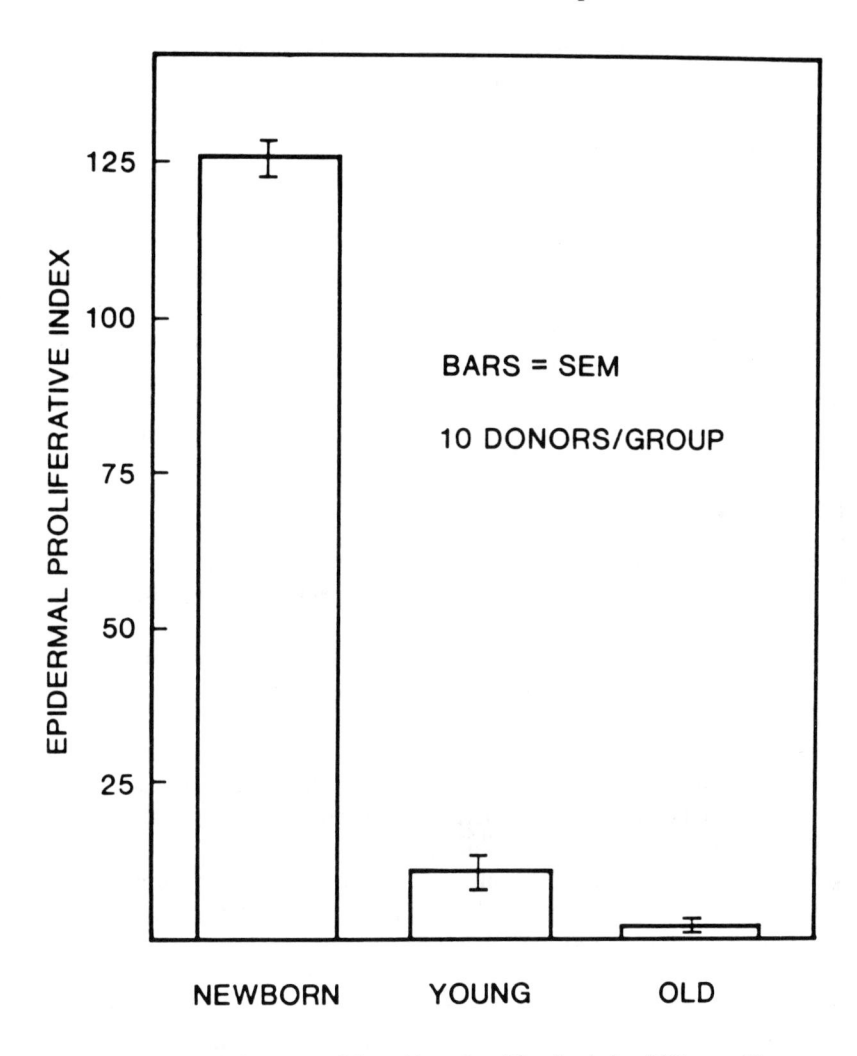

Fig. 1. Variation with donor age of the epidermal proliferation index (EPI), an arbitrary measure of growth capacity. Second passage keratinocyte cultures derived from ten donors in each age group were assayed for: 1) response to the defined mitogens epidermal growth factor and keratinocyte growth factor; 2) response to medium conditioned by confluent newborn foreskin keratinocytes; and 3) ability to condition medium for newborn donor keratinocyte cell assay lines. For each donor, the products of average cell yields in each of the three assays were multiplied together and divided by 10^{12} to yield the EPI. Young adult donors (22–27 years) and old adult donors (60–82 years) were in good to excellent general health; cultures were derived from a consistent site on the upper inner arm. Newborn cultures were derived from elective circumcision specimens of healthy 2–3-day-old infants. Differences between newborn and either young or old adult donors are highly significant ($P < 0.01$). (From Stanulis-Praeger and Gilchrest, 1986, with permission.)

its cultured keratinocyte growth [Yaar et al., 1985] and that an IFN-alpha-like protein can be identified by a variety of techniques in the proliferative basal layer of both stratified cultures in vitro and epidermal cross sections in vivo [Yaar et al., 1986], suggesting that IFN may function as a physiologic growth inhibitor for keratinocytes. Compared with 11 newborn-derived cultures that were unaffected by low-dose IFN, 5 adult-derived cultures maintained in serum-free medium and IFN alpha showed growth inhibition ($P < 0.01$) of 20% with as little as 25 units/ml and 50% with 200 units/ml [Yaar et al., 1989], approximately 5% of comparably inhibitory doses for newborn cultures [Yaar et al., 1985]. Growth inhibition in the adult cultures was accompanied by roughly proportional suppression of c-myc proto-oncogene expression, in contrast to the brisk c-myc induction observed in the non-growth-inhibited newborn cultures simultaneously exposed to fresh medium and low-dose IFN.

THE MELANOCYTE SYSTEM

Epidermal melanocytes are highly differentiated neural crest-derived cells. As a second well-defined epidermal cell type, melanocytes are also attractive for in vitro studies of skin aging. Moreover, from a gerontologic perspective, melanocytes are perhaps unique among postnatal cell populations in manifesting "phase III" senescence during the human life span.

In the Hayflick model of cell aging, cultured fibroblasts have an initial brief period of slow growth while acclimating to their in vitro environment (phase I), an extended period of rapid exponential growth during serial passages (phase II), and a final period of nonproliferation and cell death (phase III). The relevance of this model to in vivo aging has been questioned, however, on the basis that analogous phase III proliferative senescence is not observed in the epidermis, intestinal epithelium, bone marrow or other continually mitotic cell populations during old age. In contrast, greying of hair is due to total loss of the hair bulb melanocytes that normally divide and supply melanosomes to the hair during anagen (growth) cycles and then regress in number during the intervening telegen (resting) cycles. In humans, greying of the scalp hair often begins by the fourth decade, and, with large individual variation, hair is often completely white (depigmented) by the seventh decade. Thus, human hair bulb melanocytes may be considered to achieve their full proliferative life span, an unknown number of cell divisions during a crudely estimated 20 anagen cycles, in the first 50–70 years of life.

In the very first report of selective serial cultivation of human melanocytes, Eisenger and Marko [1982] noted substantially "better and faster" growth for cultures derived from newborn foreskin versus adult facial skin. A longer life span for newborn versus adult melanocyte cultures was also implied. Using a different culture system [Wilkins et al., 1982], this effect of donor age on melanocyte proliferative capacity has been quantified [Gilchrest et al., 1984]. Lines derived from three newborn foreskins or from nongenital skin of four healthy donors aged 17, 20, 50, and 60 years were serially passaged at known inoculation density in a hormone-supplemented medium containing 1% FBS until senescence, defined as failure to increase in number over a 4 week period. Newborn melanocytes underwent an estimated 3 PD in primary culture and 8.2 ± 0.6 PD in postprimary culture, approximately 11 cumulative PD, during six passages. In contrast, adult melanocytes underwent an estimated 2 PD in primary culture and only 2.4 ± 1.3 PD subsequently, a total of approximately 4.5 cumulative PD over two to three passages prior to senescence. Higher serum supplements and several minor modifications of the culture technique introduced since 1984 substantially increase culture life span for both newborn and adult melanocytes, but do not alter the marked effect of donor age [Gilchrest, unpublished observations].

In addition to the effect on proliferative capacity, aging has a subtle but distinct effect on melanocyte morphology consistent with a more advanced state of differentiation for adult-derived cells. In medium lacking phorbol esters, newborn melanocytes tend to be polygonal with large cytoplasmic area, while adult melanocytes under identical culture conditions are relatively dendritic, more closely mimicking their in vivo appearance [Gilchrest and Friedmann, 1986]. Addition of the phorbol ester 12-tetradecanoate 13-acetate, an agent known to enhance differentiation in many cell types, causes rapid transformation of both newborn and adult melanocytes into highly dendritic cells with markedly reduced perinuclear cytoplasmic area, interpretable as an exaggeration of the partially differentiated phenotype observed in adult cultures.

Further evidence of a donor age influence on the differentiated state of cultured melanocytes derives from experiments performed to examine the relationship of basement membrane (BM) to nevocellular nevus cells, melanocyte-like cells that make up the very common congenital and acquired nevi (moles). Cultured nevus cells drived from adults were found to elaborate fibronectin, laminin, BM-1 proteoglycan, and type IV collagen, suggesting they were capable of synthesizing the basement membrane noted to surround nevus cell nests in vivo [Yaar et al., 1988]. Newborn melanocytes initially selected as controls did not elaborate any of these BM components, but adult melanocytes examined as age-matched controls were identical to the nevus cells in their BM synthetic capacity. In the present context, these results again suggest that the threshold for expression of differentiated behaviors may vary at least between the newborn and adult periods and that donor age influences many cellular parameters in addition to growth potential.

CONCLUSIONS

The availability of refined culture systems for epidermal keratinocytes and melanocytes has markedly enhanced the relevance of in vitro models for studies of skin aging and indeed for gerontologic studies generally. Work with these systems to date has confirmed the important observations of early investigators utilizing fibroblast models that human diploid cell lines have a finite culture life span that is relatively constant among individuals and inversely related to donor age.

Keratinocyte cultures, now available to investigators for 15 years, have been used to demonstrate as well progressive loss of mitogen responsiveness from infancy through young adulthood to old age, compounded by a loss of cellular ability to produce and to respond to autocrine growth factors. Increased sensitivity to growth inhibitors has also been noted. Furthermore, effects of advanced *physiologic* age as well *chronological* age are demonstrable, as exemplified by studies of photoaged skin.

Melanocyte cultures, much more recently available, have similarly confirmed the donor age-associated loss of proliferative capacity. More importantly, melanocytes appear to offer an excellent opportunity to study effects of aging on cell-specific differentiated functions, an area suspected to be of great clinical importance but to date neglected for want of an adequate model system.

REFERENCES

Barrandon Y, Green H (1987): Three clonal types of keratinocyte with different capacities for multiplication. Proc Natl Acad Sci USA 84:2302–2306.

Boyce ST, Ham RG (1983): Calcium-regulated differentiation of normal human epidermal keratinocytes in chemically defined clonal culture and serum-free serial culture. J Invest Dermatol 81:33S–40S.

Eisenger M, Marko O (1982): Selective proliferation of normal human melanocytes in vitro in the presence of phorbol ester and cholera toxin. Proc Natl Acad Sci USA 79:2018–2022.

Finch PW, Rubin JS, Miki T, Ron D, Aaronson SA (1989): Human KGF is FGF-related with properties of a paracrine effector of epithelial cell growth. Science 245:752–755.

Gilchrest BA (1979): Relationship between actinic damage and chronologic aging in keratinocyte cultures in human skin. J Invest Dermatol 72:219–223.

Gilchrest BA (1980): Prior chronic sun exposure decreases the lifespan of human skin fibroblasts in vitro. J Gerontol 35:537–541.

Gilchrest BA (1983): In vitro assessment of keratinocyte aging. J Invest Dermatol 81:184s–189s.

Gilchrest BA, Friedmann PS (1986): A culture system for the study of human melanocyte physiology. In Jimbow K (ed): "Structure and Function of Melanin," Vol. 4. Sapporo, Japan: Fuji-shoin Co. Ltd., pp 1–13.

Gilchrest BA, Szabo G, Flynn E, Goldwyn RM (1983): Chronologic and actinically induced aging in human facial skin. J Invest Dermatol 80:081s–085s.

Gilchrest BA, Vrabel MA, Flynn E, Szabo G (1984): Selective cultivation of human melanocytes from newborn and adult epidermis. J Invest Dermatol 83:370–376.

Gordon PR, Gelman LK, Gilchrest BA (1988a): Demonstration of a choline requirement for optimal keratinocyte growth in a defined culture medium. J Nutr 118:1487–1494.

Gordon PR, Mawhinney TP, Gilchrest BA (1988b): Inositol is a required nutrient for keratinocyte growth. J Cell Physiol 135:416–424.

Gordon PR, Cohen SJ, Gilchrest BA (1989): Identification and isolation of a bovine brain derived growth factor for human keratinocytes. Clin Res 37:634A.

Phillips PD, Cristofalo VJ (1987): A review of recent research on cellular aging in culture. In Rothstein M (ed): "Review of Biological Research in Aging," Vol. 3. New York: Alan R. Liss, Inc. pp 385–415.

Rheinwald JR, Green H (1975): Serial cultivation of strains of human epidermal keratinocytes. Cell 6:331–344.

Sauder DN, Stanulis-Praeger BM, Gilchrest BA (1988): Autocrine growth stimulation of human keratinocytes by epidermal cell-derived thymocyte activating factor: Implications for skin aging. Arch Dermatol Res 280:71–76.

Stanulis-Praeger BM (1987): Cellular senescence revisisted: A review. Mech Ageing Dev 38:1–48.

Stanulis-Praeger BM, Gilchrest BA (1986): Growth factor responsiveness declines during adulthood for human skin-derived cells. Mech Aging Dev 35:185–198.

Stanulis-Praeger BM, Yaar M, Redziniak G, Meybeck A, Gilchrest BA (1988): An extract of bovine thymus stimulates human keratinocyte growth in vitro. J Invest Dermatol 90:749–754.

Wilkins LM, Gilchrest BA, Maciag T, Szabo G, Connell L (1982): Growth of enriched human melanocyte cultures. In Sato GH, Pardee AB, Sirbasker DA (eds): "Growth of Cells in Hormonally Defined Media," Vol. 9. Cold Spring Harbor Conferences on Cell Proliferation. Cold Spring Harbor: Cold Spring Harbor Press, pp 929–936.

Yaar M, Karassik RL, Schnipper LE, Gilchrest BA (1985): Effects of alpha and beta interferons on cultured human keratinocytes. J Invest Dermatol 85:70–74.

Yaar M, Palleroni AV, Gilchrest BA (1986): Normal human epidermis contains an interferon-like protein. J Cell Biol 103:1349–1354.

Yaar M, Woodley DT, Gilchrest BA (1988): Human nevocellular nevus cells are surrounded by basement membrane components: Immunohistologic studies of human nevus cells and melanocytes in vivo and in vitro. Lab Invest 58:157–162.

Yaar M, O'Neil CA, Genn DA, Abadi JS, Gilchrest BA (1989): Age associated increased senstivity to interferon is directly correlated with down regulation of the c-myc proto-oncogene. Clin Res 37:641A.

SECTION VI
BIOCHEMISTRY OF AGING

The Molecular Pathology of Senescence: A Comparative Analysis

Robert J. Shmookler Reis

INTRODUCTION

Recent studies of aging, in several experimental systems, have led to rather diverse conclusions regarding the molecular changes accompanying, and perhaps underlying, the aging phenotype. A review of these results indicates that proximal causes of age-dependent deterioration or loss of functionality appear to be distinct for several model systems, implying that different aberrant biochemical processes have evolved as the "weakest link" limiting longevity or intermeiotic interval in different biological systems.

GENETIC FACTORS GOVERNING SENESCENCE

All species of metazoa undergo senescence, with only a few exceptions claimed [Comfort, 1979; Beverton, 1986], suggesting that its underlying causes may also be universal among multicellular organisms; e.g., it may be an inevitable concomitant of the differentiated state, or of the cessation of cell division. The *rate* of aging, however, is clearly subject to genetic control, since life spans may differ by a factor of at least two between closely related species [Sacher, 1978; Samis, 1978; Kirkwood, 1985]. Substantial heritability of longevity is seen in crosses between subspecies or selected populations differing in life span [Johnson and Wood, 1982; Rose and Service, 1985; Lints and Bourgois, 1985]. Moreover, the inheritance of life span suggests multiple determinants [Johnson and Wood, 1982; Rose and Service, 1985; Johnson, 1988; Luckinbill et al., 1988; Kirkwood, 1989]; the paucity of mutants with markedly increased life span also argues for polygenic control [Friedman and Johnson, 1988a,b; Johnson, 1988].

Departments of Medicine and Biochemistry and Molecular Biology, University of Arkansas for Medical Sciences, and GRECC, McClellan V.A. Medical Center/151, Little Rock, Arkansas 72205

Two invertebrate systems that have well-studied genetics and that have been increasingly exploited for molecular studies of development also offer considerable potential for the genetic study of whole-animal aging: *Drosophila melanogaster* and *Caenorhabditis elegans*. In each case, the life span is conveniently brief, with nematodes (*C. elegans*) having the advantage of a shorter generation time, greater ease of inbreeding via self-fertilization of hermaphrodites, and absence of inbreeding depression. In both species, selection for increased longevity has been successful, but with outcomes that depend somewhat on the strategy employed. Attempts to breed long-lived animals, or to identify long-lived products of mutagenesis, have for the most part resulted in selection of delayed fecundity and/or reduced total reproductive yield [Lints and Bourgois, 1985; Rose and Service, 1985; Clare and Luckinbill, 1985; Johnson, 1987; Friedman and Johnson, 1988a,b]. Metabolic rate, however, appears not to be altered [Arking et al., 1988]. On the other hand, the *dunce* mutant of *Drosophila* produces a 50% reduction in longevity among sexually hyperactive females, relative to unmated controls [Bellen and Kiger, 1987], implying that it is the reproductive burden itself, rather than a separate pleiotropic effect of the mutation, that is responsible for curtailing longevity. Life span-altering mutations thus appear to exhibit "antagonistic pleiotropy" with fecundity, mating, and/or developmental processes [see Rose and Service, 1985]—which may indicate a trade-off between the energetic/entropic demands of reproduction and those of longevity.

In contrast, when recombinant lines of *C. elegans* (resulting from an interstrain cross) were inbred to homozygosity and compared, they displayed much wider variation in mean and maximal life span than the parental strains, with no correlation to developmental or reproductive timing [Johnson, 1987]. Thus the reassortment of pre-existing alleles at multiple loci can generate broad variation in longevity without affecting fertility, whereas those few single gene loci at which new mutations can extend life span (thus far limited to a single site in *C. elegans*, age-1, and an undetermined number of loci in *D. melanogaster* selection experiments) tend to be coupled to reproduction.

EXPRESSION OF SENESCENCE IN SOMATIC TISSUES

Weismann [1891] noted that the germ line is exempt from senescence, which is characteristic of somatic tissues, as indeed it must be to permit species survival. This may reflect tissue-specific expression of senescence, analogous to tissue-specific gene expression, or it may arise from some property unique to germinal cells. Most notably, reproductive cells are distinguished by cycles of meiosis, although they also undergo a greater number of mitotic divisions. The observation that meiosis also abrogates the "senescence" of unicellular organisms, sometimes termed clonal attrition or vegetative death (see below), suggests that the number of mitotic divisions, or the intermeiotic interval, may be the common limiting parameter in all such age-dependent phenomena.

It appears from serial transplantation experiments that a variety of cell types can survive considerably longer than the maximal life span of their donors/hosts [Harrison, 1985], although this has been an area of controversy [reviewed in Daniel, 1977]. It is thus possible that under ideal conditions a number of tissues would prove to be exempt from *autonomous* aging. It has been conjectured that senescence may be governed by a "pacemaker" organ, and while this notion seems at odds with its concurrent expression in multiple tissues, it cannot be strictly excluded because of the complex interactions between organs, hormonal and nutritional influences, and immune function. Indeed, these same "humoral" factors, each varying with age and each affecting the others, have the potential to complicate greatly the interpretation of any study of aging in whole organisms.

ENVIRONMENTAL CONTRIBUTIONS TO SENESCENCE

To what degree do extrinsic or environmental factors contribute to senescence? "Wear-and-tear" hypotheses, in several manifestations, argue that aging might result from a lifetime of accumulated insults to the organism. At least some types of accrued environmental damage can be excluded, however, since maximal life span is neither extended for animals raised in pathogen-free environs, nor curtailed for animals surviving recurrent trauma or disease [Curtis and Crowley, 1963; Shmookler Reis, 1976]. Nevertheless, dietary restriction studies clearly illustrate the importance of interactions between organism and environment in determining phenotype. Calorie-restricted rodents have longer life spans and appear younger for a given age by a wide variety of criteria than ad libitum fed animals [Weindruch et al., 1986; Leveille et al., 1984; Miller and Harrison, 1985; Masoro, 1988; Chatterjee et al., 1989]. In view of the very limited physical activity of these laboratory rodents, relative to the same animals in the wild, it is possible that dietary restriction might serve to offset partially the detrimental effects of sedentary existence. In any case, extrapolation to human aging is not warranted at this time; despite enormous variation in diet across the globe, differences between human populations in maximal life span are far less pronounced than those effected in rodents by caloric restriction [e.g., see Enstrom and Pauling, 1982].

Oxygen free radicals have been implicated, on the basis of correlative evidence, as causal agents in aging of both insects [Sohal and Allen, 1986] and mammals [Cutler, 1984; Adelman et al., 1988], and in replicative senescence of mammalian cells in culture [Balin et al., 1985; Hornsby and Harris, 1987]; damage due to free radicals might thus constitute a mechanism for environmental induction of senescence. The current lack of direct evidence, however, may be attributable to compensation by the multiple homeostatic pathways for control of cellular redox potential. For example, autofluorescent "lipofuscin" granules are known to result from lipid peroxidation, and to accumulate in a variety of animals as a function of metabolic activity (and thus correlate with age). Exogenous administration of antioxidants has been reported to decrease cellular lipofuscin levels, yet *not* to extend life span—results taken to imply that the cellular redox state could only be perturbed transiently, due to the efficiency of homeostatic control [Sohal and Allen, 1986; Cutler, 1984]. However, they could also be interpreted to argue against a causal role of oxidative damage in senescence.

Modification of DNA by nonenzymatic glycosylation [Lee, 1987] or by reaction with adductive mutagens [Park and Ames, 1988] and metabolic by-products [Randerath et al., 1989] all increase with in vivo age—as would be expected from cumulative exposure to the reactive species—and thus may also contribute to the senescent phenotype.

INTRINSIC LIMITATIONS ON CELL REPLICATIVE POTENTIAL

Diploid cells in serial culture undergo a progressive decline in average fitness of the cell population, as defined by the likelihood of cellular reproduction (mitosis) per interval of time, *roughly* analogous to that which occurs in aging organisms. The decrease in cell replicative probability is strictly passage-dependent [Dell'Orco et al., 1973; Goldstein and Singal, 1974; Harley and Goldstein, 1978], in contrast to the time-dependent decline in fitness (capacity for survival or reproduction) which defines senescence of intact organisms. Thus the "in vitro aging" of cells is actually observed as a collective property of the culture as a whole: increasing time between mean population doublings (MPD), or decreasing average frequency of S-phase or mitotic cells, culminating in eventual failure of cell replication to exceed cell detachment. The situation is complicated by the wide distribution of generation levels in primary cultures [Harley and Goldstein, 1978] and by the stochastic variation in proliferative potential for individual cells, even within a clone [Smith and Whitney, 1980]. Replicative senescence is accompanied by a marked increase in the ratio of cytoplasmic volume to DNA content [Sherwood et al., 1988], noted previously as in increase in protein and RNA contents per unit DNA [Schneider and Shorr, 175], which may be interpreted as a marker of terminal differentiation [Bayreuther et al., 1989].

Although the nature of the limit on cell replication potential is not known, there are several strong candidates for proximal causes. The in vitro life spans of cultured fibroblasts can be substantially extended by addition of glucocorticoids [Hosokawa et al., 1986], while those of keratinocyte cultures are increased by epidermal growth factor [Rheinwald and Green, 1977; Rockwell et al., 1987], and cultured adrenocortical cells require fibroblast factor for maximal life span [Hornsby and Harris, 1987]. The differences between cell types in factors that extend replicative life span presumably reflect tissue-specific differences in the corresponding receptors (concentration or occupancy) and may not indicate postreceptor variations in mitogenic control. Indeed, a number of growth factor receptors activate a common tyrosine phosphorylation/ PIP_2 signaling pathway [e.g., see Meisenhelder et al., 1989].

Reduction in pO_2 also increases the replicative life span of fibroblasts maintained at low density, but not when they are allowed to grow repeatedly to confluence [Balin et al., 1985], which is the usual procedure for passage in vitro. This suggests that the hyperphysiologic oxygen pressure used for routine cell culture may introduce oxidative damage, which could limit proliferative potential of at least some cultured cell types under conditions of maximal metabolic activity [Balin et al., 1985; Hornsby and Harris, 1987].

The ability to grow cells in isolation from such physiological variables as the nutritional, immunological, and hormonal state of the donor, and to follow clonal lineages by repeated subcloning, has allowed studies of cell mutation, clonal heterogeneity, gene amplification, and DNA rearrangement [Srivastava et al., Riabowol et al., 1985a,b; Stringer et al., 1985; Butner and Lo, 1986] that are believed to be relevant to mitotic cells in vivo but would have been difficult or impossible to achieve there. An inverse correlation between replicative life span of cells in culture and age of the cell donors has been demonstrated in several laboratories for both human [Hayflick, 1965; Martin et al., 1970; Schneider and Mitsui, 1976] and Syrian hamster [Bruce et al., 1986] fibroblasts, suggesting a possible mechanistic connection between replicative senescence in vitro, and organismic senescence in vivo. Other cellular growth parameters also reflect the donor's age: cloning and plating efficiency, growth rate, and growth-factor responsiveness are on average somewhat diminished in fibroblasts from aged donors [Stanulis-Praeger and Gilchrest, 1986; Papez-Zelinsky et al., 1987; E. Moerman, unpublished data].

In several instances, biochemical effects of donors age have also been demonstrated for cells in long-term culture, indicating that specific molecular markers of in vivo senescence can persevere in vitro. For example, the number of molecules of mitochondrial DNA per unit cytoplasm was observed to be twice as great in old donors' fibroblasts as in those from young donors, irrespective of in vitro passage level [Shmookler Reis and Goldstein, 1983], whereas the frequency of induced sister chromatid exchanges declines with both donor age and in vitro replicative age [Schneider, 1985].

One danger inherent in studies of "replicative senescence" in cultured cells is that cell-cycling changes can themselves produce profound alterations in cell biochemistry, which might easily be mistaken for characteristics of in vitro senescence but which actually depend on the cell-cycle phase or recent cycling status of cells, rather than on their remaining division *potential*.

Apart from the utility of studying isolated cells in culture, and the challenge to understand the nature of their replicative limit, the question remains whether that limit is germane to the senescence of organisms. The experimental data do not imply that the intrinsic limit to cell replication is responsible for aging of animals—only that it may contribute to loss of function in some continuously replicative tissues. For example, there is evidence of a senescence-dependent decrement in division potential of committed precursors, but apparently not of initial stem cells, in the lymphocytic/hematopoietic lineage [Lipschitz and Udupa, 1984; Miller and Harrison, 1985; Jones and Ennist, 1985; Miller, 1989]. These data suggest that cellular determinants of replicative capacity are adversely affected by animal age, but may not themselves be directly limiting to maximal life span. It appears likely, nevertheless, that an age-dependent diminution in reserve capacity—including cellular replicative potential—would impair the ability of older individuals to survive environmental challenges [e.g., see Lipschitz and Udupa 1986; Juckett, 1987].

CHANGES IN DNA METHYLATION WITH SENESCENCE

In following clonal lineages of human diploid fibroblasts during extended serial culture, we observed random drift of the methylation of DNA cytosines, with methylation losses predominating over gains; this drift was apparent in both total DNA and in specific gene loci [Shmookler Reis and Goldstein, 1982a,b; Goldstein and Shmookler Reis, 1985]. The observation of a progressive methylation loss from DNA cytosines [Shmookler Reis and Goldstein, 1982a] was confirmed and extended to mortal—but not immortal—cultured cells of two other mammalian species, with the rate of loss "inversely correlated" (for this small sample) to species life span [Wilson and Jones, 1983]. Similar losses of 5-methylcytosine from total DNA and from specific gene regions have been found in tissues of aging rodents [Singhal et al., 1987; Mays-Hoopes et al., 1983, 1986; Wilson et al., 1987; Slagbloom and Vijg, 1989], and in lymphocytes from old vs. young human donors [Drinkwater et al., 1989]. However, such hypomethylation is both tissue- and gene-specific, with no significant methylation change (after maturity) apparent in most cases; in a few instances, age-dependent hypermethylation predominates [Slagbloom and Vijg, 1989; Ono et al., 1989].

The above experiments point to a considerable degree of infidelity in the somatic inheritance of DNA methylation. Since the majority of methylatable cytosines (in the dinucleotide C-G) are methylated in mature tissues of vertebrates, even random drift would lead to an excess of methylation losses over gains. Nevertheless, the common observation of hypomethylation in late-passage cells has engendered speculation that loss of DNA methylation may be responsible for the limited proliferative potential of diploid cell cultures. This question has been addressed by treatment of cells with methylase inhibitors, to achieve a reduction in total DNA methylation. Results from two laboratories indicated that treatments with 5-azacytidine and/or 5-azadeoxycytidine, which are inhibitory to DNA methylation when incorporated into DNA, do indeed reduce the life span of MRC-5 fibroblasts in culture [Holliday, 1986; Fairweather et al., 1987]. It is not possible to draw firm conclusions from such experiments regarding the role of DNA methylation in cellular aging, since many perturbations of cell culture conditions have the potential to reduce replicative life span. Nonetheless, the results may be taken to suggest that either insufficient methylase activity, or the reduction in extent of DNA methylation itself, could be limiting to the replicative potential of cultured cells.

Direct assay of methylase activity in nuclear extracts, however, indicates that methylase activity is not lacking in late-passage human fibroblasts, and indeed is several-fold more abundant than at early passage, within those cells that continue to cycle [Shmookler Reis et al., 1989]. While it remains possible that methylase access to specific DNA loci may be impeded in older cells, resulting in loss of methylations, the findings of increased DNA methylation with passage in some diploid cell strains (Shmookler Reis and Goldstein 1982a), and of extensive hypomethylation in tumor cells and immortal cell lines (reviewed in Riggs and Jones 1983), argue against such alterations being directly limiting to cell replication capacity.

ALTERATIONS IN SPECIFIC GENE EXPRESSION

"Leaky" or ectopic expression of derepressed genes has been demonstrated both in cultured cells [Rosen et al., 1980; Kator et al., 1985; Milsted et al., 1982] and in animal tissues [Dean et al., 1985; Ono et al., 1985; Cattanach, 1974; Wareham et al., 1987]. In many of these studies, such leaky expression was shown to increase as a function of cell passage in culture [Kator et al., 1985] or age in vivo [Dean et al., 1985; Ono et al., 1985; Cattanach, 1974; Wareham et al., 1987; see also, however, Migeon et al., 1988]. Although there is no direct evidence at present demonstrating that hypomethylation is the mechanism of genetic derepression, in two instances [Cattanach, 1974; Wareham et al., 1987] age-dependent clonal expression was shown for genes within the inactive X chromosome, the DNA of which appears to be extensively—although not entirely—methylated [Gartler and Riggs, 1983]. It should be noted that hypomethylation per se is not normally sufficient for derepression of genes, except in cells poised for their subsequent developmentally regulated expression [reviewed in Shmookler Reis et al., 1989]; presumably other levels of control are interposed. It is certainly plausible that low-level ectopic expression of tissue-specific genes might be deleterious to cell or tissue function, but this has not yet been demonstrated.

A number of specific gene products have been shown to be differentially expressed in late-passage cells or senescent animal tissues, and these have sometimes been interpreted as evidence that such aging is governed by a genetic "program" and thus may best be considered as differentiation [Bayreuther et al., 1989]. Perhaps the most prominent example of quiescence- and senescence-induced protein expression, in cultured cells, is "statin" [Wang, 1985]—a nuclear protein with partial sequence identity to human elongation factor EF1-alpha [Wang, 1988]. The sequences diverge in the untranslated regions, and in the nucleotides encoding the C-terminal segment, which may account for the *nuclear* localization of statin [Wang, 1988]. Curiously, EF1-alpha mRNA and protein have been reported to decrease during in vivo aging [e.g., Webster and Webster, 1984], but this could be unrelated to statin expression, or perhaps competitive with it.

Other well-documented examples of protein alterations specific to noncycling cells include an increase of histone H_1^0 [Medvedev et al., 1978; Houde et al., 1989] and diminished activity and fidelity of DNA polymerases alpha and beta [Fry et al., 1981, 1984; Krauss and Linn, 1986]. It is interesting that both the histone and DNA-polymerase changes—which, like statin synthesis, are associated in vitro with cessation of cell division rather than with replicative aging per se—have also been observed to accompany senescence in vivo for tissues that are predominantly nondividing [Medvedev et al., 1978; Webster and Webster, 1984; Fry et al., 1981, 1984; Krauss and Linn, 1986]. They may thus reflect the phenotype of cells following prolonged absence of cycling, however induced.

Attempts to define the extent of senescence-dependent alteration of gene expression have generally led to the conclusion that rather few of the highly expressed and/or stable gene products detectable on two-dimensional protein gels are altered significantly during aging of cells in culture [Harley et al., 1980; Lincoln et al., 1984; Bayreuther et al., 1989], or of *Drosophila* in vivo [Fleming et al., 1986, 1988], although in the latter case there was an increase in the number and amount of heat shock-inducible polypeptides—interpreted to arise as a consequence of increased concentrations of modified or partially denatured proteins with aging [reviewed in Adelman and Roth, 1983]. During mammalian senescence, a wide range of age-dependent changes have been observed in the expression of specific genes, many remaining static, some increasing, and some declining [reviewed in Slagbloom and Vijg, 1989]. Many of these, however, may be secondary to systemic humoral changes accompanying aging—e.g., alterations in hormonal, nutrient, and growth factor levels.

DNA LOSS DURING AGING

Chromosome fragments or whole chromosomes are lost from nuclei of somatic cells during senescence in vivo [Staino-Coico et al., 1983; Marlhens et al., 1986; Martin et al., 1986], and to a lesser extent during replicative aging in vitro [Shmookler Reis and Goldstein, 1980; Sherwood et al., 1988]. Moreover, repetitive DNA sequences (in particular centromeric tandem repeats) become under-represented in many or all chromosomes of terminally differentiated cells in *Drosophila, Ascaris,* and *Cucumis* [reviewed in Harley et al., 1982], and of human diploid fibroblasts undergoing serial passage [Shmookler Reis and Goldstein, 1980]. The mechanism appears to be unequal recombination [Harley et al., 1982], associated primarily with developmental cell divisions, whereas further change during senescence—either in vivo or in vitro—could be explicable, at least in part, by cell selection rather than continued recombination.

Among the tandemly repeated sequence families, telomeric repeats may be unique in possessing a known, essential function: stabilizing linear chromosomes and allowing complete replication of their termini. Telomeres from human somatic cells are several kilobase pairs shorter than those from germ cells [Cooke and Smith, 1986; Allshire et al., 1988], suggesting a somatic cell deficiency in telomere terminal transferase [Cooke and Smith, 1986], which could conceivably limit the replicative potential of mitotic cells. In this regard it is intriguing that mutations at a locus required for telomere elongation in *S. cerevisiae* lead to a senescence-like phenotype, characterized by progressive truncation of telomeric DNA, chromosomal instability, and cell death [Lundblad and Szostack, 1989].

Hemizygosity or complete loss of chromosomal regions may be responsible for the "senescent" phenotype (clonal attrition) observed in a variety of ciliated protozoa. Following sexual fertilization, by either conjugation or autogamy, mitotic growth of clones deteriorates over a vegetative life span ranging from less than 200 to over 1,000 divisions in various species of *Paramecium* or *Euplotes* [Smith-Sonneborn, 1985]. Clonal senescence is also observed in some *Tetrahymena* strains, or can be induced by inbreeding [Nanney, 1959]; it also arises spontaneously in clonal *Amoeba* cultures, especially when grown on a maintenance diet [Muggleton and Danielli, 1958]. In ciliates, of which *Paramecium* has been studied most extensively with respect to clonal attrition [reviewed in Smith-Sonneborn, 1985, 1897], a transcriptionally active macronucleus derives from the micronucleus—which serves as the "germ line," generating gametes by meiosis. Sexual conjugation consists of the exchange of gametic (micro-)nuclei, which fuse to form diploid zygotes, after which a new macronucleus is formed by chromosome fragmentation and reduplication.

Senescence of a clonal culture is characterized by decreased rates of fission, and of DNA and RNA synthesis; conjugation results in "rejuvenation," but even this may be compromised as parental clones grow older [Smith-Sonneborn, 1985]. The process of rejuvenation clearly cannot be due solely to the elimination of homozygosity, since fusion of identical gametes (in autogamy) is also effective. Macronuclear function is apparently impaired in aged cultures [Aufderheide, 1987], perhaps reflecting the accrual of DNA damage, or the unequal partition of chromosome fragments at mitosis [Aufderheide and Schneller, 1985]. Because most of these fragments lack centromeres, only their presence in multiple copies prevents a much earlier loss of essential genes by unbalanced chromosome partition, and clearly the probability of such a lethal event must approach one (certainty) as the number of generations increases. Thus, whether or not asymmetric chromosome distribution actually causes clonal attrition, it is a sufficient cause for limited macronuclear replication potential. Sequestration of the relatively inactive micronucleus may protect it from DNA damage or other factors limiting mitotic propagation, but also would ensure a safe repository of complete genomes.

DNA BREAKAGE, REARRANGEMENT, AND RECOMBINATION

Karyotypic abnormalities have been reported to increase with passage level of diploid human cells in culture [Saksela and Moorhead, 1963; Kadanka et al., 1973; Smith and Hayflick, 1974; Thompson and Holliday, 1975; Sherwood et al., 1988]; these comprise predominantly aneuploidies, but also include polyploidy, chromosome breakage, translocation, and the appearance of chromatid gaps, dicentrics, and ring chromosomes. Similarly, senescence-associated increases have been reported for a wide variety of chromosomal aberrations in mouse kidney cells [Martin et al., 1986] and in human lymphocytes [Staino-Coico et al., 1983; Marlhens et al., 1986]. The donor age dependence of chromosome breakage in lymphocytes is also significant following treatment with clastogens [Esposito et al., 1989].

The most remarkable increases in aneuploidy of late-passage fibroblasts, predominantly hypodiploidies, ranged from 26% at early passage to 43% at late passage [Thompson and Holliday, 1975]. It should be noted that much lower levels of hypodiploidy, increasing from 2.5% at early passage to 4.5% at late phase III, were obtained [Shmookler Reis and Goldstein, 1980] by use of a novel in situ lysis method to avoid fragmentation of the lobulated nuclei seen in late-passage cells. This study confirmed, however, the more modest passage-dependent increase in tetraploidy and endotetraploidy noted previously [Thompson and Holliday, 1975].

Apart from chromosome loss and breakage, a number of other DNA structural changes have been observed to accompany aging in vivo and replicative senescence of cells in culture: 1) DNA adducts and alkali-labile sites accumulate during aging of postmitotic cells [Hanawalt, 1987; Lee, 1987; Mullaart et al., 1988; Park and Ames, 1988]; 2) repetitive DNA sequences associated with centromeric heterochromatin are diminished in late-passage fibroblasts [Shmookler Reis and Goldstein, 1980; Macieira-Coelho and Puvion-Dutilleul, 1989], possibly by underreplication or unequal recombination; 3) moderate amplification (up to fourfold) of the c-H-*ras* proto-oncogene occurs during the in vitro life span of human fibroblasts, in seven strains examined [Srivastava et al., 1985]; and 4) extrachromosomal circular DNAs, which contain an overrepresentation (relative to total genomic DNA) of interspersed-repetitive DNA elements, accumulate in human diploid fibroblasts at late passage [Riabowol et al., 1985a,b; Kunisada et al., 1985] and in rodent tissues during in vivo aging [Kunisada et al., 1985; Flores et al., 1988].

The above studies suggest an increase in chromosomal instability with both in vitro and in vivo aging of mammals. In contrast, mutagen-induced sister chromatid exchange (SCE) is reported to decrease in frequency, both in late-passage cultured cells and in senescent rodents relative to mature young animals—whereas spontaneous SCE remained constant in each case [reviewed in Schneider, 1985]. Spontaneous SCE frequencies may reflect primarily rate-limiting initiating events such as oxidative damage to DNA [Emerit, 1984; Schafer, 1984]; however, induced SCE would appear to monitor a specific outcome of a repair pathway that is not fully understood. The consensus of current experimental evidence is that neither "senescent" in vitro replicative arrest nor in vivo aging entails any clear deficit in the repair of UV-induced dimers [Cleaver, 1984] or of single- or double-strand breaks in DNA [Little, 1976; Mayer et al., 1986], although the difference in repair efficacy between coding and noncoding regions of DNA may necessitate a reevaluation of this issue [Hanawalt, 1987].

The evidence summarized above indicates a general senescence-dependent increase in the frequency of phenomena related to DNA and chromosomal instability, all presumably involving DNA strand rejoining. We have recently used a plasmid-borne substrate for homologous recombination, in a transient assay, to evaluate the efficiency with which normal diploid human cells carry out this process. No significant alteration in recombination rate was seen as a function of cell-culture passage level, although there was a modest suggestion of a donor age effect (unpublished data). A marked and consistent increase was observed, however, in recombination mediated by five immortally transformed cell lines, compared with any of three mortal cell strains [Finn et al., 1989]—suggesting that enhanced DNA recombination might underlie both the chromosome instability and genetic heterogeneity associated with immortally transformed cells. Indeed, an elevated level of recombination might favor the co-occurrence of the multiple, rare mutational events required for cellular immortalization.

STRUCTURAL ALTERATIONS TO MITOCHONDRIAL DNA

Vegetative propagation of the filamentous fungi is limited by eventual "senescent" deterioration and cessation of mycelial growth, at rates that differ between fungal isolates (races) and that are maternally inherited. In *Podospora anserina*, senescence is associated with a variety of *petite*-like mitochondrial (mt) DNA deletions wherein circular DNAs, comprising multimers of small (2.5–10 kbp [kilo-base-pair]) fragments of the mtDNA genome, gradually displace the normal complement of 94 kbp mtDNA molecules [Belcour and Begel, 1978; Kuck et al., 1981; Cummings et al., 1985; Jamet-Vierny, 1988]. A causal connection to senescence is implied by studies of a temperature-sensitive mutant for the senescent phenotype, in which several species of small circular mtDNA—including the previously characterized alpha-senDNA—were detected consistently and uniquely at the nonpermissive (senescing) temperature [Turker et al., 1987]. Similar deletion mutants of mitochondrial DNA have been described in *S. cerevisiae (petite* mutants*), Aspergillus,* and *Neurospora* [reviewed in Cummings et al., 1985].

Senescence in some strains of *Neurospora crassa* and *N. intermedia* is characterized by proliferation of a mitochondrial plasmid DNA, which is unrelated to the mtDNA genome, and its insertion into mtDNA [Stohl et al., 1982; Bertrand et al., 1986]. If there is a common mechanism for these "senescent" fungal growth limitations, it may be that the amplifying elements can all function as transposable elements [Cummings et al., 1985; Bertrand et al., 1986; Osiewacz et al., 1989] that result in disruption of the mitochondrial genome and/or inactivation of mitochondrial genes. In each case, as for protozoan clonal attenuation, the senescent phenotype is seen only during prolonged vegetative growth and is reversed by sexual propogation.

The question of whether similar events occur in mammalian cells has been addressed for human diploid fibroblasts, but no subgenomic-size fragments of mtDNA could be detected at any passage level, except for a small, 0.6 kbp fragment interpreted as D-loop primer, which *decreased* during prolonged mitotic arrest but was otherwise unaffected by passage level [Shmookler Reis and Goldstein, 1983]. Nonetheless, nondeletional inactivation of mitochondrial genes could arise relatively frequently in somatic cells during normal aging, especially in view of the extreme vulnerability of mitochondrial DNA to mutagenic and oxidative damage [Shay and Werbin, 1987; Richter et al., 1988; Linnane et al., 1989]. The observed increase in mtDNA copy number, per cytoplasmic protein or volume, for human diploid fibroblasts from older donors [Shmookler Reis and Goldstein, 1983] might then be understood as a compensatory adjustment to inactive mtDNA genomes.

DISCUSSION: MECHANISMS OF SENESCENCE

The universality of aging, as most broadly construed, may be inextricably tied to the evolution of sex. The stochastic reassortment of gene clusters, achieved through meiotic reduction and zygotic fusion, clearly serves a valuable function in evolution: the rapid creation of new combinations of mutationally altered genes. Since natural selection acts on phenotypes governed by the complex interactions of many genes, new combinations of alleles constitute phenotypic variants, which may prove advantageous in unstable selective environments [Maynard Smith, 1972]. In principle, vegetative growth (or adult life of the parent) might otherwise continue without limit, in competition with progeny created via gametogenesis. This is precisely what aging precludes, in effect determining the maximal *intermeiotic span*, and ensuring that a successful parental phenotype does not impede survival of, or the accumulation of genetic diversity among, its offspring—diversity that would be essential to species survival should the environment alter [see Maynard Smith, 1972]. There might also be advantages to senescence that would act at the level of the individual (with respect to propagation of its genes), rather than the group: faithful mitotic expansion and/or maintenance of germ line cells could be impaired progressively over time, as may the ability to "reset the switches" for germ line totipotency.

Quite apart from the evolutionary advantages of senescence, a number of biochemical/genetic factors may impose limits on mitotic or postmitotic life spans of cells and organisms: DNA damage, oxidative or adductive; chromosome breakage, rearrangement or loss; random alterations to the DNA methylation patterns; or increasing homozygosity introduced through recombinational repair of loci. The data summarized briefly above indicate that any of these processes can occur in some genera, effectively limiting mitotic propagation of cells and organismal longevity. Indeed, in the two systems subjected to most intensive study, mammalian senescence in vivo and human fibroblast aging in vitro, most or all of the above deteriorative processes have been documented. This suggests that whatever *can* go wrong eventually will go wrong, provided that selection does not eliminate the least fit cells. The diversity of molecular pathologies observed in different cells or organisms, however, as they approach senescence or a mitotic limit, implies that these are unlikely candidates for an *initiating* event in the evolution of senescence, but probably arose independently as secondary factors reinforcing an extant limit on life span.

It is clear that once a life span limit were imposed, whether due to molecular disabilities as discussed above, or more simply arising from the cumulative effect of mortality by accident or predation [Medawar, 1952], the accrual of new mutations should act to consolidate that boundary, since there would be no advantage to possessing other features of physiology or cellular metabolism that favor survival well beyond the expected life span [Medawar, 1952; Maynard Smith, 1966; see also Kirkwood, 1985]. There thus would be no point in expending metabolic energy to reduce oxidative damage, or to maintain DNA information, extracellular matrix, or cell replicative capacity, beyond the "warranty period." This may indeed provide the simplest explanation for correlations reported between species life span and a wide variety of disparate parameters: relative antioxidant levels [Cutler, 1984; Sohal and Allen, 1986], DNA repair capacity [Hart and Setlow, 1974; Tice and Setlow, 1985], rate of DNA methylation loss [Wilson and Jones, 1983], collagen cross-linking [reviewed in Kligman et al., 1985], or the replicative potential of cultured cells [Rohme, 1981]. While different biochemical processes have evolved as the "weakest links" limiting longevity or maximal intermeiotic interval in different systems, these *proximal* causes of senescent deterioration of function may reflect cellular economies made in the face of other, prior factors limiting life span—thereby accounting for the diversity of apparent causes of aging.

Presumably all molecular and organ systems, in a given organism, would have evolved to postpone deterioration to roughly the same age [Medawar, 1952; Maynard Smith, 1966], but external changes (e.g., due to the laboratory environment) may have advanced individual factors to life span-limiting status. Thus, for example, the sedentary lifestyle of laboratory rodents may be responsible for the life extension effects of caloric restriction and exercise; long-term selection exclusively for maximal reproduction, in a protected environment, may have resulted in sexual hyperactivity of *Drosophila* strains and the observed coselection of extended life span with reduced or delayed fecundity, while growth factor depletion and/or oxidative stress may have become limiting factors in cells grown under nonphysiological culture conditions—continuous replication in near-atmospheric O_2 tension.

Common features of aging may exist, but at present we can only say that multicellular organisms have a limited capacity to replicate or survive between meioses. What *first* goes wrong appears to vary widely, but for each source of organism failure, meiotic cells are somehow exempted. Each experimental system for the study of aging may illuminate a distinctive way in which complex molecular systems can become vulnerable to time-dependent or replication-dependent failure. It will be an intriguing challenge to learn how, for each situation, such failure is obviated or corrected in the germ line.

ACKNOWLEDGMENTS

This paper is dedicated to the memory of Dr. James M. Gilliam, who contributed through many helpful discussions. The work was supported by grants from the National Institutes of Health (AG-03787) and the Veterans Administration.

REFERENCES

Adelman RC, Roth GS (eds) (1983): "Altered Proteins and Aging." Boca Raton, FL: CRC Press.

Adelman R, Saul RL, Ames BN (1988): Oxidative damage to DNA: Relation to species metabolic rate and life span. Proc Natl Acad Sci USA 85:2706–2708.

Allshire RC, Gosden JR, Cross SH, Cranston G, Rout D, Sugawara N, Szostak JW, Fantes PA, Hastie ND (1988): Telomeric repeat from *T. thermophila* cross hybridizes with human telomeres. Nature 332:656–659.

Arking R, Buck S, Wells RA, Pretzlaff R (1988): Metabolic rates in genetically based long-lived strains of *Drosophila*. Exp Gerontol 21:59–76.

Aufderheide K (1987): Clonal aging in *Paramecium tetraurelia*. II. Evidence of functional changes in the macronucleus with age. Mech Ageing Dev 37:265–279.

Aufderheide K, Schneller MV (1985): Phenotypes associated with early clonal death in *Paramecium tetraurelia*. Mech Ageing Dev 32:299–309.

Balin AK, Lustberg S, Leong I, Carter DM (1985): Oxygen modulates the proliferative lifespan of human fibroblasts. J Invest Dermatol 84:315.

Bayreuther K, Rodemann HP, Hommel R, Dittmann K, Albiez M, Francz PI (1989): Human skin fibroblasts in vitro differentiate along a terminal cell lineage. Proc Natl Acad Sci USA 85:5112–5116.

Belcour L, Begel O (1978): Lethal mitochondrial genotypes in *Podospora anserina:* A model for senescence. Mol Gen Genet 163:113–123.

Bellen HJ, Kiger JA (1987): Sexual hyperactivity and reduced longevity of *dunce* females of *D. melanogaster*. Genetics 115:153–160.

Bertrand H, Griffiths AJ, Court DA, Cheng CK (1986): An extrachromosomal plasmid is the etiological precursor of kalDNA insertion sequences in the mitochondrial chromosome of senescent neurospora. Cell 47:829–837.

Beverton RJH (1986): Longevity in fish: Some ecological and evolutionary considerations. Basic Life Sci 42:161–185.

Bruce SA, Deamond SF, Ts'o POP (1986): In vitro senescence of Syrian hamster mesenchymal cells of fetal to aged adult origin. Inverse relationship between in vivo donor age and in vitro proliferative capacity. Mech Ageing Dev 34:151–173.

Butner KA, Lo CW (1986): High frequency DNA rearrangements associated with mouse centromeric satellite DNA. J Mol Biol 187:547–556.

Cattanach BM (1974): Age-related reactivation of autosomal genes inserted in an X-chromosome. Genet Res 23:291–306.

Chatterjee B, Fernandes G, Yu BP, Song C, Kim JM, Demyan W, Roy AK (1989): Calorie restriction delays age-dependent loss of androgen responsiveness of the rat liver. FASEB J 3:169–173.

Clare MJ, Luckinbill LS (1985): The effects of gene-environment interaction on the expression of longevity. Heredity 55:19–29.

Cleaver JE (1984): DNA repair deficiencies and cellular senescence are unrelated in *Xeroderma pigmentosum* cell line. Mech Aging Dev 27:189–196.

Comfort A (1979): "The Biology of Senescence." New York: Elsevier.

Cooke HJ, Smith BA (1986): Variability at the telomeres of the human S/Y pseudoautosomal region. Cold Spring Harbor Symp Quant Biol 51:213–219.

Cummings DJ, MacNeil IA, Domenico J, Matsuura ET (1985): Excision-amplification of mitochondrial DNA during senescence in *Podospora anserina*. J Mol Biol 185:659–680.

Curtis HJ, Crowley C (1963): Life-span shortening from various tissue insults. In "Cellular Basis and Aetiology of Late Somatic Effects of Ionizing Radiation." London: Academic Press, pp 267–271.

Cutler RG (1984): Evolutionary biology of aging and longevity in mammalian species. In Johnson JE Jr (ed): "Aging and Cell Function." New York: Plenum Press, pp 1–147.
Daniel CW (1977): Cell longevity. In Finch CE, Hayflick L (eds): "Handbook of the Biology of Aging." New York: Van Nostrand Reinhold, pp 122–158.
Dean RG, Socher SH, Cutler RG (1985): Dysdifferentiative nature of aging: Age-dependent expression of mouse mammary tumor virus and casein genes in brain and liver tissues of the C57BL/6J mouse strain. Arch Gerontol Geriatr 4:43–41.
Dell'Orco RT, Mertens JG, Kruse PF (1973): Doubling potential, calender time, and donor age of human diploid cells in culture. Exp Cell Res 77:356–360.
Drinkwater RD, Blake TJ, Morley AA, Turner DR (1989): Human lymphocytes aged in vivo have reduced levels of methylation in transcriptionally active and inactive DNA. Mutat Res 219:29–37.
Emerit I (1984): Activated oxygen species at the origin of sister chromatid exchange repair. In Tice RR, Hollaender A (eds): "Sister Chromatid Exchange." New York: Plenum Press, pp 127–140.
Enstrom JE, Pauling L (1982): Mortality among health-conscious elderly Californians. Proc Natl Acad Sci USA 79:6023–6027.
Esposito D, Fassina G, Szabo P, DeAngelis P, Rodgers L, Weksler M, Siniscalco M (1989): Chromosomes of older humans are more prone to aminopterine-induced breakage. Proc Natl Acad Sci USA 86:1302–1306.
Fairweather DS, Fox M, Margison GP (1987): The in vitro lifespan of MRC-5 cells is shortened by 5-azacytidine-induced demethylation. Exp Cell Res 168:153–159.
Finn GK, Kurz BW, Cheng RZ, Shmookler Reis RJ (1989): Homologous plasmid recombination is elevated in immortally transformed cells. Mol Cell Biol 9:4009–4017.
Fleming JE, Quattrocki E, Latter G, Miquel J, Marcuson R, Zuckerkandl E, Bensch KG (1986): Age-dependent changes in proteins of *Drosophila melanogaster*. Science 231:1157–1159.
Fleming JE, Walton JK, Dubitsky R, Bensch KG (1988): Aging results in an unusual expression of *Drosophila* heat shock proteins. Proc Natl Acad Sci USA 85:4099–4103.
Flores SC, Sunnerhagen P, Moore TK, Gaubatz JW (1988): Characterization of repetitive sequence families in mouse heart small polydisperse circular DNAs: Age-related studies. Nucleic Acids Res 16:3889–3906.
Friedman DB, Johnson TE (1988a): A mutation in the age-1 gene in *Caenorhabditis elegans* lengthens life and reduces hermaphrodite fertility. Genetics 118:75–86.
Friedman DB, Johnson TE (1988b): Three mutants that extend both mean and maximum lifespan of the nematode *Caenorhabditis elegans* define the age-1 gene. J Gerontol 43:B102–B109.
Fry M, Loeb LA, Martin GM (1981): On the activity and fidelity of chromatin-associated hepatic DNA polymerase-beta in aging murine species of different life spans. J Cell Physiol 106:435–444.
Fry M, Silber J, Loeb LA, Martin GM (1984): Delayed and reduced cell replication and diminishing levels of DNA polymerase-alpha in regenerating liver of aging mice. J Cell Physiol 118:225–232.
Gartler SM, Riggs AD (1983): Mammalian X-chromosome inactivation. Annu Rev Genet 17:155–190.
Goldstein S, Singal DP (1974): Senescence of cultured human fibroblasts: Mitotic versus metabolic time. Exp Cell Res 88:359–364.
Goldstein S, Shmookler Reis RJ (1985): Methylation patterns in the gene for the alpha subunit of chorionic gonadotropin are inherited with variable fidelity in clonal lineages of human fibroblasts. Nucleic Acids Res 13:7055–7065.
Hanawalt PC (1987): On the role of DNA damage and repair processes in aging: Evidence for and against. In Warner HR, Butler RN, Sprott RL, Schneider EL (eds): "Modern Biological Theories of Aging." New York: Raven Press, pp 183–198.

Harley CB, Goldstein S (1978): Cultured human fibroblasts: Distribution of cell generations and a critical limit. J Cell Physiol 97:509–516.

Harley CB, Shmookler Reis RJ, Goldstein S (1982): Loss of repetitious DNA in proliferating somatic cells may be due to unequal recombination. J Theor Biol 94:1–12.

Harrison DE (1985): Cell and tissue transplantation: A means of studying the aging process. In Finch CE, Schneider EL (eds): "Handbook of the Biology of Aging." New York: Van Nostrand Reinhold, pp 322–356.

Hart RW, Setlow RB (1974): Correlation between DNA excision-repair and life-span in a number of mammalian species. Proc Natl Acad Sci USA 71:2169–2172.

Hayflick L (1965): The limited in vitro lifespan of human diploid cell strains. Exp Cell Res 37:614–636.

Holliday R (1986) Strong effects of 5-azacytidine on the in vitro lifespan of human diploid fibroblasts. Exp Cell Res 166:543–552.

Hornsby PJ, Harris SE (1987): Oxidative damage to DNA and replicative lifespan in cultured adrenocortical cells. Exp Cell Res 168:203–217.

Hosokawa M, Philips PD, Cristophalo VJ (1986): The effect of dexamethasone on epidermal growth factor binding and stimulation of proliferation in young and senescent WI-38 cells. Exp Cell Res 164:408–414.

Houde M, Shmookler Reis RJ, Goldstein S (1989): Proportions of H1 histone subspecies in human fibroblasts shift during density-dependent growth arrest independent of replicative senescence. Exp Cell Res 184:256–261.

Jamet-Vierny C (1988): Senescence in *Podospora anserina:* A possible role for nucleic acid interacting proteins suggested by the sequence of a mtDNA region specifically amplified in senescent cultures. Gene 74:387–398.

Johnson TE (1987): Aging can be genetically dissected into component processes using long-lived lines of *Caenorhabditis elegans*. Proc Natl Acad Sci USA 84:3777–3781.

Johnson TE (1988): Genetic specification of life span: Processes, problems, and potentials. J Gerontol 43:B87–B92.

Johnson TE, Wood WB (1982): Genetic analysis of life span in *Caenorhabditis elegans*. Proc Natl Acad Sci USA 79:6603–6607.

Jones KH, Ennist DL (1985): Mechanisms of age-related changes in cell-mediated immunity. In Rothstein M (ed): "Review of Biological Research in Aging," Vol. 2. New York: Alan R. Liss, Inc., pp 155–177.

Juckett DA (1987): Cellular aging (the Hayflick limit) and species longevity: A unification model based on clonal succession. Mech Aging Dev 38:49–71.

Kadanka ZK, Sparks JD, MacMorine HG (1973): A study of the cytogenetics of the human cell strain WI-38. In Vitro 8:353–361.

Kator K, Cristofalo V, Charpentier R, Cutler RG (1985): Dysdifferentiative nature of aging: Passage number dependency of globin gene expression in normal human diploid cells grown in tissue culture. Gerontology 31:355–361.

Kirkwood TBL (1985): Comparative and evolutionary aspects of longevity. In Finch CE, Schneider EL (eds): "Handbook of the Biology of Aging." New York: Van Nostrand Reinhold, pp 27–44.

Kirkwood TBL (1989): DNA, mutations and aging. Mutat Res 219:1–7.

Kligman AM, Grove GL, Balin AK (1985): Aging of human skin. In Finch CE, Schneider EL (eds): "Handbook of the Biology of Aging." New York: Van Nostrand Reinhold, pp 820–841.

Krauss SW, Linn S (1986): Studies of DNA polymerases alpha and beta from cultured human cells in various replicative states. J Cell Physiol 126:99–106.

Kuck U, Stahl U, Esser K (1981): Plasmid-like DNA is part of mitochondrial DNA in *Podospora anserina*. Curr Genet 3:151–156.

Kunisada T, Yamagishi H, Ogita Z, Hirakawa T, Mitsui Y (1985): Appearance of extrachromosomal circular DNAs during in vivo and in vitro ageing of mammalian cells. Mech Ageing Dev 29:89–99.

Lee AT (1987): The nonenzymatic glycosylation of DNA by reducing sugars in vivo may contribute to DNA damage associated with aging. Age 10:150–155.

Leveille PJ, Weindruch R, Walford RL, Bok D, Horwitz J (1984): Dietary restriction retards age-related loss of gamma crystallins in the mouse lens. Science 224:1247–1249.

Lincoln DW II, Braunschweiger KI, Braunschweiger WR, Smith JR (1984): The two-dimensional polypeptidè profile of terminally non-dividing human diploid cells. Exp Cell Res 154:136–146.

Linnane AW, Marzuki S, Ozawa T, Tanaka M (1989): Mitochondrial DNA mutations as an important contributor to ageing and degenerative diseases. Lancet 25:642–645.

Lints FA, Bourgois M (1985): Aging and lifespan in insects with special regard to *Drosophila*. In Rothstein M (ed): "Review of Biological Research in Aging," Vol. 2. New York: Alan R. Liss, Inc., pp 61–84.

Lipschitz DA, Udupa KB (1984): Effect of donor age on long term culture of bone marrow in vitro. Mech Aging Dev 26:119–127.

Lipschitz DA, Udupa KB (1986): Influence of aging and protein deficiency on neutrophil function. J Gerontol 41:690–694.

Little JB (1976): Relationship between DNA repair capacity and cellular aging. Gerontology 22:28–55.

Luckinbill LS, Graves JL, Reed AH, Koetsawang S (1988): Localizing genes that defer senescence in *Drosophila melanogaster*. Heredity (Edinburgh) 60:367–374.

Lundblad V, Szostack JW (1989): A mutant with a defect in telomere elongation leads to senescence in yeast. Cell 57:633–643.

Macieira-Coelho A, Puvion-Dutilleul F (1989): Evaluation of the reorganization in the high-order structure of DNA occurring during cell senescence. Mutat Res 219:179–185.

Marlhens F, Achkar WA, Aurias A, Couturier J, Dutrillaux AM, Gerbault-Sereau M, Hoffschir F, Lamoliatte E, Lefrancois D, Lombard M, Mularis M, Prieur M, Prod'homme M, Sabatier L, Viegas-Pequignot E, Volobouev V, Dutrillaux B (1986): The rate of chromosome breakage is age dependent in lymphocytes of adult controls. Hum Genet 73:290–297.

Martin GM, Sprague CA, Epstein CJ (1970): Replicative life-span of cultivated human cells. Effects of donor's age, tissue, and genotype. Lab Invest 23:86–92.

Martin GM, Smith AC, Ketterer DJ, Ogburn CE, Disteche CM (1986): Increased chromosomal aberrations in first metaphases of cells isolated from the kidneys of aged mice. Isr J Med Sci 21:296–301.

Masoro EJ (1988): Food restriction in rodents: An evaluation of its role in the study of aging. J Gerontol 43:B59–B64.

Mayer PJ, Bradley MO, Nichols WW (1986): No change in DNA damage or repair of single- and double-strand breaks as human diploid fibroblasts age in vitro. Exp Cell Res 166:497–509.

Maynard Smith J (1966): Theories of aging. In Krohn PL (ed): "Topics in the Biology of Aging." New York: Interscience/John Wiley, pp 1–35.

Maynard Smith J (1972): "On Evolution." Edinburgh: Edinburgh University Press.

Mays-Hoopes LL, Brown A, Huang RCC (1983): Methylation and rearrangement of mouse intracisternal A particle genes in development, aging, and myeloma. Mol Cell Biol 3:1371–1380.

Mays-Hoopes LL, Chao W, Butcher HC, Huang RCC (1986): Decreased methylation of the major mouse long interspersed repeated DNA during aging and in myeloma cells. Dev Genet 7:65–73.

Medawar PB (1952): "An Unsolved Problem in Biology." London: H.K. Lewis.

Medvedev ZA, Medvedeva RN, Robson L (1978): Tissue-specificity and age changes of the pattern of the H1 group of histones in chromatin from mouse tissues. Gerontology 24:286–292.

Meisenhelder J, Suh P-G, Rhee SG, Hunter T (1989): Phospholipase C-gamma is a substrate for the PDGF and EGF receptor protein-tyrosine kinases in vivo and in vitro. Cell 57:1109–1122.

Migeon BR, Axelman J, Beggs AH (1988): Effect of ageing on reactivation of the human X-linked HPRT locus. Nature 335:93–96.

Miller RA, Harrison DE (1985): Delayed reduction in T cell precursor frequencies accompanies diet-induced lifespan extension. J Immunol 85:1343–1426.

Miller RA (1989): The cell biology of aging: Immunological models. J Gerontol 44:B4–B8.

Milsted A, Day DL, Cox RP (1982): Glycopeptide hormone production by cultured human diploid fibroblasts. J Cell Physiol 113:420–429.

Muggleton A, Danielli JF (1958): Aging of *Amoeba proteus* and *A. discoides* cells. Nature 181:1783.

Mullaart E, Boerrigter METI, Brouver A, Berends F, Vijg G (1988): Age-related accumulation of alkali-labile sites in DNA of post-mitotic but not in that of mitotic rat liver cells. Mech Aging Dev 45:41–49.

Nanney DL (1959): Vegetative mutants and clonal senility in *Tetrahymena*. J Protozool 6:171–177.

Ono T, Dean RG, Chattopadhyay SK, Cutler RG (1985): Dysdifferentiative nature of aging: Age-dependent expression of MuLV and globin genes in thymus, liver and brain in the AKR mouse strain. Gerontology 31:362–372.

Ono T, Takahashi N, Okada S (1989): Age-associated changes in DNA methylation and mRNA level of the c-myc gene in spleen and liver of mice. Mutat Res 219:39–50.

Osiewacz HD, Hermanns J, Marcou D, Triffi M, Esser K (1989): Mitochondrial DNA rearrangements are correlated with a delayed amplification of the mobile intron (p1DNA) in a long-lived mutant of *Podospora anserina*. Mutat Res 219:9–15.

Papez-Zelinsky K, Carter TH, Zimmerman JA (1987): Isolation and characterization of chemically transformed pancreatic acinar cell lines from young and old mice. In Vitro Cell Dev Biol 23:118–122.

Park J, Ames BN (1988): 7-methylguanine adducts in DNA are normally present at high levels and increase on aging: Analysis by HPLC with electrochemical detection. Proc Natl Acad Sci USA 85:7467–7470.

Randerath K, Liehr JG, Gladek A, Randerath E (1989): Age-dependent covalent DNA alterations (I-compounds) in rodent tissues: species, tissue and sex specificities. Mutat Res 219:121–133.

Rheinwald JG, Green H (1977): Epidermal growth factor and the multiplication of cultured human epidermal keratinocytes. Nature 265:421–424.

Riabowol KT, Shmookler Reis RJ, Goldstein S (1985a): Tandemly repetitive and interspersed repetitive sequences are differentially represented in extrachromosomal covalently closed circular DNA of human diploid fibroblasts. Nucleic Acids Res 13:5563–5584.

Riabowol KT, Shmookler Reis RJ, Goldstein S (1985b): Properties of extrachromosomal covalently closed circular DNA isolated and cloned from aged human fibroblasts. Age 8:114–121.

Richter C, Park J, Ames BN (1988): Normal oxidative damage to mitochondrial and nuclear DNA is extensive. Proc Natl Acad Sci USA 85:6465–6467.

Rockwell GA, Johnson G, Sibatani A (1987): In vitro senescence of human keratinocyte cultures. Cell Struct Funct 12:539–548.

Rohme D (1981): Evidence for a relationship between longevity of mammalian species and life spans of normal fibroblasts in vitro and erythrocytes in vivo. Proc Natl Acad Sci USA 78:5009–5013.

Rose MR, Service PM (1985): Evolution of aging. In Rothstein M (ed): "Review of Biological Research in Aging," Vol. 2. New York: Alan R. Liss, Inc., pp 85–98.

Rosen SW, Weintraub BD, Aaronson SA (1980): Nonrandom ectopic protein production by malignant cells: Direct evidence in vitro. J Clin Endocrinol Metab 50:834–847.

Saksela E, Moorhead PS (1963): Aneuploidy in the degenerative phase of serial cultivation of human cell strains. Proc Natl Acad Sci USA 50:390–395.

Sacher GA (1978): Evolution of longevity and survival characteristics in mammals. In Schneider EL (ed): "The Genetics of Aging." New York: Plenum Press, pp 151–168.

Samis HV (1978): Molecular genetics of aging. In Schneider EL (ed): "The Genetics of Aging." New York: Plenum Press, pp 7–25.

Schafer DE (1984): Replication bypass SCE mechanisms and the induction of SCE by single-strand adducts or lesions of DNA. In Tice RR, Hollaender A (ed): "Sister Chromatid Exchange." New York: Plenum Press, pp 245–265.

Schneider EL (1985): Cytogenetics of aging. In Finch CE, Schneider EL (eds): "Handbook of the Biology of Aging." New York: Van Nostrand Reinhold, pp 357–373.

Schneider EL, Shorr SS (1975): Alteration in cellular RNAs during the in vitro lifespan of cultured human diploid fibroblasts. Cell 6:179–184.

Schneider EL, Mitsui Y (1976): The relationship between in vitro cellular aging and in vivo human age. Proc Natl Acad Sci USA 73:3584–3588.

Shay JW, Werbin H (1987): Are mitochondrial DNA mutations involved in the carcinogenic process? Mutat Res 186:149–160.

Sherwood SW, Rush D, Ellsworth JL, Schimke RT (1988): Defining cellular senescence in IMR-90 cells: A flow cytometric analysis. Proc Natl Acad Sci USA 85:9086–9090.

Shmookler Reis RJ (1976): Enzyme fidelity and metazoan aging. Interdisc Topics Gerontol 10:11–23.

Shmookler Reis RJ, Goldstein S (1980): Loss of reiterated DNA sequences during serial passage of human diploid fibroblasts. Cell 21:739–749.

Shmookler Reis RJ, Goldstein S (1982a): Variability of DNA methylation patterns during serial passage of human diploid fibroblasts. Proc Natl Acad Sci USA 79:3949–3953.

Shmookler Reis RJ, Goldstein S (1982b): Interclonal variation in methylation patterns for expressed and non-expressed genes. Nucleic Acids Res 10:4293–4304.

Shmookler Reis RJ, Goldstein S (1983): Mitochondrial DNA in mortal and immortal human cells: Genome number, integrity and methylation. J Biol Chem 258:9078–9085.

Shmookler Reis RJ, Moerman EJ, Goldstein S (1989): DNA methylation, maintenance methylase, and senescence. In Warner HR, Wang E (eds): "Growth Control During Cellular Aging." Boca Raton, FL: CRC Press, pp 191–201.

Singhal RP, Mays-Hoopes LL, Eichhorn GL (1987): DNA methylation in aging of mice. Mech Ageing Dev 41:199–210.

Slagbloom PE, Vijg J (1989): Genetic instability and aging: Theories, facts and future perspectives. Genome 31:373–385.

Smith JR, Hayflick L (1974): Variation in the life-span of clones derived from human diploid cell strains. J Cell Biol 62:48–53.

Smith JR, Whitney RG (1980): Intraclonal variation in proliferative potential of human diploid fibroblasts: Stochastic mechanism for cellular aging. Science 207:82–84.

Smith-Sonneborn J (1985): Aging in unicellular organisms. In Finch CE, Schneider EL (eds): "Handbook of the Biology of Aging." New York: Van Nostrand Reinhold, pp 79–104.

Smith-Sonneborn J (1987): Longevity in the protozoa. Basic Life Sci 42:101–109.

Sohal RS, Allen RG (1986): Relationship between oxygen metabolism, aging and development. Adv Free Radical Biol Med 2:117–160.

Srivastava A, Norris JS, Shmookler Reis RJ, Goldstein S (1985): C-Ha-ras-1 proto-oncogene amplification and overexpression during the limited replicative life span of normal fibroblasts. J Biol Chem 260:6404–6409.

Staino-Coico L, Darzynkiewicz Z, Hefton JM, Dutkowski R, Darlington GJ, Weksler ME (1983): Increased sensitivity of lymphocytes from people over 65 to cell cycle arrest and chromosomal damage. Science 219:1335–1337.

Stanulis-Praeger BM, Gilchrest BA (1986): Growth factor responsiveness declines during adulthood for human skin-derived cells. Mech Ageing Dev 35:185–198.

Stohl LL, Collins RA, Cole MD, Lambowitz AM (1982): Characterization of two new plasmid

DNAs found in mitochondria of wild-type *Neurospora intermedia* strains. Nucleic Acids Res 10:1439–1458.

Stringer J, Kuhn R, Newman J, Meade J (1985): Unequal homologous recombination between tandemly arranged sequences stably incorporated into cultured rat cells. Mol Cell Biol 5:2613–2622.

Thompson KVA, Holliday R (1975): Chromosome changes during the in vitro aging of MRC-5 human fibroblasts. Exp Cell Res 96:1–6.

Tice RR, Setlow RB (1985): DNA repair and replication in aging organisms and cells. In Finch CE, Schneider EL (eds): "Handbook of the Biology of Aging. New York: Van Nostrand Reinhold, pp 173–224.

Turker MS, Nelson JG, Cummings DJ (1987): A *Podospora anserina* longevity mutant with a temperature-sensitive phenotype for senescence. Mol Cell Biol 7:3199–3204.

Wang E (1985): Rapid disappearance of statin, a nonproliferating and senescent cell-specific protein, upon reentering the process of cell cycling. J Cell Biol 101:1695–1701.

Wang E (1988): Molecular and biochemical characterization of statin, a marker for nonproliferative and quiescent state. In "Proceedings of the Fourth International Congress of Cell Biology." Ottawa: NRC of Canada, p 31.

Wareham KA, Lyon MF, Glenister PH, Williams ED (1987): Age related reactivation of an X-linked gene. Nature 327:725–727.

Webster GC, Webster SL (1984): Specific disappearance of translatable messenger RNA for elongation factor one in aging *Drosophila melanogaster*. Mech Ageing Dev 24:335–342.

Weindruch R, Walford RL, Fligiel S, Guthrie D (1986): The retardation of aging in mice by dietary restriction: Longevity, cancer, immunity and lifetime energy intake. J Nutr 116:641–654.

Weismann A (1891): "Essays Upon Heredity and Kindred Biological Problems." London: Oxford University Press.

Wilson VL, Jones PA (1983): DNA methylation decreases in aging but not in immortal cells. Science 220:1055–1057.

Wilson VL, Smith RA, Ma S, Cutler RG (1987): Genomic 5-methyldeoxycytidine decreases with age. J Biol Chem 262:9984–9951.

Age-Related Effects in Enzyme Metabolism and Catalysis

Ari Gafni

INTRODUCTION

The effects of biological aging on enzyme levels and properties have been previously reviewed in this series by Rothstein [1983,1985]. The present review, therefore, focuses on the more recent developments in this field. While an ever increasing number of old enzymes (i.e., enzymes from old animals) have been found to display modified properties compared with their young counterparts, detailed examinations of the structural modifications responsible for these effects have only been made in a small number of systems. In all cases thus far studied the aging effects were found to be introduced postsynthetically, although the nature of the modifications and their mechanism of development may vary from one enzyme to the other.

At the time of the previous review in this series [Rothstein, 1985], it was noted that much of the work in the area of enzyme aging was mainly concerned with the documentation of changes in various enzyme activities, and did not provide significant insight into the molecular mechanisms of aging. While such descriptive work is still typical of many of the publications in the recent literature, it is clear that the emphasis has begun to shift, during the period covered by this review, to more focused studies in which single enzymes, or small groups of enzymes, are targeted in an attempt to characterize the structural modifications and to elucidate the underlying biological mechanism. Several possible mechanisms of enzyme aging have been put forward and tested. Their experimental support, as well as their relative merits and limitations, will be discussed below.

While the progress being made in the elucidation of enzyme aging mechanisms is significant, it is noteworthy that information connecting these changes at the molecular level with the effects of aging at the cellular and organ level

Institute of Gerontology and Department of Biological Chemistry, The University of Michigan, Ann Arbor, Michigan 48109

is still lacking. Thus, in spite of its potential importance in some specific cases (i.e., collagen crosslinking, red blood cell metabolism, eye-lens opacification, etc.), the general biological importance of enzyme aging is still to be demonstrated and documented.

RECENT WORK DOCUMENTING AGING EFFECTS IN ENZYMES
Antioxidant Protecting Enzymes

The deleterious effects of oxygen free radicals on cells and cellular organelles are well recognized and their potential role in the aging process has continued to be evaluated extensively [Harman, 1986, 1988; Nohl, 1986; Gutteridge et al., 1986; Kay et al., 1987]. It is therefore not surprising that several recent studies addressed age-related changes in enzymes that act as antioxidant defense systems. Hazelton and Lang [1985] assayed the activities of glutathione peroxidase and glutathione reductase in liver, kidney, and heart tissues of mature (10 months) and very old (36 months) mice. The activities of glutathione peroxidase in all three organs were significantly (27–53%) lower in old animals than in their young counterparts. Glutathione reductase activities were also reduced with age (by 25–28%) in liver and kidney, while the level in heart was unaffected. These changes apparently reflect declines in the specific activities of the enzymes since organ weights and total protein contents were found to be constant from 10 to 36 months of age. It was noted that due to these decreases in enzyme activities the detoxification via glutathione peroxidase and glutathione reductase could be impaired in old age.

Croute et al. [1985] investigated the activities of several antioxidant enzymes (catalase, glutathione peroxidase, superoxide-dismutase) in *Paramecium tetraurelia* as a function of clonal age. They found that while catalase activity slightly increased during aging, no significant changes in the activities of the other two enzymes could be detected. The increase in catalase was interpreted to be a response to an age-related stimulation of the steady-state formation of hydrogen peroxide.

A significant decrease in enzymic protection against oxidative damage in rat erythrocytes as a function of both cell and donor age has been documented by Glass and Gershon [1984]. The specific activities per cell were determined for glucose-6-phosphate dehydrogenase, glutathione peroxidase, glutathione reductase, and catalase. All four enzymes displayed a decline in activity with increasing cell age. More interestingly, significant differences in the activities of these enzymes were found between cells of the same age-fraction but from young and old donors. This demonstrates that the level of protection against oxidative damage in cells of old rats is greatly reduced already early in the erythrocyte life in circulation. A shift in the redox status of old rat muscle cells, towards a higher oxidation potential, was reported by Noy et al. [1985], and although a direct connection with a reduction in activities of protective enzymes was not made, this appears to be a very plausible explanation.

Somville et al. [1985] compared the effects of age on several properties of superoxide dismutase in human fibroblasts in culture. Young cells (after 29 passages), and old cells (52 passages) were harvested, homogenized, and separated into a particulate fraction, containing most of the mitochondrial enzyme, and a supernatant fraction that contained cytoplasmic superoxide dismutase. Corresponding fractions from young and old cells were compared by immunotitration of superoxide dismutase and by heat inactivation at 70°C. These tests revealed no age-related changes in mitochondrial superoxide dismutase, while the old cytoplasmic enzyme displayed a significantly reduced heat stability and a loss of activity per antigenic site as compared with its young counterpart. These age-related alterations could be induced in young superoxide dismutase by its incubation in the supernatant from old cells and, more interestingly, could be eliminated in samples of the old enzyme upon its incubation in young supernatant. The modifications thus appear to be induced by the environment, are fully reversible, and hence must develop postsynthetically. While the detailed nature of the modifications and their mechanism of development was not revealed, a molecular origin was suggested based on a shift in the equilibrium between a dimeric (young) and a tetrameric form of the enzyme, the latter form subsequently undergoing some (as yet undefined) conformational changes to yield old superoxide dismutase. More detailed information on this process is obviously needed.

Enzymes Involved in Protein Synthesis

Changes, with age, in the properties of enzymes that play a role in protein synthesis have attracted special attention ever since Orgel [1963] developed the error catastrophe theory of aging. It was, therefore, an interesting observation when Gabius et al. [1982] reported large (up to 70%) declines in the specific activities of 17 aminoacyl tRNA synthetases in several tissues of 39-month-old mice compared with activities of the same enzymes in 2-month-old controls. In his review of the field Rothstein [1985] commented that at 39 months of age mice are very old and may have turned off practically all protein synthesis while at 2 months the animals are still growing and their protein synthesis activity is enhanced compared with young, but fully grown, animals. It was therefore recommended that intermediate ages be examined. Indeed, this approach was taken in a study addressing the age-dependent alteration in heat stability in each of six aminoacyl tRNA synthetases from liver, kidney, and brain tissues of male Wistar rats ranging in age between 3 and 24 months [Takahashi et al., 1985]. The enzymes were only partially purified to avoid possible loss of modified forms during purification. No significant change, with age, in the specific activity of any of the aminioacyl tRNA synthetases could be detected. In contrast, heat denaturation experiments clearly showed a continuous increase in the level of heat-labile forms of the enzymes from the age of 15 months on and their presence in large amounts (up to 50% total enzyme) in old animals. The authors concluded from these findings that the decrease in functional activities in tissues of old animals may be related to the presence of altered aminoacyl tRNA synthetases. This suggestion, while appealing, should be regarded with caution in view of the fact that the biological activities of the enzymes were not affected by age and that the differences in stability reported by Takahashi et al. [1985] were only revealed under nonphysiological conditions (incubation at 48°C). It is possible, though not demonstrated, that the labile but still active forms of aminoacyl tRNA synthetases observed at 24 months become inactivated at older ages. Experiments with 28–30-month-old rats should help in answering this important question.

Enzymes From Cultured Cells

Aging of enzymes in human fibroblasts in culture with increasing passage number has been well documented [Holliday and Tarrant, 1972]. Several recent studies have focused on alterations in glucose-6-phosphate dehydrogenase (G6PD) [Houben et al., 1984a,b] and it was shown that the altered form of this enzyme is an intermediate state between the active dimer and an inactive monomeric subunit. Nbemba Fufu et al. [1985] have successfully simulated the aging of G6PD by incubating a cell supernatant from young human fibroblasts under nonreducing conditions. Addition of reduced glutathione had a stabilizing effect on the enzyme, while substances that deplete the cells of glutathione were found to increase the inactivation rate dramatically. The aging mechanism of G6PD, as suggested based on these experiments, involves an initial oxidation of cysteine residues in the active dimeric enzyme, leading to the formation of a thermolabile (but still active) dimer. Conformational changes then transform this species into an inactive dimer that is unstable and dissociates into monomeric subunits. The coenzyme $NADP^+$ has a protecting effect, presumably by stabilizing the active dimeric form (i.e., slowing the conformational transitions to the inactive dimer). The aging of G6PD is thus clearly postsynthetic. More significantly, it follows the same two-step mechanism responsible for the aging of glyceraldehyde-3-phosphate-dehydrogenase (GPDH) [Gafni and Noy, 1984; Gafni, 1985] and of phosphoglycerate kinase (PGK) [Yuh and Gafni, 1987; Zuniga and Gafni, 1988; Cook and Gafni, 1988]. A general pattern of aging of sulfhydryl-dependent enzymes is thus emerging: the process involves oxidation of reactive cysteine residues, an event that induces conformational transitions in the enzyme. Old cells are more susceptible to this process due to their reduced protection against oxidation [Noy et al., 1985]. A reduction, with aging, in the rate of protein synthesis [Sojar and Rothstein, 1986; Klaude and Von Der Decken, 1986] and protein degradation

[Karey and Rothstein, 1986] leading to a longer dwell-time of proteins in old cells promotes the development of these postsynthetic modifications by giving them more time to occur [Rothstein, 1982 Goldspink et al., 1986; Reff, 1985; Sarkis et al., 1988; Ward, 1988].

In line with the studies cited above Jahani and Gracy [1985] found a significant decrease in the activity of cathepsin B in human skin fibroblasts from old donors (73–84 years) as compared with cells from young individuals (17–32 years). Fibroblasts from donors with the premature aging syndromes, progeria, and Werner's syndrome were especially deficient in cathepsin B activity.

Membrane-Bound Enzymes

Several recent studies focused on membrane bound proteins. In these systems effects of aging on both protein and membrane component may potentially contribute to the observed age-related modification in functional properties. Velez et al. [1985] examined the origin of the 1.7-fold increase in the activity of rat heart succinate dehydrogenase in old animals. This membrane-bound mitochondrial enzyme is present in interconvertible active and inactive conformations. The increase in succinate dehydrogenase activity thus could either reflect a change in the fraction of active enzyme in response to modifications of metabolic conditions in the old heart, or be due to an age-related increase in the total amount of enzyme. The conclusion that the increase in activity is due to an increase in enzyme protein was reached, based on the observation that the activation state of the enzyme is independent of age, and on the finding that the increase in activity is accompanied by a similar increase in immunoprecipitable succinate dehydrogenase in old heart mitochondria. In addition to the quantity of enzyme, however, some properties of old succinate dehydrogenase were also found to become modified. These include an increased K_m for succinate, a reduced thermal stability, and a change in the reactivity of cysteine residues toward the reagent N-ethylmaleimide.

Velez et al. [1985] interpreted the results described above as reflecting a change in the configuration of succinate dehydrogenase that is induced by age-related modifications in the fluidity and phospholipid composition of the mitochondrial membrane. A critical test of this conclusion, by a comparison of purified succinate dehydrogenase from young and old animals, either in detergent-solubilized state or when reconstituted into identical membrane preparations, has not been performed. Changes in the enzyme molecule itself may therefore have gone undetected.

A decrease, with aging, in the rate of calcium uptake by cardiac sarcoplasmic reticulum-enriched fractions has been reported by several groups of investigators [Froehlich et al., 1978; Narayanan, 1981; Lakatta, 1987; Narayanan, 1987]. In a detailed discussion by Orchard and Lakatta [1985], it was pointed out that this decline in the rate of removal of calcium is responsible for the prolonged duration of the myoplasmic calcium transient observed in senescent muscle and may thus be a major contributor to the prolongation of the duration of contraction of old cardiac muscle, the latter being a significant, well-documented aging phenomenon [Lakatta et al., 1975]. A detailed study of the cardiac calcium pump [Narayanan, 1987] revealed a progressive decline, with aging, in the rate of ATP-dependent calcium accumulation by rat cardiac sarcoplasmic reticulum. This decline reached 50% by 24 months of age (compared with 3- or 6-month-old controls). In contrast, sarcolemma preparations displayed an age-related increase in the rate of ATP-dependent calcium accumulation. With either membrane it was demonstrated that the change in the velocity of calcium transport was not associated with changes in the affinity of the transport system for this cation. An interesting observation was made that the calcium-stimulated ATPase activities of the two membranes are practically unaffected by age. This finding is corroborated by results obtained using rat skeletal muscle [Nakano and Tauchi, 1986], which showed no change, with aging, in Ca^{2+} stimulated ATPase activity. An activation, by Tris:HCl, of the Ca^{2+}-ATPase was more pronounced in preparations from young animals, indicating that some age-related modifications do occur in the pump system. A decline in the rate of Ca^{2+} transport with no change in ATPase activity clearly reflects the fact that the pumping stoichiometry (i.e., number of calcium ions pumped per ATP cleaved) is modulated (reduced) by aging. The molecular origin of this change, i.e., whether it reflects an uncoupling of ATPase activity from ion translocation, it yet to be determined.

Gafni and Yuh [1988] compared several properties of the sarcoplasmic reticulum ATPase in membrane fragments prepared from skeletal muscles of rats of several age groups. The quantity of membrane-bound protein, per gram of muscle tissue, was found to decline markedly with age. The capacity of the sarcoplasmic reticulum vesicles to store calcium was also significantly reduced in older animals, and the rate of heat inactivation of the ATPase in these vesicles was greatly increased at old age. Some differences in the rate of ATPase inactivation persisted when the enzyme was solubilized in Triton X-100 and incubated at several pH values between 5.5 and 8.5. These results indicate that age-related modifications may exist in the ATPase protein, in addition to changes in the membrane component; however, purification of the solubilized enzyme followed by a comparative study of young and old ATPases is needed before any final conclusions can be drawn.

That at least some of the aging effects in the sarcoplasmic reticulum calcium pump may be a result of oxidative damage was pointed out by Luo and Hutlin [1986]. An in vitro incubation of vesicles, isolated from 2–8-year-old winter flounder, with the peroxidizing cofactors NADH, ADP, and Fe^{3+}, revealed a higher rate of lipid peroxidation in the old membrane. Also, two high molecular weight, crosslinked, proteins appeared in young sarcoplasmic reticulum vesicles during this incubation; they resembled protein species that increase in native membranes with aging. However, a partial loss of calcium-dependent ATPase activity also occured upon oxidation, in contrast to aging in vivo, where the activity is unaffected. This reduction may be due to loss of active ATPase protein by crosslinking, to the destruction of essential cysteines by the strongly oxidizing conditions or, as pointed out by Luo and Hutlin [1986], by captive stress in the fish.

Significant age-related effects have also been found in the calcium-translocating system of rat parotid basolateral membrane. Ambudkar et al. [1988] reported a 30–50% decrease in ATP-dependent calcium transport in membrane vesicles isolated from 24-month-old rats as compared with that in vesicles from 4-month-old animals. An interesting observation, from kinetic studies, was that the maximal velocity of calcium transport (i.e., at saturating concentrations of the metal ion) was practically independent of age. The aging effect mentioned above was due to an increase, by about 50%, in the K_m for calcium in the old membrane preparations. The differences between the rates of calcium transport by young and old pump units is therefore dependent on the concentration of this ion and may become very marked when this concentration is small (i.e., when the calcium gradient across the membrane is large). Old vesicles may thus not be able to attain the same gradients of Ca^{2+} across their membranes. Again, addressing the important question of whether this is due to changes in the membrane or in the protein would be very rewarding.

MECHANISMS OF ENZYME AGING

Although the error catastrophe theory [Orgel, 1963] played an important role in stimulating the search for age-modified proteins and their study, the bulk of the experimental evidence accumulated since then strongly points to postsynthetic modifications as the major cause for enzyme aging. Any proposed mechanism for this phenomenon must, therefore, rely on some inherent difference between the environments in young and old tissues that would promote the development of modifications in enzymes in the latter. Several such differences have indeed been observed. One significant change in old tissues is the shift toward a higher oxidation status [Noy et al., 1985] and the reduced protection against oxidative damage [Gutteridge et al., 1986; Harman, 1988]. This may lead to an increase, with aging, in the level of oxidized, damaged, proteins—a phenomenon to be reviewed in more detail below. Alternatively, modifications may occur more readily in old tissue if protein turnover is slowed down with age, so that enzymes in aged tissues are provided more time to become altered. This hypothesis, originally proposed by Reiss and Rothstein [1974] and further developed by Rothstein [1975, 1982] has found significant experimental support. Indeed, declines with aging have been documented in the rates of both protein synthesis [Richardson and Semsei, 1987] and degradation [Karey and Rothstein, 1986; Sarkis et al., 1988].

The hypothesis that enzymes in old tissues undergo a stepwise conversion from newly synthesized, fully active, forms to old, less stable forms that further change to partially denatured and easily degradable products was tested theoretically by Sadana and Henley [1985a,b]. These authors conducted a mathematical analysis of the deactivation kinetics of several enzymes from young and old animals assuming the series-type enzyme deactivation sequence mentioned above. While the model is useful in demonstrating the existence of intermediate unfolding states of enzymes and makes it possible to obtain their relative concentrations and rates of interconversion, it is very general and could possibly even be compatible with the notion that old enzymes are predisposed to the stepwise postsynthetic alterations due to errors in synthesis (i.e., that enzymes in old tissue become modified more easily not because of a longer dwell-time but due to some synthetic defect). The usefulness of such models in deciding the basic questions of enzyme aging is therefore limited.

Aging of Long-Lived Proteins

Evidence that a longer dwell-time in the cell increases posttranslational protein modification may be provided by studies involving tissues in which protein turnover is very slow or nonexistent. Two such tissues, i.e., red blood cells and the eye-lens, have indeed been addressed in a number of studies.

The lens is apparently devoid of active protein degradation and is, therefore, a particularly suitable organ for studying age-related posttranslational modifications in proteins in vivo. Dovrat et al. [1984, 1986] studied the fate of several lens enzymes as a function of age in humans and in rats. The specific activities of superoxide dismutase, aldolase, G6PD, and GPDH were all significantly reduced in the old lens. Immunotitration studies indicated that the loss of activity was in each case associated with an accumulation of enzyme forms totally devoid of catalytic activity and at least partially denatured. Thus, antibodies against native G6PD failed to cross-react with the enzyme form in old lenses, while antibody prepared against denatured G6PD (which did not react with the native young enzyme) was active against G6PD molecules from old lenses.

Evidence that significant, albeit slow, protein degradation does occur in the lens has been recently presented by FitzGerald [1988], who reported on the progressive, stepwise, posttranslational degradation of an extrinsic membrane protein in lens fiber cells. Proteolysis of lens crystallins during aging has also been documented, resulting in a gradual increase of cleaved polypeptides in the lens [van Kleef et al., 1974]. A connection between the proteolytic product and cataract development has been suggested but not as yet demonstrated. Srivastava [1988a,b] has recently determined the levels of degraded polypeptides in human lenses of various ages and found these levels to increase in both water-soluble and -insoluble protein fractions. The degraded proteins were also found to aggregate, by an unknown mechanism, to produce high molecular weight products (possibly related to cataract formation). Among nonenzymatic degradation mechanisms that affect lens proteins, photooxidative insult has been recently shown to be more effective in old lens tissue. Hibino et al. [1988] reported that the susceptibility of alpha crystallin to photolytic damage by UV irradiation significantly increases with lens age and suggested that this increase is due to an age-related formation of a photosensitizer in alpha crystallin.

Asparagine Modification Processes

To the class of nonenzymatic, spontaneous, reactions that lead to the covalent modification of proteins also belongs the deamidation of asparaginyl residues. Gracy et al. [1985] compared several enzymes from young and old fibroblasts, lymphocytes, and eye-lens cells. Accumulation, with aging, of labile forms was observed in all cases. In a detailed study of triose phosphate isomerase (TPI) from bovine eye lens these investigators found an age-related accumulation of four, more acidic (and labile) isoenzymic forms and traced their origin to a sequential deamidation of two asparagine residues in each subunit of the dimeric enzyme. This conclusion found strong support in the demonstration by Yuksel and Gracy [1986] that identical TPI isozymes can be produced from the native enzyme in vitro, an observation that also confirms that the observed effects develop postsynthetically. A similar molecular basis has been recently proposed for the presence of isozymes of glucose-6-phosphate isomerase (GPI) in bovine eye lens [Cini and Gracy, 1986] and heart [Cini et al., 1988]. In both TPI and GPI the deamidation reactions give rise to isozymes that are less stable and more susceptible to proteolytic digestion. The higher prevalence of deamidated isozymes in old and nonproliferating tissue is quite certainly due to the reduced protein degradation rates in this tissue. It was noted in the studies mentioned above that not all asparagine residues in TPI and GPI undergo deamidation at the same rate but that certain sequences, especially Asn–Gly, are particularly prone to deamidation and are indeed involved in this modification in both enzymes.

A detailed mechanism for deamidation (as well as isomerization and racemization) of asparagine residues in proteins has been proposed by Geiger and Clarke [1987] based on their studies of model peptides. A peptide containing the Asn–Gly combination underwent deamidation rapidly (with a half-life of 1.4 days at 37°C, pH 7.4) yielding an aspartyl succinimide derivative that further reacted by hydrolysis to yield a mixture of aspartyl and isoaspartyl residues. Some racemization was found to accompany the deamidation, most of it at the succinimide intermediate; hence each of the two hydrolytic products contained a small fraction of the D-isomer. When the glycyl residue following the Asn was replaced by Pro or Leu residues, the degradation rate was dramatically reduced, confirming the great susceptibility of the Asn–Gly sequence to deamidation. In contrast to these results with synthetic peptides, a subsequent study [Clarke, 1987] in which the rates of succinimide formation in a group of proteins with known sequences and three-dimensional structures were compared, revealed that the conformational restrictions on succinimide formation are also very important and hence the prediction of the rate of this reaction from amino acid sequence alone is not reliable.

Strong evidence that the tertiary structure of a protein is a major determinant for asparaginyl deamidation has been recently provided by Kossiakoff [1988] from the neutron crystallographic structure of trypsin. Three of 13 asparagine residues were found to be deamidated following incubation of the enzyme in vitro for about 18 months during crystallization and sample preparation. Detailed analysis of the experimental data revealed no correlation between the tendency of Asn residues to deamidate and the nature of the amino acid residues flanking the reactive Asn. Here again it was concluded that the rate of the deamidation reaction, which is known to be catalyzed by the peptide nitrogen of the succeeding residue, depends more on this residue's favorable configuration than on its identity.

The age-dependent accumulation of D-Asp and D- and L-iso Asp in proteins in eye-lens and erythrocytes has been documented and investigated in several recent studies [Brunauer and Clarke, 1986; McFadden and Clarke, 1986; Ota et al., 1987; Lowenson and Clarke, 1988]. It is pertinent to note that the formation of D-Asp residues is a slow process—in eye-lens the rate is about 0.14% per year—hence even in old tissue only a relatively small fraction of Asp residues are modified. Indeed, studies with erythrocytes revealed no accumulation of D-Asp, D-Asn, or D-iso Asp in cytosolic proteins, and only a small increase was noted in the erythrocyte membrane proteins over the life span of these cells both in vivo and in vitro [Brunauer and Clarke, 1986]. Thus, while isomerization and racemization of Asn do occur and result in protein modification, the contribution of these reactions to the pool of altered proteins is small even in cells where enzyme turnover is very slow. In most living cells, where protein turnover proceeds with higher rates, Asn racemization and isomerization is probably not very significant. This may explain why in his comparative examination of the sequences of several abundant and stable proteins Clarke [1987] found little sequence selectivity, i.e., there is no tendency for Asx residues in proteins to be followed in the sequence by a large amino acid residue so as to be more protected against modification. Moreover, Asx residues are not underrepresented in the sequence—as could be achieved by their substitution by the much more stable Glx residue. Thus, there is no evolutionary selection against Asn residues, indicating that they usually cause no problem. It has also not yet been established what effect a racemization of an Asp residue has on a protein activity or stability, although it is reasonable to assume that it should [Lowenson and Clarke, 1988]. Although the level of Asp isomerization and racemization is small in most cells, a possible cellular response mechanism has been described based on carboxymethylation of these derivatives by a specific carboxymethyl–transferase [Clarke, 1987; Johnson et al., 1987; Ota et al., 1987; McFadden and Clarke, 1986; Sellinger et al., 1988].

Protein Glycation

One nonenzymatic protein modification reaction that may contribute to the pool of modified proteins in old cells involves protein glycation. This reaction begins with the formation of a Schiff base between the carbonyl group of glucose and an alpha or epsilon amino group in a protein. This derivative is quickly transformed, by an Amadori rearrangement, to a relatively stable ketoamine. The reaction then proceeds via a complex pathway to eventually form crosslinked, fluorescent, brown pigments [Mauron, 1981]. The idea that this reaction sequence may play a role in aging of long-lived proteins, especially in uncontrolled diabetics (due to the abnormally high level of blood glucose) was expressed by Brownlee et al. [1984] and supported by the finding that the typical brown, fluorescent end products are indeed found in lens crystallins and collagen in old or diabetic people [Kohn et al., 1984; Monnier et al., 1984]. Ulrich et al. [1985] identified the key product of advanced glycosylation, and the mechanism of the process was reviewed by Cerami et al. [1987]. The effects of glycation on the activity of RNase were reported by Ahmed et al. [1986], who found that the incorporation of glucose into the enzyme resulted in reduction of activity.

The glycation of Cu–Zn superoxide dismutase from human erythrocytes was studied in detail by Arai et al. [1987a,b] and was found to increase with the age of these cells in circulation, especially in diabetics. Purified superoxide dismutase was found to consist of a mixture of glycated and nonglycated forms, the modified enzyme being largely inactivated. The glycosylation sites were identified by incubating unmodified superoxide dismutase with tritiated glucose followed by trapping of the Schiff-base adducts by reduction with $NaBH_4$ and by HPLC separation and analysis of tryptic fragments. Six lysine residues (of 11) were found to become glycated, the remaining 5 being relatively resistant toward this modification. Interestingly, the inactivation of the enzyme was reported to be due mainly to the glycation of Lys-122 and Lys-128, both located in an active-site ligand loop.

It is clear, from the reports mentioned above, that glycation can modify proteins by forming conjugates and crosslinks. Modifications that occur in residues in, or near, the active sites of an enzyme may lead to inactivation.

While the process is significant in long-lived proteins, it probably has a much smaller effect on proteins that turnover rapidly relative to the rate of glycation. Since the latter process develops over a period of many days it is probably safe to assume that glycation contributes little to the pool of modified proteins in most tissues, where protein synthesis and degradation processes are active.

Enzyme Oxidation

Several different mechanisms of protein damage by oxygen radicals have been recently studied by Davies and coworkers [Davies, 1987; Davies et al., 1987; Davies and Delsignore, 1987]. These studies revealed oxidation-related modifications at all levels of protein structure (primary, secondary, and tertiary). The reduced protection against oxidation characteristic of aged tissues [Noy et al., 1985; Glass and Gershon, 1984; Lavie and Gershon, 1988], along with the longer average dwell-time of enzymes in old cells, makes these enzymes particularly good targets for posttranslational modifications by oxidation of susceptible amino acid residues. Indeed, the role of oxygen free radical damage in the aging process has been the basis for the free radical theory of aging [Harman, 1986 and references therein]. A considerable number of reports highlighting the potential significance of protein oxidation in aging have recently appeared. The role of both enzymatic and nonenzymatic mixed-function oxidation (MFO) systems in enzyme modifications has been studied by Stadtman and his coworkers in a large number of enzymes. Several MFO systems were identified and described [Fucci et al., 1983; Kim et al., 1985; Stadtman, 1986; Oliver et al., 1987a,b], all employing a suitable electron donor, molecular oxygen, and ferric ions to generate H_2O_2 and Fe^{2+} as key intermediates that react with metal binding sites on the enzyme to yield active oxygen species, which in turn oxidize amino acids. A detailed account of this process as it relates to the modification of glutamine synthetase was given by Farber and Levine [1986]. Lysine, histidine, arginine proline, and serine residues were found to be most susceptible to this oxidative damage [Oliver et al., 1985]. Gordillo et al. [1988] found that the aging of rat liver malic enzyme, which is reflected by a 36% reduction in catalytic activity, is caused by the oxidation of a single histidine residue. The aging effects could be simulated in young malic enzyme by exposure to a MFO system, and a loss of one histidine was indeed found here too. A convenient way to monitor the MFO-mediated oxidation of proteins was described [Ahn et al., 1987; Starke et al., 1987]; it involves the detection and quantitation of the protein carbonyl derivatives that are a product of the reaction. The oxidation was invariably found to lead to a significant loss of enzyme activity [Stadtman, 1988] and in most cases to a much enhanced susceptibility of the modified enzyme towards proteolysis [Davies and Goldberg, 1987; Rivett and Levine, 1987]. Moreover, several rat liver proteases were found by Rivett [1985a] to exhibit a high degree of selectivity for the oxidized forms of enzymes, suggesting that oxidation may mark proteins for intracellular proteolysis and protein turnover. The fact that many of the enzymes that have been reported to accumulate as modified forms in aged tissues were also effectively inactivated by the MFO systems led Oliver et al. [1987b] and Stadtman [1988] to propose that the modifications of proteins during aging may be due, at least partially, to oxidative damage catalyzed by MFO systems. Indeed, Oliver et al. [1987a] found that the levels of oxidatively modified proteins increase in erythrocytes with their age in circulation, and in cultured fibroblasts with the age of their human donor. The levels of oxidized proteins in fibroblasts from patients with progeria or Werner's syndrome were found to be markedly more elevated, even in cells from young individuals. Starke et al. [1987] reported that elevated levels of oxidized proteins, as observed in aging, can be induced by short exposure of young rats to a 100% oxygen environment. As in aging, the activities of the oxidized enzymes were significantly reduced.

While it is clear from the published data that enzyme oxidation does increase with aging and leads to modification and inactivation of many enzymes, the proposition that this is a significant and common mechanism responsible for the observed age-related effects in enzymes in vivo needs direct experimental support, as it appears to be in variance with some of the reported aging effects in enzymes. Several apparent shortcomings of the enzyme-oxidation model are listed below.

1. From the published work [Oliver et al., 1987a,b] it is clear that the increase with age in carbonyl content is relatively small (both in erythrocytes and in fibroblasts the differences in carbonyl levels between the youngest and oldest populations tested is about 3.5 nmole carbonyl/mg protein, i.e., about 0.1 carbonyl per protein molecule of MW 30,000), and the oxidation is therefore very substoichiometric. Such modification may account for at most a small fraction of the aged protein, while the majority of enzyme molecules in the old tissue must be altered by other mechanisms. Indeed, the data presented by Oliver et al. [1987a,b] for erythrocyte fractions of increasing density (age) show that for all enzymes tested the degree of carbonylation initially increases rapidly with density and then levels off, while the enzyme activities continue to decline. Thus the data show that in some enzymes (GPDH, PGK) about half of the activity loss takes place among fractions that do not differ at all in their carbonyl content, and this loss must arise by other mechanisms.

2. In two enzymes, GPDH and PGK, in which age-related modifications have been studied in great detail [see Gafni, 1985; Rothstein, 1985 for reviews with numerous references], the alterations introduced by oxidation differ from the ones found in vivo at old age. Thus native old PGK is known to be significantly more heat-stable than its young counterpart, while the opposite was found for the oxidized enzyme [Oliver et al., 1987b]. Moreover, while

the activity of native old PGK is practically identical to that of the young enzyme, the oxidation by MFO systems causes rapid inhibition [Fucci et al., 1983]. Dulic and Gafni [1987] compared GPDH forms produced by selective oxidation with hydrogen peroxide, superoxide radicals, and atmospheric oxygen. They concluded that although modified, partially inactivated, forms of the enzyme were produced by the oxidation, these GPDH species differed in several of their properties from the native old enzyme.

3. The enhancement in the susceptibility to proteolysis that accompanies enzyme oxidation [Rivett, 1985a,b; Davies and Goldberg, 1987] is sometimes much larger than the one found in old enzymes. Thus Rivett [1985b] found the oxidative inactivation of enolase to cause a dramatic (20-fold) increase in the rate of degradation of this enzyme by trypsin, while the rates of proteolysis of native young and old enolase forms were reported by Sharma and Rothstein [1978] to differ by less than a factor of 2.

4. At least some enzymes (i.e., fructose 1,6 diphosphatase, and G6PD) that are known to undergo age-related modifications are not affected by the MFO systems tried by Fucci et al. [1983], proving that in these enzymes the aging effects are not introduced by this oxidation reaction and hence other mechanisms must be responsible for their aging.

There is considerable evidence that oxidation can modify and inactivate enzymes both in vitro and in vivo. Moreover, old tissues have been shown to lose some of the protection against this damaging reaction and hence become more susceptible to it. It is, therefore, very reasonable to assume that protein oxidation contributes to the pool of modified enzymes in old cells. However, the level and significance of this contribution to in vivo enzyme aging is still not clear. Even when oxidation does occur, the important demonstration that the age-related effects in an enzyme are indeed identical to the effects caused by its oxidation has yet to be made.

Conformational Isomerization of Enzymes

It is a common phenomenon for enzymes in solution (even when kept under conditions that prohibit any significant development of covalent modifications) to lose activity with time. One major mechanism responsible for this process involves spontaneous conformational transitions by which the correctly folded, active, form of the enzyme is converted to a conformational isomer. The proposition that altered enzymes in old cells result from conformational changes, without covalent modifications, was put forward by Rothstein [1975, 1979, 1982, 1985], who explained this phenomenon as being due to the longer dwell-time of enzymes in the tissues of old animals. Direct evidence supporting this scheme has now been obtained for several enzymes, making conformational isomerization the only proposed mechanism of enzyme aging capable of explaining these effects at the levels observed in vivo. Sharma and Rothstein [1980] found that when samples of pure young and old nematode enolase were unfolded in 2 M guanidine:HCl and refolded under controlled conditions, the refolded products were identical, and very similar (but not identical) to native old enolase. This experimentation clearly proves that young and old forms of enolase do not differ in their sequence (and hence that the aging effects are indeed postsynthetic). However, the fact that the refolded products were simlar to old enolase rather than to the young enzyme, could be argued to indicate the presence of covalent modifications in the old enzyme, these being introduced into the young enzyme during the unfolding–refolding procedure. Deamidation (and isomerization) of asparagines is one such possible change since this reaction is strongly enhanced in unfolded peptides [Clarke, 1987]. While highly unlikely, the possibility that old nematode enolase is covalently modified cannot, therefore, be ruled out.

Reiss and Sacktor [1983], by using monoclonal antibodies, found that old rat kidney maltase contains a higher fraction of inactive forms than the enzyme in young animals. A two-dimentional chromatography of cyanogen bromide fragments revealed no sequence, or covalent, changes in the inactive enzyme, indicating that the age-related modifications in maltase are purely conformational.

Several studies by Gafni and coworkers, reviewed by Gafni [1985], provided strong evidence that the age-related modifications in rat muscle GPDH are also purely conformational. These investigators found that all the observed aging effects may be introduced into samples of young enzyme by selective oxidation of the reactive cysteine 149 residue in the active site followed by a 45 minute incubation at 0°C and subsequently by complete reduction of the oxidized cysteines with excess 2-mercaptoethanol. Since this procedure is highly unlikely to introduce covalent modifications in the enzyme, it was concluded that young and old GPDHs are conformational isomers. It is pertinent to note, however, that these results were obtained when iodine was used as the oxidizing agent, while attempts to simulate the aging effects using hydrogen peroxide or superoxide radicals produced GPDH species different from the native old enzyme, and possibly covalenty modified [Dulic and Gafni, 1987].

Undoubtedly the best documented example of conformationally altered old enzyme is phosphoglycerate kinase. Rothstein and his coworkers conducted a number of studies and found this enzyme to become modified in rat muscle [Sharma et al., 1980] liver [Hiremath and Rothstein, 1982], and brain [Sharma and Rothstein, 1984]. More recently PGK in rat cardiac muscle was also found to become altered with aging [Zuniga and Gafni, 1988]. While the specific activity in all cases was not affected by the animal's age, other properties were altered. Thus, old PGK from any of the tissues listed above was found to be significantly more heat-stable than its young counterpart. Strong evidence that the age-related alterations in PGK do not involve covalent modifications was provided by Hardt and Rothstein [1985] from peptide mapping by HPLC. Three matched pairs of pure young and old PGKs were subjected to digestion each with a different proteinase. A fourth pair of PGKs was treated sequentially by two of the proteinases. In all cases the resulting peptides were resolved by reverse-phase HPLC; the peptide patterns obtained from young and old PGKs were compared and always found to be identical. These results strongly suggest that the age-related modifications in PGK involve no sequence changes and no covalent modifications and are, therefore, purely conformational.

In an effort to test the above hypothesis critically Yuh and Gafni [1987] extensively unfolded samples of PGK purified from skeletal muscles of young and old rats by an 18 hour incubation in a 2 M guanidine hydrochloride solution. A dilution of these unfolded enzyme solutions into a large excess of denaturant-free buffer, followed by a 4 hour incubation at 25°C resulted in a complete reactivation in both cases. The refolding and reactivation kinetics of the young and old enzymes were identical. Moreover, the refolded products were found to have identical heat inactivation kinetics, which coincided with that of native young PGK and differed substantially from the inactivation rate of the old enzyme. Since it is highly unlikely that any covalent modification in old PGK has been removed by the unfolding–refolding procedure, the results can only be explained by assuming that the age-related modifications in rat muscle PGK are purely conformational.

In a subsequent study [Zuniga and Gafni, 1988] the reversibility of aging effects in phosphoglycerate kinase from rat cardiac tissue was also demonstrated by an unfolding–refolding procedure similar to the one used for the skeletal muscle enzyme. It is interesting to note that immunotitration did not reveal any difference between the phosphoglycerate kinase forms in crude extracts from young and old rat hearts [Sharma et al., 1980], and the presence of differences was only established from heat inactivation rates in samples of partially purified enzyme [Zuniga and Gafni, 1988]. It seems likely, therefore, from these findings that the modifications in heart PGK reside in a domain that is not involved in antibody binding—possibly in the interior of the enzyme. While the detailed nature of the age-related structural modifications is yet to be determined, the unfolding–refolding experiments clearly demonstrate that only conformational modifications are involved.

Yuh and Gafni [1987] as well as Zuniga and Gafni [1988] reported that the aging effects in PGK could be simulated in vitro by prolonged incubation of young enzyme under nonreducing conditions with subsequent reduction by excess 2-mercaptoethanol. This simulated-old enzyme displayed heat inactivation kinetics as native old PGK and, like the latter, could be rejuvenated by unfolding–refolding. Based on these observations a two-step mechanism for PGK aging was proposed whereby an initial, and reversible, oxidation of the enzyme is followed by conformational transitions that persist even after subsequent enzyme reduction. Cook and Gafni [1988] tested the involvement of reactive cysteine residues in the initial oxidation of PGK by selectively blocking a fraction of these residues, by methylation, and exposing the modified enzyme to the in vitro aging protocol described above. This methylated enzyme was found to be fully active and was resistant to aging by the procedure that effectively transformed (unprotected) young PGK into its old counterpart. These results give strong support to the hypothesis that a reversible cysteine oxidation is an essential step in the aging process of rat muscle PGK and possibly of other enzymes that possess reactive sulfhydryl groups.

CONCLUSIONS

About 18 years have passed since the first demonstration of the occurrence of age-related modifications in enzymes. During this period it has become clear that aging affects some, but by no means all, enzymes, and that various properties—most notably enzymatic activity—may be altered. The idea that aging at the molecular level may have deleterious effects at the level of cells and tissues has been expressed frequently and indeed is one of the strong motivations for research in this field.

All the experimental evidence so far compiled points to postsynthetic modifications as the origin of the observed aging effects in enzymes. Several plausible, more detailed, mechanisms that have been developed in an effort to account for the observed modifications are described in this review. While it is very possible that no one mechanism is responsible for all the observed aging effects, and that several modification-generating reactions contribute to the pool of altered enzymes in old tissues, it is clear that any significant postsynthetic modification has to be introduced during the dwell-time of the affected protein in the cell. While many pathways may account for the aging of enzymes in tissues in which protein turnover is nonexistent or very slow (eye-lens, erythrocytes), the above requirement severely limits the number of viable mechanisms for enzyme aging in cells in which protein synthesis and degradation are active processes—albeit slower at old age. In this context, the spontaneous change in enzyme conformation deserves special attention. This mechanism, which was originally supported by negative evidence—the failure of attempts to detect covalent changes in amino acid composition—has been recently unambiguously demonstrated to be responsible for the aging of PGK in both heart and skeletal muscle tissues of rats. This demonstration was based on the successful rejuvenation of the old enzyme by an unfolding–refolding protocol. It is pertinent to note that old PGK is not composed of native and modified forms but is rather uniformly in the old conformation, and is fully rejuvenated by unfolding–refolding. To date, conformational isomerization is, therefore, the only mechanism that satisfactorily accounts for enzyme aging at the level at which this effect is observed in vivo in active, nucleated cells.

In his conclusion of the previous review of enzyme aging in this series Rothstein [1985] addressed several areas in which more information would particularly enhance our understanding of the role of enzyme modification in the aging process. Information was deemed necessary on the age of onset of aging effects in various enzymes, on whether they develop gradually or abruptly, and on whether any one enzyme in different tissues is altered at the same age and by the same mechanism. Most of these questions are beginning to be tackled and indeed yield significant information. The most intriguing question—the physiological effects of enzyme modifications in old tissues, especially under conditions whereby the old tissue is challenged or stressed—still has to be answered.

REFERENCES

Ahmed MU, Thorpe SR, Baynes JW (1986): Identification of N-carboxymethyl lysine as a degradation product of fructoselysine in glycated protein. J Biol Chem 261:4889–4894.

Ahn B, Rhee SG, Stadtman ER (1987): Use of fluorescein hydrazide and fluorescein thiosemicarbazide reagents for the fluorometric determination of protein carbonyl groups and for the detection of oxidized protein on polyacrylamide gels. Anal Biochem 161:245–257.

Ambudkar IS, Kuyatt BL, Roth GS, Baum BJ (1988): Modification of ATP-dependent Ca^{2+} transport in rat parotid basolateral membranes during aging. Mech Ageing Dev 43:45–60.

Arai K, Iizuka S, Tada Y, Oikawa K, Taniguchi N (1987a): Increase in the glycosylated form of erythrocyte Cu–Zn superoxide dismutase in diabetes and close association of the nonenzymatic glucosylation with the enzyme activity. Biochim Biophys Acta 924:292–296.

Arai K, Maguchi S, Fujii S, Ishibashi H, Oikawa K, Taniguchi N (1987b): Glycation and inactivation of human Cu–Zn-superoxide dismutase. J Biol Chem 262:16969–16972.

Brownlee M, Vlassara H, Cerami A (1984): Nonenzymatic glycosylation and the pathogenesis of diabetic complications. Ann Intern Med 101:527–537.

Brunauer LS, Clarke S (1986): Age-related accumulation of protein residues which can be hydrolyzed to D-aspartic acid in human erythrocytes. J Biol Chem 261:12538–12543.

Cerami A, Vlassara H, Brownlee M (1987): Glucose and aging. Sci Am 256:90–96.

Cini JK, Gracy RW (1986): Molecular basis of the isozymes of bovine glucose-6-phosphate isomerase. Arch Biochem Biophys 249:500–505.

Cini JK, Cook PF, Gracy RW (1988): Molecular basis for the isozymes of bovine glucose-6-phosphate isomerase. Arch Biochem Biophys 263:96–106.

Clarke S (1987): Propensity for spontaneous succinimide formation from aspartyl and asparaginyl residues in cellular proteins. Int J Pept Protein Res 30:808–821.

Cook LL, Gafni A (1988): Protection of phosphoglycerate kinase against in vitro aging by selective cysteine methylation. J Biol Chem 263:13991–13993.

Croute F, Vidal S, Dupony D, Soleilhavoup JP, Serre G (1985): Studies on catalase, glutathione peroxidase and superoxide dismutase activities in aging cells of paramecium tetraurelia. Mech Ageing Dev 29:53–62.

Davies KJA (1987): Protein damage and degradation by oxygen radicals. I. General aspects. J Biol Chem 262:9895–9901.

Davies KJA, Delsignore ME (1987): Protein damage and degradation by oxygen radicals. III. Modification of secondary and tertiary structure. J Biol Chem 262:9908–9913.

Davies KJA, Goldberg AL (1987): Proteins damaged by oxygen radicals are rapidly degraded in extracts of red blood cells. J Biol Chem 262:8227–8234.

Davies KJA, Delsignore ME, Lin SW (1987): Protein damage and degradation by oxygen radicals. II. Modification of amino acids. J Biol Chem 262:9902–9907.

Dovrat A, Scharf J, Gershon D (1984): Glyceraldehyde 3-phosphate dehydrogenase activity in rat and human lenses and the fate of enzyme molecules in the aging lens. Mech Ageing Dev 28:187–191.

Dovrat A, Scharf J, Eisenbach L, Gershon D (1986): G6PD molecules devoid of catalytic activity are present in the nucleus of the rat lens. Exp Eye Res 42:489–496.

Dulic V, Gafni A (1987): Mechanism of aging of rat muscle glyceraldehyde-3-phosphate dehydrogenase studied by selective enzyme oxidation. Mech Ageing Dev 40:289–306.

Farber JM, Levine RL (1986): Sequence of a peptide susceptible to mixed-function oxidation: Probable cation binding site in glutamine synthetase. J Biol Chem 261:4574–4578.

FitzGerald P (1988): Age-related changes in a fiber-cell-specific extrinsic membrane protein. Curr Eye Res 7:1255–1261.

Froehlich JP, Lakatta EG, Beard E, Spurgeon HA, Weisfeldt ML, Gerstenblith G (1978): Studies of sarcoplasmic reticulum function and contraction duration in young adult and aged rat myocardium. J Mol Cell Cardiol 10:427–438.

Fucci L, Oliver CN, Coon MJ, Stadtman ER (1983): Inactivation of key metabolic enzymes by mixed-function oxidation reactions: Possible implication in protein turnover and ageing. Proc Natl Acad Sci USA 80:1521–1525.

Gabius HJ, Goldbach S, Graupner G, Rehm S, Cramer F (1982): Organ pattern of age-related changes in the aminoacyl tRNA synthetase activity of the mouse. Mech Ageing Dev 20:305–313.

Gafni A (1985): Age-related modifications in a muscle enzyme. In Adelman RC, Dekker EE (eds): "Modification of Proteins During Aging". New York: Alan R. Liss, Inc., pp 19–38.

Gafni A, Noy N (1984): Age related effects in enzyme catalysis. Mol Cell Biochem 59:113–129.

Gafni A, Yuh KCM (1988): Age-related molecular changes in skeletal muscle. In Snyder DL (ed): "Dietary Restriction and Aging." New York: Alan R. Liss, Inc., pp 277–282.

Geiger T, Clarke S (1987): Deamidation, isomerization and racemization at asparaginyl and aspartyl residues in proteins. J Biol Chem 262:785–794.

Glass AG, Gershon D (1984): Decreased enzymic protection and increased sensitivity to oxidative damage in erythrocytes as a function of cell and donor aging. Biochem J 218:531–537.

Goldspink DF, Lewis SEM, Merry BJ (1986): Effects of aging and long term dietary intervention on protein turnover and growth of ventricular muscle in the rat heart. Cardiovasc Res 20:672–678.

Gordillo E, Ayala A, F-Lobato M, Bautista J, Machado A (1988): Possible involvement of histidine residues in the loss of enzymatic activity of rat liver malic enzyme during aging. J Biol Chem 263:8053–8057.

Gracy RW, Yuksel KU, Chapman ML, Cini JK, Jahani M, Lu HS, Oray B, Talent JM (1985): Impaired protein degradation may account for the accumulation of "abnormal" proteins in aging cells. In Adelman RC, Dekker EE (eds): "Modification of Proteins During Aging.": New York: Alan R. Liss, Inc., pp 1–18.

Gutteridge JMC, Westermarck T, Halliwell B (1986): Oxygen radical damage in biological systems. In Johnson JE, Walford R, Harman D, Miquel T (eds): "Free Radicals, Aging and Degenerative Diseases." New York: Alan R. Liss, Inc., pp 99–140.

Hardt H, Rothstein M (1985): Altered phosphoglycerate kinase from old rat muscle shows no change in primary structure. Biochim Biophys Acta 831:13–21.

Harman D (1986): Free radical theory of aging: Role of free radicals in the origination and evolution of life, aging and disease processes. In Johnson JE, Walford R, Harman D, Miquel J (eds): "Free Radicals, Aging and Degenerative Diseases." New York: Alan R. Liss, Inc., pp 3–50.

Harman D (1988): Free radicals in aging. Molec Cell Biochem 84:155–161.
Hazelton GA, Lang CA (1985): Glutathione peroxidase and reductase activities in the aging mouse. Mech Ageing Dev 29:71–81.
Hibino K, Du J, Dillon J, Malinowski K (1988): Age-dependent effect of UV light in abnormal alpha neoprotein formation in the lens. Curr Eye Res 7:1113–1124.
Hiremath LS, Rothstein M (1982): The effect of aging on rat liver phosphoglycerate kinase and comparison with the muscle enzyme. Biochim Biophys Acta 705:200–209.
Holliday R, Tarrant GM (1972): Altered enzymes in ageing human fibroblasts. Nature 238:26–30.
Houben A, Raes M, Houbion A, Remacle J (1984a): Alteration of enzymes in aging human fibroblasts in culture. I. Conditions of the appearance of an alteration in glucose-6-phosphate dehydrogenase. Mech Ageing Dev 25:23–34.
Houben A, Raes M, Houbion A, Remacle J (1984b): Alteration of enzymes in aging human fibroblasts in culture. II. Conditions for the reversibility and the mechanism of the alteration of glucose-6-phosphate dehydrogenase. Mech Ageing Dev 25:35–45.
Jahani M, Gracy RW (1985): Cathepsin B activity in human skin fibroblasts from young, old and premature aging syndromes. Fed Proc 44:876.
Johnson BA, Langmack EL, Aswad DW (1987): Partial repair of deamidation-damaged calmodulin by protein carboxyl methyltransferase. J Biol Chem 262:12283–12287.
Karey KP, Rothstein M (1986): Evidence for the lack of lysosomal involvement in the age-related slowing of protein breakdown in Turbatrix aceti. Mech Ageing Dev 35:169–178.
Kay MMB, Bosman GJCGM, Shapiro SS (1987): Oxidation as a possible mechanism of cellular aging: Vitamin E deficiency causes premature aging and IgG binding to erythrocytes. Proc Natl Acad Sci USA 83:2463–2467.
Klaude M, Von Der Decken A (1986): Protein synthesis activity in vitro in mouse tissues in relation to methionine-cysteine deficiency and age. Mech Ageing Dev 34:261–272.
Kohn RR, Monnier VM, Cerami A (1984): Collagen aging in vitro by nonenzymatic glycosylation and browning. Diabetes 33:57–62.
Kossiakoff AA (1988): Tertiary structure is a principal determinant to protein deamidation. Science 240:191–194.
Kim K, Rhee SG, Stadtman ER (1985): Nonenzymatic cleavage of proteins by reactive oxygen species generated by dithiothreitol and iron. J Biol Chem 260:15394–15397.
Lakatta EG (1987): Cardiac muscle changes in senescence. Annu Rev Physiol 49:519–531.
Lakatta EG, Gerstenblith G, Angell CS, Shock NW, Weisfeldt ML (1975): Prolonged contraction duration in aged myocardium. J Clin Invest 55:61–68.
Lavie L, Gershon D (1988): Oxygen free radical production by mouse peritoneal macrophages as a function of age. Mech Ageing Dev 45:177–189.
Lowenson J, Clarke S (1988): Does the chemical instability of aspartyl and asparaginyl residues in proteins contribute to erythrocyte aging? Blood Cells 14:103–117.
Luo SW, Hutlin HO (1986): Effect of age of winter flounder on some properties of the sarcoplasmic reticulum. Mech Ageing Dev 35:275–289.
Mauron J (1981): The Maillard reaction in food; a critical review from the nutritional standpoint. Prog Food Nutr Sci 5:5–35.
McFadden PN, Clarke S (1986): Protein carboxyl methyltransferase and methyl acceptor proteins in aging and cataractous tissue of the human eye lens. Mech Ageing Dev 34:91–105.
Monnier VM, Kohn RR, Cerami A (1984): Accelerated age-related browning of human collagen in diabetes mellitus. Proc Natl Acad Sci USA 81:583–587.
Nakano M, Tauchi H (1986): Difference in activation by tris (hydroxymethyl) aminomethane of Ca, Mg-ATPase activity between young and old rat skeletal muscle. Mech Ageing Dev 36:287–294.
Narayanan N (1981): Differential alterations in ATP-supported calcium transport activities of

sarcoplasmic reticulum and sarcolemma of aging myocardium. Biochim Biophys Acta 678:442–459.
Narayanan N (1987): Comparison of ATP dependent calcium transport and calcium activated ATPase activities of cardiac sarcoplasmic reticulum and sarcolemma from rats of various ages. Mech Ageing Dev 38:127–143.
Nbemba Fufu DL, Houbion A, Remacle J (1985): Alteration of enzymes in aging human fibroblasts in culture. IV. Effect of glutathione on the alteration of glucose-6-phosphate dehydrogenase. Mech Ageing Dev 32:249–266.
Nohl H (1986): Oxygen radical release in mitochondria: Influence of age. In Johnson JE, Walford R, Harman D, Miquel J (eds): "Free Radicals, Aging and Degenerative Diseases." New York: Alan R. Liss, Inc., pp 77–98.
Noy N, Schwartz H, Gafni A (1985): Age-related changes in the redox status of rat muscle cells and their role in enzyme aging. Mech Ageing Dev 29:63–69.
Oliver CN, Ahn BW, Wittenberger ME, Stadtman ER (1985): Oxidative inactivation of enzymes: Implication in protein turnover and aging. In Ebashi S (ed): "Cellular Regulation and Malignant Growth." Tokyo: Japan Scientific Societies Press Berlin: Springer-Verlag, pp 320–331.
Oliver CN, Ahn BW, Moerman EJ, Goldstein S, Stadtman ER (1987a): Age-related changes in oxidized proteins. J Biol Chem 262:5488–5491.
Oliver CN, Levine RL, Stadtman ER (1987b): A role of mixed-function oxidation reactions in the accumulation of altered enzymes forms during aging. J Am Geriatr Soc 35:947–956.
Orchard CH, Lakatta EG (1985): Intracellular calcium transients and developed tensions in rat heart muscle. A mechanism for the negative interval-strength relationship. J Gen Physiol 86:637–651.
Orgel LE (1963): The maintenance of the accuracy of protein synthesis and its relevance to ageing. Proc Natl Acad Sci USA 49:517–521.
Ota IM, Ding L, Clarke S (1987): Methylation at specific altered aspartyl and asparaginyl residues in glucagon by the erythrocyte protein carboxyl methyltransferase. J Biol Chem 262:8522–8531.
Reff ME (1985): In Schneider EL and Finch CE (eds): "Handbook of the Biology of Aging." New York: Van Nostrand-Reinhold, pp 225–254.
Reiss U, Rothstein M (1974): Heat labile isozymes of isocitrate lyase from aging Turbatrix aceti. Biochim Biophys Res Commun 61:1012–1016.
Reiss U, Sacktor B (1983): Monoclonal antibodies to renal brush border membrane maltase: Age-related antigenic alterations. Proc Natl Acad Sci USA 80:3255–3260.
Richardson A, Semsei I (1987): Effect of aging on translation and transcription. In Rothstein M (ed): "Review of Biological Research in Aging." Vol. 3. New York: Alan R. Liss, Inc., pp 467–483.
Rivett AJ (1985a): Preferential degradation of the oxidatively modified form of glutamine synthetase by intracellular mammalian proteases. J Biol Chem 260:300–305.
Rivett AJ (1985b): The effect of mixed-function oxidation of enzymes on their susceptibility to degradation by a nonlysosomal cysteine proteinase. Arch Biochem Biophys 243:624–632.
Rivett AJ, Levine RL (1987): Enhanced proteolytic susceptibility of oxidized proteins. Biochem Soc Trans 15:816–818.
Rothstein M (1975): Aging and the alteration of enzymes: A review. Mech Aging Dev 4:325–338.
Rothstein M (1979): The formation of altered enzymes in aging animals. Mech Ageing Dev 9:197–202.
Rothstein M (1982): "Biochemical Approaches to Aging." New York: Academic Press, pp 213–255.
Rothstein M (1983): Enzymes, enzyme alteration, and protein turnover. In Rothstein M (ed): "Review of Biological Research in Aging," Vol. 1. New York: Alan R. Liss, Inc., pp 305–314.

Rothstein M (1985): Age-related changes in enzyme levels and enzyme properties. In Rothstein M (ed) "Review of Biological Research in Aging," Vol. 2. New York: Alan R. Liss, Inc., pp 421–433.
Sadana A, Henley JP (1985a): A mathematical analysis of aging influences on enzyme deactivation/activation kinetics. Examples of the influence of regional brain development and drugs in rats. Mech Ageing Dev 30:201–219.
Sadana A, Henley JP (1985b): A mathematical analysis of the influence of aging on enzyme deactivation kinetics. Mech Ageing Dev 32:113–130.
Sarkis GJ, Ashcom JD, Hawdon JM, Jacobson LA (1988): Decline in protease activities with age in the nematode caenorhabditis elegans. Mech Ageing Dev 45:191–201.
Sellinger OZ, Kramer CM, Conger A, Duboff GS (1988): The carboxymethylation of cerebral membrane-bound proteins increase with age. Mech Ageing Dev 43:161–173.
Sharma HK, Rothstein M (1978): Age-related changes in the properties of enolase from Turbatrix aceti. Biochemistry 17:2869–2876.
Sharma HK, Rothstein M (1980): Altered enolase in aged Turbatrix aceti results from conformational changes in the enzyme. Proc Natl Acad Sci USA 77:5865–5868.
Sharma HK, Rothstein M (1984): Altered brain phosphoglycerate kinase from aging rats. Mech Ageing Dev 25:285–296.
Sharma HK, Prasanna HR, Rothstein M (1980): Altered phosphoglycerate kinase in aging rats. J Biol Chem 255:5043–5050.
Sojar HT, Rothstein M (1986): Protein synthesis by liver ribosomes from aged rats. Mech Ageing Dev 35:47–57.
Somville M, Houben A, Raes M, Houbion A, Henin V, Remacle J (1985): Alteration of enzymes in aging human fibroblasts in culture III. Modification of superoxide dismutase as an environmental and reversible process. Mech Ageing Dev 29:35–51.
Srivastava OP (1988a): Purification and characterization of a membrane proteinase from bovine lens. Exp Eye Res 46:269–283.
Srivastava OP (1988b): Age-related increase in concentration and aggregation of degraded polypeptides in human lenses. Exp Eye Res 47:525–543.
Stadtman ER (1986): Oxidation of proteins by mixed-function oxidation systems: Implication in protein turnover, ageing and nentrophil function. Trends Biochem Sci 11:11–12.
Stadtman ER (1988): Protein modification in aging. J Gerontol 43:B112–120.
Starke PE, Oliver CN, Stadtman ER (1987): Modification of hepatic proteins in rats exposed to high oxygen concentration. FASEB J 1:36–39.
Takahashi R, Mori N, Goto S (1985): Alteration of amino-acyl tRNA synthetases with age: Accumulation of heat-labile enzyme molecules in rat liver, kidney and brain. Mech Ageing Dev 33:67–75.
Ulrich P, Pongor S, Chang JCF, Bencsath FA, Cerami A (1985): Aging of proteins. The furoyl furanyl imidazole crosslink as a key advanced glycosylation event. In: Adelman RC, Dekker EE (eds): "Modifications of Proteins During Aging." New York: Alan R. Liss, Inc., pp 83–92.
Van Kleef FSM, Nijink-Mass MJCM, Hoenders HJ (1974): Intracellular degradation of alpha-crystallin. Eur J Biochem 48:563–570.
Velez M, Machado A, Satrustegui J (1985): Age-dependent modifications in rat heart succinate dehydrogenase. Mech Ageing Dev 32:131–140.
Ward WF (1988): Enhancement by food restriction of liver protein synthesis in the aging Fischer 344 rat. J Gerontol 43:B50–53.
Yuh KCM, Gafni A (1987): Reversal of age-related effects in rat muscle phosphoglycerate kinase. Proc Natl Acad Sci USA 84:7458–7462.
Yuksel KU, Gracy RW (1986): In vitro deamidation of human triose phosphate isomerase. Arch Biochem Biophys 248:452–459.
Zuniga A, Gafni A (1988): Age-related modifications in rat cardiac phosphoglycerate kinase. Rejuvenation of the old enzyme by unfolding-refolding. Biochim Biophys Acta 955:50–57.

Alzheimer's Disease: A View Toward the Neurites

Kenneth S. Kosik

INTRODUCTION

Whether one chooses a classical neuropathological definition of Alzheimer's disease or a clinical definition, the separation of phenomena related to normal aging from those related to the disease process remains a conceptual problem. Indeed, whether early Alzheimer's disease is an entity distinct from normal cognitive function or is a continuum with normal senescence has not yet been answered by neuropsychological research. Some of the most common chronic diseases, such as benign prostatic hypertrophy, chronic obstructive lung disease and hypertension, are not clearly separated from normality. For a dementing process the problems are many times more complex since they involve constructing sensitive cognitive instruments for the detection of impaired function and cognitive norms among various cultural, geographic, economic, and ethnic populations. Furthermore, unlike most systemic diseases we don't have serial access to tissue to study the temporal process, but must rely on end-stage analysis. Despite these uncertainties, we do have a remarkably rich body of data consisting of descriptive neuropathology, detailed quantitative and descriptive biochemistry, and an approach toward a molecular genetic analysis.

Since most investigators agree that regardless of the parameter studied the only distinction between Alzheimer's disease and normal senescene is quantitative, it would clearly be of interest to evaluate the Alzheimer data set with regard to the relative value of each datum point as a marker of the disease versus normal senescence. However, multivariate analyses comparing, for example, various neurotransmitter markers taken from a variety of tissue sources, e.g., brain or cerebrospinal fluid, with peripheral markers, such as membrane fluidity or fibroblast calcium, represent formidable statistical and data collection tasks. Perhaps nowhere among the various putative disease markers is the

Center for Neurologic Diseases, Brigham and Women's Hospital, Harvard Medical School, Boston, Massachusetts 02115

question more pressing than with regard to the relationship of the senile plaques to the neurofibrillary tangles. How these two structures covary [DeWolfe Miller et al., 1984] with the disease has led to the somewhat unsatisfactory view that dementia can occur with tangles only, particularly in younger individuals [Ulrich, 1985], or with plaques only, particularly in older individuals [Terry et al., 1987]. While the question of the relationship between these structures is long standing, several new dimensions of the question have appeared as a result of more sensitive probes for these structures. These more sensitive probes are based upon an expanded knowledge of the composition of the senile plaques and neurofibrillary tangles.

DESCRIPTIVE NEUROPATHOLOGY

Senile plaques consist of an amyloid core that contains aggregated filaments derived from the β-amyloid prescursor protein, about which there is now considerable information [recently reviewed in Selkoe, 1989]. As "cotravelers" with the β-amyloid, there are at least two other components, α_1-antichymotrypsin [Abraham et al., 1988] and complement factors [Ishii and Haga, 1984; Pouplard and Emile, 1985; Eikelenboom and Stam, et al., 1982; Alafuzoff et al., 1987]. Other proteins may also be present [Shirahama et al., 1982; Ishii and Haga, 1976]. While it is not believed that these associated proteins are integral components of the amyloid filament, their quantitative and mechanistic contribution to the senile plaque are unknown. The β-amyloid protein is also deposited in a number of other conditions including normal aging, hereditary Dutch disease [Prelli et al., 1988; van Duinen et al., 1987], Down syndrome (in patients who are not necessarily demented), adjacent to arteriovenous malformations [Hart et al., 1988], and in the lipofuscinoses [Maslinska et al., 1989]. It has been shown for some of these conditions and considered probable in all of them that the amyloid-associated proteins are always present. Thus the presence of β-amyloid is not a specific marker of the Alzheimer disease process, and in Alzheimer's disease the β-amyloid protein alone is not a specific marker of the senile plaque. The most confounding condition in which these proteins are found is normal aging. Several highly sensitive techniques including pretreatment of slides with formic acid [Kitamoto et al., 1987; Yamaguchi et al., 1988a], the use of specific antibodies, and sensitive silver stains all reveal amyloid deposition in brain sites such as the cerebellum, subcortical white matter, and spinal cord [Pro et al., 1980; Ogomori et al., 1989], where clinical pathology is not observed. β-amyloid protein is frequently observed around cerebral blood vessels in both Alzheimer's disease and normal aging, in which it also does not have a very good correlation with dementia. Finally, a subgroup with numerous senile plaques in the neocortex has been described in individuals with preserved mental status [Katzman et al., 1988; Crystal et al., 1988].

Even less disease-specific than the presence of the β-amyloid protein is the presence of the neurofibrillary tangle. This structure, the composition of which remains controversial, is seen in a great many neuropathological entities of diverse etiology [Wisniewski et al., 1979; Halper et al., 1986; Hirano et al., 1966; Ishii and Nakamura, 1981; Eidelberg et al., 1987; Horoupian and Yang, 1978; Harada et al., 1988]; the tangles seen in other diseases share immunochemical characteristics with Alzheimer neurofibrillary tangles [Gambetti et al., 1983; Dickson et al., 1985]. The presence of neurofibrillary tangles is, however, less well correlated with normal aging than is the presence of β-amyloid protein. As such, the neurofibrillary tangle probably represents a somewhat nonspecific lesion among the limited repertoire of neuronal responses to a brain insult. One attempt to unify these distinct pathological entities was the report that the neurofibrillary tangle consists of the β-amyloid precursor protein [Guiroy et al., 1987; Masters et al., 1985]. However, these findings could not be corroborated in the hands of other investigators [Roher et al., 1988] and probably result from the inherent difficulties in purifying neurofibrillary tangles—amyloid fibrils may be present as a common contaminant [Gorevic et al., 1986]. By studying the progression of neurofibrillary tangles and senile plaques in Alzheimer patients for 3–7 years following a diagnostic craniotomy, Mann and coworkers [1988] concluded that neither of these markers shows useful correlates with the clinical deficits.

A possible link between the β-amyloid protein at the core of the senile plaque and the neurofibrillary tangle is the dystrophic neurite. Dystrophic neurites are observed around (and often within) the senile plaque amyloid, and they are best visualized by those antibodies that label neurofibrillary tangles. Dystrophic neurites can also be observed at sites remote from senile plaques; these neurites scattered within the neuropil have been referred to as curly fibers [Kowall and Kosik, 1987] or neuropil threads [Braak et al., 1986]. It should be made clear, however, that with any given technique, there is not a perfect correlation between tangle labeling and the labeling of the neurites. For instance, tubulin antibodies label a subset of dystrophic neurites, but do not label neurofibrillary tangles. Ubiquitin antibodies also label a subset of dystrophic neurites [Shaw and Chau, 1988]. Tau antibodies, which are particularly good at labeling large numbers of dystrophic neurites, fail to label a small subset of them that are labeled by NADPH [Kowall and Kosik, 1987]. Alz 50 appears to label neurites in a pattern similar to that of tau, which is not surprising given that the epitope recognized by Alz 50 has been localized to the carboxy terminus of tau [Kosik et al., submitted]. Another class of antibodies that recognizes neurites is that directed against phosphorylated neurofilaments. While this staining has been attributed to the cross-reactivity of these antibodies with tau [Ksiezak-Reding et al., 1987; Nukina et al., 1987], several neurofilament antibodies have now been described that recognize neurofibrillary tangles but fail to label tau on immunoblots [Gambetti et al., 1986; Lee et al., 1988].

This class of antibodies was found to be particularly good at recognizing the neuritic dystrophy in the neural epithelium of the nose and as such was considered to have some potential diagnostic utility [Talamo et al., 1989]. What exactly dystrophic neurites are has not been established, and a pervasive biochemical marker for these structures remains a worthy pursuit.

Double-labeling studies demonstrate every combination of neurites and amyloid protein [Dickson et al., 1988]. Some amyloid is deposited without any obvious association with neurites. This is almost always the case with diffuse plaques, which rarely, if ever, include any neuritic response [Yamaguchi et al., 1988a,b]. Some compacted plaques may also lack a neuritic response. On occasion dystrophic neurites can be observed around a central core that lacks any detectable β-amyloid by light microscopic immunohistochemistry.

In sorting out this closely intertwined process of the neuritic pathology from the deposition of β-amyloid protein, one is led to conclude that β-amyloid is not a direct cause of the clinical dementia and may bear a closer association with the aging process than the disease process. The neuritic pathology, on the other hand, may bear a closer direct relationship to the clinical dementia, but it remains unclear to what extent this pathology is fundamental to the disease process. Dickson et al. [1988] reached a similar conclusion because they observed extensive neuritic degeneration in Alzheimer brain, but not in nondemented elderly individuals with senile plaques or in non-Alzheimer dementia cases. It would be of interest to know to what extent the reported 30% of elderly demented patients that lack neurofibrillary tangles [Terry et al., 1987] have neurites around their senile plaques. The hypothetical scenario in which neuritic pathology occurs in vulnerable brain regions in response to a certain amyloid burden is supported by some temporal studies in Down syndrome [Giaconne et al., 1989; Mann, 1988; Wisniewski et al., 1985]. A conflicting conclusion was reached in the non-human primate regarding the chronological sequence of amyloid and the neuritic pathology [Cork and Price, personal communication]. A caveat in these studies, however, is that while amyloid deposition may precede neurite formation, that fact alone is not sufficient to establish causation of neurites. An idealized experimental approach toward these problems would be to separate those biochemical aspects of Alzheimer's disease that are more likely related to normal aging from those that are more closely associated with the clinical disease process.

CELL BIOLOGICAL APPROACHES TO THE NEURITES

Our knowledge of the composition of the neurites around amyloid plaques and those within the neuropil is based mostly upon immunocytochemical studies (described above) and to a lesser degree upon ultrastructural analyses. Neurites around senile plaques have been observed to contain paired helical filaments. These ultrastructural studies are difficult due to the poor preservation of post mortem brain tissue; however, one report suggested that neuropil neurites may consist of large numbers of densely packed straight filaments [Perry et al., 1987]. None of the available studies is sufficient to decide the cell biological context in which neurites occur, that is, whether they represent degenerative or regenerative phenomena or both. This question has been of some recent interest.

McKee et al. [1989a] have suggested that tangle-bearing neurons are undergoing simultaneous degeneration and regeneration. This suggestion was based on the observation that tau immunocytochemistry revealed filopodial-like neurites extending from pyramidal cell bodies and their proximal processes [McKee et al., 1989a; Ihara, 1988]. Furthermore, immunocytochemical labels revealed supernumerary basilar dendrites on pyramidal cells in Alzheimer's disease [McKee et al., 1989a]. Previous studies that did not reveal expansions of dendritic fields [Buell and Coleman, 1981; Flood et al., 1987] may have been limited by the problem of selective filling inherent in the Golgi technique. However, other investigators have observed sprouting with similar techniques [Probst et al., 1983; Scheibel and Tomiyasu, 1978; Ferrer et al., 1983; Arendt et al., 1986; Paula-Barbosa et al., 1980]. The possibility has been raised that the dystrophic neurites in the neuropil themselves represent a regenerative phenomenon [Ihara, 1988], particularly since they are studded with "microspikes" that are characteristic of sprouts [McKee et al., 1989a]. These observations were enhanced by the use of very different techniques and probes demonstrating that a growth response occurs in the Alzheimer brain [Geddes et al., 1985; Uchida et al., 1988] and that the hippocampus is capable of synaptic reorganization [Sutula et al., 1988].

Neural regeneration has been associated with a fetal pattern of gene expression [Miller et al., 1989; Hoffman and Cleveland, 1988]. For this reason it was of interest when one of the tau sequences directly obtained from paired helical filaments corresponded to a fetal isoform [Kosik et al., 1989a]. This work focused upon a splice junction in the carboxy terminal microtubule binding domain of tau where a fourth tandemly repeated sequence is inserted among the three trandem repeats that are present in the immature brain [Kosik et al., 1989a]. If sprouting is responsible for the switch to a fetal tau isoform, this reversion to a fetal splicing pattern is more likely linked to the development of tangles and neurites than to a direct increase in tau synthesis. To the limited extent that measurements can be made there has been no evidence for increased tau synthesis in Alzheimer's disease, and in experimental models of regeneration tau mRNA actually decreases after axotomy [Argasinski et al., 1989]. This latter observation is likely related to the microtubule-stabilizing function of tau and the possibility of increased plasticity in the immediate postaxotomy period. Another immature microtubule-associated protein that has recently been associated with neurites is MAP 1b [McKee et al., 1989b]. On the other hand, several antibodies that should label growth cone structures, such as GAP 43 and actin, have not been revealing [Kowall and Kosik, unpublished observations].

An alternative way of viewing the dystrophic neurites in the neuropil is that they represent a dendritic dystrophy. At sites where it is possible to make such observations it appears that the neurites tend to be present in dendritic fields. This appearance is best studied in regions immediately surrounding the layer II entorhinal star cell clusters, which normally are labeled by the somatodendritic microtubule-associated protein MAP2; in Alzheimer's disease this pattern is replaced by tau-immunoreactive dystrophic neurites [McKee et al., 1989a]. Braak and Braak [1988] have directly observed that dystrophic neurites occur in dendrites and are contiguous with tangle-bearing neuronal cell bodies. Recently we have reported that the location of tau mRNA within neurons is coincident with the location of the neurofibrillary tangles [Kosik et al., 1989b]. Thus, while the normal tau protein is axonal, the message is in the soma and proximal dendrite of the neuron. Since cortical axons are generally considered to be devoid of ribosomes beyond the axon hillock, it is not surprising that the tau mRNA does not colocalize with the protein; however, the problem remains as to how tau protein is selectively targeted to the axon. The discovery of anterograde and retrograde motor proteins [for review see Vale, 1987 and Vallee et al., 1989] has provided a potential mechanism for selective sorting of vesicle-bound proteins [Black and Baas, 1989]. The problem of sorting, however, remains for microtubule-associated proteins. Some degree of localization for cytoskeletal proteins may be conferred by the topography of the ribosomal array [Lawrence aand Singer, 1986]. The distribution of the ribosomes for MAP2 [Garner et al., 1988] is quite distinct from the distribution we reported for tau [Kosik et al., 1989]. The ribosomal population for MAP2, a somatodendritic protein, extends to the distal portions of the pyramidal cell apical dendrite. One mechanism by which tau could accumulate in neuronal somata might relate to an impairment in the correct targeting of tau from the ribosome to its axonal destination.

Some degree of plasticity in tau localization has been observed during the establishment of neuronal polarity in culture [Kosik and Finch, 1987]. Specifically when rat cerebrocortical neurons are first plated and when they regenerate their sheared neurites, tau is observed in all of the neurites. Tau protein continues to be observed in all the neurites even when the neurites begin to assume their identity as axons or dendrites. Only relatively late in culture does one observe the segregation of tau into the axonal compartment. Alzheimer's disease might present cues that bear some similarity to the injury of neurites during neuronal dissociation for culture which result in the loss of discrimination in targeting of MAPs to previously polarized neurites. We have hypothesized that part of the mechanism by which tau is targeted to the axon relates to its association with the microtubule. When tau is bound to microtubules it moves in the slow transport system [Tytell et al., 1984]; however, when it is free of microtubules it very likely moves rapidly by diffusion. Thus, under normal conditions, immediately after the synthesis of tau, it may remain free of microtubules until it reaches the axon, where it associates with microtubules and increases in concentration as part of the slow transport system. This mechanism might explain the failure to detect tau protein at its site of synthesis since it rapidly diffuses. Whether its somatodendritic accumulation in Alzheimer's disease represents a premature association with microtubules or a failure to associate with axonal microtubules correctly is of interest. The lack of microtubules in neurofibrillary tangles favors the latter speculation.

The presence of tau in Alzheimer neurites, located in distal dendritic fields, suggests that tau is displaced some distance from its normal locus. Recently a model system using the lamprey has been described in which following a close axotomy, ectopic axonal sprouting occurs from dendritic tips and the dendritic processes assume an axonal appearance ultrastructurally [Hall and Cohen, 1983; Hall and Cohen, 1988a,b; Hall et al., 1989]. The development and characterization of model regenerative and degenerative neuronal systems, as well as detailed quantitative and correlative analysis of the neuritic dystrophy in the Alzheimer brain, should permit additional insights into this aspect of the pathology.

REFERENCES

Abraham CA, Selkoe DJ, Potter H (1988): Immunochemical identification of the serine protease inhibitor α_1-antichymotrypsin in the brain amyloid deposits of Alzheimer's disease. Cell 52:487–501.

Alafuzoff I, Adolfsson R, Grundke-Iqbal I, Winblad B (1987): Blood–brain barrier in Alzheimer's dementia and in non-demented elderly. Acta Neuropathol (Berl) 73:160–166.

Argasinski A, Wong J, Kosik KS, Oblinger MM (1989): Changes in tan gene expression in DRG neurons during development and regeneration. J Cell Biol 109:79a.

Arendt T, Zvegintseva HG, Leontovich TA (1986): Dendritic changes in the basal nucleus of Meynert and in the diagonal band nucleus in Alzheimer's disease—a quantitative Golgi investigation. Neuroscience 19:1265–1278.

Black MM, Baas PW (1989): The basis of polarity in neurons. Trends Neurosci 12:211–214.

Braak H, Braak E (1988): Neuropil threads occur in dendrites of tangle-bearing nerve cells. Neuropathol Applied Neurobiol 14:39–44.

Braak H, Braak E, Grundke-Iqbal I, Iqbal K (1986): Occurrence of neuropil threads in the senile human brain and in Alzheimer's disease: A third location of paired helical filaments outside of neurofibrillary tangles and neuritic plaques. Neurosci Lett 65:351–355.

Buell SJ, Coleman PD (1981): Quantitative evidence for selective dendritic growth in normal human aging but not in senile dementia. Brain Res 214:23–41.

Crystal H, Dickson D, Fuld P, Masur D, Scott R, Mehler M, Masdeu J, Kawas C, Aronson M, Wolfson L (1988): Clinico-pathologic studies in dementia: Nondemented subjects with pathologically confirmed Alzheimer's disease. Neurology 38:1682–1687.

DeWolfe Miller F, Hicks SP, D'Amato CJ, Landis JR (1984): A descriptive study of neuritic plaques and neurofibrillary tangles in an autopsy population. Am J Epidemiol 120:331–341.
Dickson DW, Kress Y, Crowe A, Yen S-H (1985): Monoclonal antibodies to Alzheimer neurofibrillary tangles (ANT): 2. Demonstration of a common antigenic determinant between ANT and neurofibrillary degeneration in progressive supranuclear palsy. Am J Pathol 120:292–303.
Dickson DW, Farlo J, Davies P, Crystal H, Fuld P, Yen S-HC (1988): Alzheimer's disease: A double labeling immunohistochemical study of senile plaques. Am J Pathol 132:86–101.
Eidelberg D, Sotrel A, Joachim C, Selkoe D, Forman A, Pendlebury WW, Perl DP (1987): Adult onset Hallervorden–Spatz disease with neurofibrillary pathology. Brain 110:993–1013.
Eikelenboom P, Stam FC (1982): Immunoglobulins and complement factors in senile plaques: An immunohistoperoxidase study. Acta Neuropathol (Berl) 57:239–242.
Ferrer I, Aymami A, Rovira A, Grau Veciana JM (1983): Growth of abnormal neurites in atypical Alzheimer's disease. Acta Neuropathol 59:167–170.
Flood DG, Buell SJ, Horwitz GJ, Coleman PD (1987): Dendritic extent in human dentate gyrus granule cells in normal aging and senile dementia. Brain Res 402:205–216.
Gambetti P, Shecket G, Ghetti B, et al. (1983): Neurofibrillary change in human brain: An immunocytochemical study with neurofilament antiserum. J Neuropathol Exp Neurol 42:69–79.
Gambetti P, Perry G, Autilio-Gambetti L (1986): Paired helical filaments: Do they recognize contain neurofilament epitopes? Neurobiol Aging 7:451–452.
Garner CC, Tucker RP, Matus A (1988): Selective localization of messenger RNA for cytoskeletal protein MAP2 in dendrites. Nature 336:674–677.
Geddes JW, Monaghan DT, Cotman CW, Lott IT, Kim RC, Chui HC (1985): Plasticity of hippocampal circuitry in Alzheimer's disease. Science 230:1179–1181.
Giaccone G, Tagliavini F, Linoli G, Bouras C, Frigerio L, Frangione B, Bugiani O (1989): Down patients: Extracellular preamyloid deposits precede neuritic degeneration and senile plaques. Neurosci Lett 97:232–238.
Gorevic PD, Goni F, Pons-Estel B, Alvarez F, Peress NS, Frangione B (1986): Isolation and partial characterization of neurofibrillary tangles and amyloid plaque core in Alzheimer's disease: Immunohistological studies. J Neuropathol Exp Neurol 45:647–664.
Guiroy DC, Miyazaki M, Multhaup G, Fischer P, Garruto RM, Beyreuther K, Masters C, Simms G, Gibbs CJ, Gajdusek DC (1987): Amyloid of neurofibrillary tangles of Guamanian Parkinson-dementia and Alzheimer disease share identical amino acid sequence. Proc Natl Acad Sci USA 84:2073–2077.
Hall GF, Cohen MJ (1983): Extensive dendritic sprouting induced by close axotomy of central neurons in the lamprey. Science 222:518–521.
Hall GF, Cohen MJ (1988a): The pattern of dendritic sprouting and retraction induced by axotomy of lamprey central neurons. J Neurosci 8:3584–3597.
Hall GF, Cohen MJ (1988b): Dendritic amputation redistributes sprouting evoked by axotomy in lamprey central neurons. J Neurosci 8:3598–3606.
Hall GF, Poulos A, Cohen MJ (1989): Sprouts emerging from the dendrites of axotomized lamprey central neurons have axonlike ultrastructure. J Neurosci 9:588–599.
Halper J, Scheithauer BW, Okazaki H, Laws ER Jr (1986): Meningio-angiomatosis: A report of six cases with special reference to the occurrence of neurofiibrillary tangles. J Neuropathol Exp Neurol 45:426–46.
Harada K, Krucke W, Mancardi JL, Mandybur TI (1988): Alzheimer's tangles in sudanophilic leukodystrophy. Neurology 38:55–59.
Hart MN, Merz P, Bennett-Gray J, Menezes AH, Goeken JA, Schelper RL, Wisniewski HM (1988): Beta-amyloid protein of Alzheimer's disease is found in cerebral and spinal cord vascular malformations. Am J Pathol 132:167–172.
Hirano A, Malamud N, Elizan TS, Kurland LT (1966): Amyotrophic lateral sclerosis and Parkinson-dementia complex on Guam. Arch Neurol 15:35–51.

Hoffman PN, Cleveland DW (1988): Neurofilament and tubulin expression recapitulates the developmental program during axonal regeneration: Induction of a specific beta-tubulin isotype. Proc Natl Acad Sci USA 85:4530–4533.

Horoupian DS, Yang SS (1978): Paired helical filaments in neurovisceral lipidosis (juvenile dystonic Lipidosis). Ann Neurol 4:404–411.

Ihara Y (1988): Massive somatodendritic sprouting of cortical neurons in Alzheimer's disease. Brain Res 459:138–144.

Ishii T, Haga S (1976): Immuno-electron microscopic localization of immunoglobulins in amyloid fibrils of senile plaques. Acta Neuropathol 36:243–249.

Ishii T, Nakamura Y (1981): Distribution and ultrastructure of Alzheimer's neurofibrillary tangles in postencephalitis parkinsonism of Economo type. Acta Neuropathol (Berl) 55:59–62.

Ishii T, Haga S (1984): Immuno-electron-microscopic localization of complements in amyloid fibrils of senile plaques. Acta Neuropathol 63:296–300.

Katzman R, Terry R, DeTeresa R, Brown T, Davies P, Fuld P, Renbing X, Peck A (1988): Clinical, pathological, and neurochemical changes in dementia: A subgroup with preserved mental status and numerous neocortical plaques. Ann Neurol 23:138–144.

Kitamoto T, Ogomori K, Tateishi J, Prusiner SB (1987): Formic acid pretreatment enhances immunostaining of cerebral and systemic amyloids. Lab Invest 57:230–236.

Kosik KS, Finch EA (1987): MAP2 and tau segregate into axonal and dendritic domains after the elaboration of morphologically distinct neurites: An immunocytochemical study of cultured rat cerebrum. J Neurosci 7:3142–3153.

Kosik KS, Orecchio LD, Bakalis S, Neve RL (1989a): Developmentally regulated expression of specific tau sequences. Neuron 2:1389–1397.

Kosik KS, Crandall JE, Mufson E, Neve RL (1989b): Tau in situ hybridization in normal and Alzheimer brain: A predominant localization in the neuronal somatodendritic compartment. Ann Neurol 26: (in press).

Kosik KS, Bakalis S, Scoble H: An *in vivo* phosphate and an adjacent conformational epitope in the carboxy terminus of bovine tau epitope (submitted).

Kowall NW, Kosik KS (1987): Axonal disruption and aberrant localization of tau protein characterize the neuropil pathology of Alzheimer's disease. Ann Neurol 22:639–643.

Ksiezak-Reding H, Dickson DW, Davies P, Yen S-H (1987): Recognition of tau epitopes by anti-neurofilament antibodies that bind to Alzheimer neurofibrillary tangles. Proc Natl Acad Sci USA 84:3410–3414.

Lawrence JB, Singer RH (1986): Intracellular localization of messenger RNAs for cytoskeletal proteins. Cell 45:407–415.

Lee VM-Y, Otvos L, Schmidt ML, Trojanowski JQ (1988): Immunological similarities between multi-phosphorylation repeats in the large neurofilament (NF) proteins and neurofibrillary tangles (NFT). J Neuropathol Exp Neurol 47:319.

Mann DMA (1988): The pathological association between Down syndrome and Alzheimer disease. Mech Aging Dev 43:99–136.

Mann DMA, Marcyniuk B, Yates PO, Neary D, Snowden JS (1988): The progression of the pathological changes of Alzheimer's disease in frontal and temporal neocortex examined both at biopsy and at autopsy. Neuropathol Appl Neurobiol 14:177–195.

Maslinska D, Wisniewski KE, Goebel HH, Kim KS (1989): Amyloid-beta-protein epitopes in brains with different forms of neuronal ceroid-lipofuscinosis. J Neuropathol Exp Neurol 48:332.

Masters CL, Multhaup G, Simms G, Pottigiesser J, Martins RN, Beyreuther K (1985): Neuronal origin of a cerebral amyloid: Neurofibrillary tangles of Alzheimer's disease contain the same protein as the amyloid of plaque cores and blood vessels. EMBO J 4:2757–2763.

McKee AC, Kowall NW, Kosik KS (1989a): Microtubular reorganization and growth response in Alzheimer's disease. Ann Neurol; in press.

McKee A, Binder L, Kowall N (1989b): [Abstract]. Am Neurol Acad.
Miller FD, Tetzlaff W, Bisby MA, Fawcett JW, Milner RJ (1989): Rapid induction of the major embryonic alpha-tubulin mRNA, T alpha 1, during nerve regeneration in adult rats. J Neurosci 9:1452–1463.
Nukina N, Kosik KS, Selkoe DJ (1987): Recognition of Alzheimer paired helical filaments by monoclonal neurofilament antibodies is due to cross-reaction with tau protein. Proc Natl Acad Sci USA 84:3415–3419.
Ogomori K, Kitamoto T, Tateishi J, Sato Y, Suetsugu M, Abe M (1989): Beta-protein is widely distributed in the central nervous system of patients with Alzheimer's disease. Am J Pathol 134:243–251.
Paula-Barbosa MM, Mota Cardoso R, Guimaraes ML, Cruz C (1980): Dendritic degeneration and regrowth in the cerebral cortex of patients with Alzheimer's disease. J Neurol Sci 45:129–134.
Perry G, Mulvihill P, Manetto V, Autilio-Gambetti L, Gambetti P (1987): Immunocytochemical properties of Alzheimer straight filaments. J Neurosci 7:3736–3738.
Pouplard A, Emile J (1985): New immunological findings in senile dementia. Interdiscipl Topics Gerontol 19:62–71.
Prelli F, Castano EM, van Duinen SG, Bots GThAM, Luyendij W, et al. (1988): Different processing of Alzheimer's β-protein precursor in the vessel wall of patients with hereditary cerebral hemorrhage with amyloidosis—Dutch type. Biochem Biophys Res Commun 151:1150–1155.
Pro JD, Smith CH, Sumi SM (1980): Presenile Alzheimer disease: Amyloid plaques in the cerebellum. Neurology 30:820–825.
Probst A, Basler V, Bron B, Ulrich J (1983): Neuritic plaques in senile dementia of Alzheimer type: A Golgi analysis in the hippocampal region Brain Res 268:249–254.
Roher AE, Palmer KC, Chau V, Ball MJ (1988): Isolation and chemical characterization of Alzheimer's disease paired helical filament cytoskeleton: Differentiation from amyloid plaque core protein. J Cell Biol 107:2703–2716.
Scheibel AB, Tomiyasu U (1978): Dendritic sprouting in Alzheimer's presenile dementia. Exp Neurol 60:1–8.
Selkoe DJ (1989): Biochemistry of altered brain proteins in Alzheimer's disease. Annu Rev Neurosci 12:463–490.
Shaw G, Chau V (1988): Ubiquitin and microtubule-associated protein tau immunoreactivity each define distinct structures with differing distributions and solubility properties in Alzheimer brain. Proc Natl Acad Sci USA 85:2854–2858.
Shirahama T, Skinner M, Westermark P, Rubinow A, Cohen AS, Brun A, Kemper TL (1982): Senile cerebral amyloid: Prealbumin as a common constituent in the neuritic plaque, in the neurofibrillary tangle, and in the microangiopathic lesion. Am J Pathol 107:41–50.
Sutula T, Xiao-Xian H, Cavazos J, Scott G (1988): Synaptic reorganization in the hippocampus induced by abnormal functional activity. Science 239:1147–1150.
Talamo BR, Rudel RA, Kosik KS, Lee VM-Y, Neff S, Adelman L, Kauer JS (1989): Pathological changes in olfactory neurons in patients with Alzheimer's disease. Nature 337:736–739.
Terry RD, Hansen LA, DeTeresa R, Davies P, Tobias H, Katzman R (1987): Senile dementia of the Alzheimer type without neocortical neurofibrillary tangles. J Neuropathol Exp Neurol 46:262–268.
Tytell M, Brady ST, Lasek RJ (1984): Axonal transport of a subclass of tau proteins: Evidence for the regional differentiation of microtubules in neurons. Proc Natl Acad Sci USA 81:1570–1574.
Uchida Y, Ihara Y, Tomonaga M (1988): Alzheimer's disease brain extract stimulates the survival of cerebral cortical neurons from neonatal rats. Biochem Biophys Res Commun 150:1263–1267.
Ulrich J (1985): Alzheimer Changes in nondemented patients younger than sixty-five: Possible

early stages of Alzheimer's disease and senile dementia of Alzheimer type. Ann Neurol 17:273–277.
Vale RD (1987): Intracellular transport using microtubule-based motors. Annu Rev Cell Biol 3:347–378.
Vallee RB, Shpetner HS, Paschal BM (1989): The role of dynein in retrograde axonal transport. Trends Neurosci 12:66–70.
van Duinen SG, Castano EM, Prelli F, Bots GTAB, Luyendijk W, Frangione B (1987): Hereditary cerebral hemorrhage with amyloidosis in patients of Dutch origin is related to Alzheimer disease. Proc Natl Acad Sci USA 84:5991.
Wisniewski K, Jervis GA, Moretz RC, Wisniewski HM (1979): Alzheimer neurofibrillary tangles in diseases other than senile and presenile dementia. Ann Neurol 5:288–294.
Wisniewski KE, Wisniewski HM, Wen GY (1985): Occurrence of neuropathological changes and dementia of Alzheimer's disease in Down's syndrome. Ann Neurol 17:278–282.
Yamaguchi H, Hirai S, Morimatsu M, Shoji M, Ihara Y (1988a): A variety of cerebral amyloid deposits in the brains of the Alzheimer-type dementia demonstrated by beta protein staining. Acta Neuropathol 76:541–549.
Yamaguchi H, Hirai S, Morimatsu M, Shoji M, Harigaya Y (1988b): Diffuse type of senile plaques in the brains of Alzheimer-type dementia. Acta Neuropathol 538:1–6.

Food Restriction Research: Past and Present Status

Byung P. Yu

INTRODUCTION

The popularity of food restriction as the most efficient and convenient means to intervene in aging and disease processes has led to the accumulation of a large amount of information during the last decade. This situation poses researchers in the field great difficulty in keeping abreast of the progress being made on this important phenomenon.

To alleviate some of these difficulties, in the first section of this review, a somewhat unorthodox tabulated format of cataloging the data is used to provide the reader with quick and easy access to the original source. It should be noted that the data included in the table deal mainly with recent studies. The impressive numbers of recent studies listed in the table only signify the usefulness of food restriction in aging research, with diversified action on all levels of physiological systems including subcellular and genomic activities.

The second part of this review attempts to cover the most interesting findings on food restriction that have appeared during the past few years. These findings focus on the effects of modulating dietary components, on lipid peroxidation of membranes, on cytosolic antioxidant systems, on malondialdehyde oxidation, on gene expression activity, and on repair of DNA damage. On aggregate these data bring us one step closer to finding the cellular mechanisms for the action of food restriction.

Department of Physiology, Health Science Center at San Antonio, University of Texas, San Antonio, Texas 78284-7756

TABULATION OF STUDIES ON THE EFFECTS OF FOOD RESTRICTION

TABLE I. Studies on the Effects of Food Restriction

Physiological Parameters Studied	Tissues or Procedures Used	Species or Strain	Age Effect	Types of Restriction	Dietary Effect	Reference
1. Blood components:						
triglyceride	serum	male Fischer 344 rats	increased	calorie	decreased	Masoro et al. (1983); Yu et al. (1984)
rate of VLDL triglyceride clearance	serum	male Fischer 344 rats	not affected	calorie	no effect	Masoro et al. (1983); Yu et al. (1984)
ketone bodies	serum	male Fischer 344 rats	increased but late in life	calorie	no late-life increase	Masoro et al. (1983)
glucose	serum	male Fischer 344 rats	elevated	calorie	attenuated	Masoro et al. (1989)
cholesterol	serum	male Fischer 344 rats	increased	calorie	decreased	Masoro et al. (1983); Yu et al. (1984)
phospholipids	serum	male Fischer 344 rats	increased	calorie	decreased	Liepa et al. (1980)
vitamin E	serum	male Fischer 344 rats	elevated	calorie	lowered	Laganiere and Yu (1989)
	plasma	male Lobund-Wistar rats	increased	calorie	reduced	Chen and Lowry (1989)
phosphorus	serum	male Fischer 344 rats	decreased	calorie	no dietary differences	Kalu et al. (1988b)
creatinine	serum	male Fischer 344 rats	increased	calorie	lower	Kalu et al. (1988b)
Ca (ionized)	serum	male Fischer 344 rats	no change	calorie	same as control	Kalu et al. (1988b)
Ca (total)	serum	male Fischer 344 rats	elevated	calorie	lower	Kalu et al. (1988b)
25 (OH) Vitamin D	serum	male Fischer 344 rats	decreased	calorie	increased	Kalu et al. (1988b)
2. Hormones:						
insulin	serum	male Sprague-Dawley rats	increased	calorie	decreased	Reaven and Reaven (1981)
	serum	male Fischer 344 rats	increased	calorie	slight reduction	Masoro et al. (1983)
glucagon	serum	male Fischer 344 rats	increased	calorie & protein	no effect	Masoro et al. (1983)
calcitonin	serum and thyroid	male Fischer 344 rats	increased	calorie	decreased	Kalu et al. (1983)
	plasma clearance	male Fischer 344 rats	not tested	calorie	no effect	Kalu et al. (1988)

Continued

TABLE I. Studies on the Effects of Food Restriction (*Continued*)

Physiological Parameters Studied	Tissues or Procedures Used	Species or Strain	Age Effect	Types of Restriction	Dietary Effect	Reference
T_3,						
	serum	male Lobund-Wistar rats	no change	calorie	no effect	Snyder and Towne (1989)
	plasma	male Sprague-Dawley rats	no change	calorie	reduced	Merry and Holehan (1984)
	plasma	male Fischer 344 rats	not tested	calorie	reduced 24 hr. mean and eliminated diurnal rhythm	Herlihy et al. (1989)
T_4						
	plasma	male Fischer 344 rats	not tested	calorie	reduced 24 hr. mean and altered diurnal rhythm	Herlihy et al. (1989)
	serum	male Lobund-Wistar rats	decreased	calorie	not affected	Snyder and Towne (1989)
	blood	male Sprague-Dawley rats	no change	calorie	reduced (up to 457 days)	Merry and Holehan (1984)
growth hormone						
	plasma and pituitary	male Sprague-Dawley rats	not tested	calorie	depressed	Merry and Holehan (1984)
FSH						
	serum	female Sprague-Dawley rats	no change (after prepubertal peak)	calorie	decreased (by 30-40 days)	Holehan and Merry (1985)
LH						
	serum	female Sprague-Dawley rats	marginal change	calorie	decreased (ovulatory peak)	Holehan and Merry (1985)
progesterone						
	serum	female Sprague-Dawley rats	increased	calorie	decreased	Holehan and Merry (1985)
prolactin						
	serum	male Lobund-Wistar rats	increased	calorie	suppressed	Snyder and Towne (1989)
parathyroid						
	serum and thyroid	male Fischer 344 rats	increased	calorie	suppressed	Kalu et al. (1984); Kalu et al. (1988b)
testosterone						
	serum	male Sprague-Dawley rats	fall after 100 days	calorie	suppressed peak height	Merry and Holehan (1984)
	serum	male Lobund-Wistar rats	decreased	calorie	prevented	Snyder and Towne (1989)
	serum	male Fischer 344 rats	decreased	calorie	prevented	Chatterjee et al. (1989)
3. Receptor Activity:						
number						
	lung, β-adrenergic receptor	male Fischer 344 rats	decreased	calorie	increased	Scarpace and Yu (1987)
	striatal dopamine receptor	male Wistar rats	decreased	calorie	sparing effect	Roth et al. (1984)
binding						
	lung, β-adrenergic receptor	male Fischer 344 rats	no change	calorie	no change	Scarpace and Yu (1987)
adenylate cyclase						
	liver, fluoride-stimulated	male Fischer 344 rats	no effect	calorie	no effect	Yu et al. (1984)

Continued

TABLE I. Studies on the Effects of Food Restriction (*Continued*)

Physiological Parameters Studied	Tissues or Procedures Used	Species or Strain	Age Effect	Types of Restriction	Dietary Effect	Reference
β-adrenergic responsiveness	lung, isoproterenol-stimulated	male Fischer 344 rats	decreased	calorie	increased	Scarpace and Yu (1987)
	whole blood vessels	Sprague-Dawley rats	decreased	calorie	partially prevented	Volicer et al. (1983)
4. Collagen:						
amount	lung, liver, kidney, gastrocnemius	male Fischer 344 rats	increased	calorie	decreased	Yu et al. (1982)
cross-linking	tail tendon, solubility	male B6CBAF1 mice	increased	calorie	slowed	Harrison and Archer (1987)
	liver, lung, kidney	male Wister rats	increased	calorie	delayed increase	Deyl et al. (1971)
	skeletal muscle	male Fischer 344 rats	increased	calorie	decreased	McCarter and McGee (1987)
5. Protein:						
amount	liver	male Wistar-derived rats	decreased	calorie	increased	Bale et al. (1988)
	liver homogenates	male Fischer 344 rats	decreased	calorie	increased	Laganiere and Yu (1989)
	liver, kidney, heart	male Sprague-Dawley rats	decreased during 2nd yr	calorie	increased (decreased in 3rd yr)	Merry et al. (1987)
	liver	male Sprague-Dawley rats	increased until 6 mo	calorie	lower	Merry & Holehan (1985)
	hepatocytes	male Fischer 344 rats	increased	calorie	lower	Birchenall-Sparks et al. (1985)
synthesis	hepatocytes	male Fischer 344 rats	55% decrease	calorie	higher	Birchenall-Sparks et al. (1985)
	liver, valine incorporation	male Fischer 344 rats	decreased	calorie	enhanced	Ward (1988a)
	cell free, testis, spleen, kidney, lymphocytes	male Fischer 344 rats	decreased	calorie	increased	Ricketts et al. (1985)
	whole body fractional rate	male Sprague-Dawley rats	decreased	calorie	no significant effect	Lewis et al. (1985)
degradation	liver, proteolytic capacity	male Fischer 344 rats	decreased	calorie	increased	Ward (1988b)
	whole body, phenylalanine incorporation	male Sprague-Dawley rats	decreased	calorie	enhanced	Lewis et al. (1985)
translational efficiency	liver, protein synthesized gRNA	male Sprague-Dawley rats	progressive loss	calorie	delayed loss	Merry et al. (1987)
6. Immune System:						
mitogen-response	T-cell mitogen	(NZBXNZW)F$_1$	decline	calorie	retarded	Jung et al. (1982)
IL-2 production	spleen cells	(NZBXNZW) F$_1$ (B/W) mice	slight decline	calorie	increase at 5, 11 months	Jung et al. (1982)

Continued

TABLE I. Studies on the Effects of Food Restriction (*Continued*)

Physiological Parameters Studied	Tissues or Procedures Used	Species or Strain	Age Effect	Types of Restriction	Dietary Effect	Reference
thymus involution	3H-thymidine incorporated into DNA	male Sprague-Dawley rats	increased	calorie	decreased	Merry and Holehan (1985a)
lymphocyte proliferation	spleen lymphocytes	male Fischer 344 rats	decline	calorie	increased	Richardson and Cheung (1982)
	response to T-cell antigens	male C57Bl/6/J mice	not tested	calorie	reduction	Christadoss et al. (1984)
thymosin α₁	serum	female CSBIDRF₁, mice	decrease in early life	calorie	29% lower at 19 months	Weindruch et al. (1988)
7. Reproductive activity:						
estrous cyclicity	vaginal lavage	female C57BL/6J mice	decreased	calorie	delayed	Nelson et al. (1985)
follicular depletion	primordial follicles	female C57BL/6J mice	increased	calorie	retarded	Nelson et al. (1985)
8. Bone:						
loss	femur	male Fischer 344 rats	increased	calorie	prevented	Kalu et al. (1988b)
strength	femur, transverse fracture	male Fischer 344 rats	decreased	calorie	increased	Kalu et al. (1984)
maturation	femur	male Fischer 344 rats	increased	calorie	delayed	Kalu et al. (1984)
mineral content	femur, tibia and fibula	male Fischer 344 rats	not tested	calorie (16 weeks of restriction)	lowered	Lee et al. (1986)
lipid content	femur	male Fischer 344 rats	increased	calorie	higher than control	Kalu et al. (1984)
9. Mitochondria:						
amount	liver, homogenates	male Fischer 344 rats	decreased	calorie	prevented	Laganiere and Yu (1987)
number	liver, electron microscopy	male Fischer 344 rats	decreased after 27 mos	calorie	prevented	Iwasaki et al. (1988c)
size	liver, electron microscopy	male C57BL/6J mice (for age effect); C5B10F, female mice (for restriction study)	not determined	calorie	larger	Weindruch et al. (1980)
	liver, electron microscopy	male Fischer 344 rats	no change	calorie	no effect	Iwasaki et al. (1988c)
substrate oxidation	liver, pyruvate, glutamate, acetate, palmitoyl-1-carnitine	male Fischer 344 rats	no change	calorie (starting at 10 months)	enhanced	Rumsey et al. (1987)

Continued

TABLE I. Studies on the Effects of Food Restriction (*Continued*)

Physiological Parameters Studied	Tissues or Procedures Used	Species or Strain	Age Effect	Types of Restriction	Dietary Effect	Reference
composition	phospholipid / protein	male Fischer 344 rats	decreased	calorie	higher	Laganiere and Yu (1989)
cytochrome c oxidase	liver homogenates	male Fischer 344 rats (retired breeders)	decreased (after 24 mos)	calorie (starting at 10 months)	increased	Rumsey et al. (1987)
citrate synthase	liver homogenates	male Fischer 344 rats (retired breeders)	decreased	calorie (starting at 10 months)	attenuated	Rumsey et al. (1987)
state 3	liver, β-hydroxy butryate supported	female C3B10F₁ mice (for restriction study)	decreased	calorie	increase	Weindruch et al. (1980)
state 4	liver, glutamate supported	male C57BL/6J mice (for age study)	no effect	calorie	no difference	Weindruch et al. (1980)
uncoupling	liver, dinitrophenol	male C57BL/6J mice (for age study); female C3B10F₁ mice (for restriction study)	decreased	calorie	increase	Weindruch et al. (1980)
10. Nucleic Acids: DNA	liver	male Wistar-derived rats	increased	protein or calorie	elevated (in young rats)	Bale et al. (1988)
	kidney	male Wistar-derived rats	not affected	protein or calorie	not affected	Bale et al. (1988)
	liver	female C57BL/6J mice	not tested	protein	increased	Barrows and Kokkonen (1978)
	hepatocytes	male Fischer 344 rats	increased	calorie	attenuated	Birchenall-Sparks et al. (1985)
	liver, kidney, heart	male Sprague-Dawley rats	increased	protein	increased	Barrows and Kokkonen (1978)
	kidney	female C57BL/6J mice	not changed	protein	reduced	Leto et al. (1976)
	liver	male Sprague-Dawley rats	increased	calorie	reduced	Merry et al. (1987)
	aorta	male Fischer 344 rats	decreased	calorie	attenuation of loss	Herlihy and Yu (1984)
DNA synthesis	heart, ³H-thymidine incorporation	male Sprague-Dawley rats	increased to 100 days	calorie	reduced	Merry and Holehan (1985)
DNA repair	splenocytes, UV irradiation	male & female C3BIORF₁ hybrid mouse	decreased	calorie	decelerated	Licastro et al. (1988)
	hepatocytes, kidney cells	male Fischer 344 rats	decreased	calorie	increased	Weraarchakul and Richardson (1988)
RNA	liver	male Wistar-derived rats	increased	calorie	no effect	Bale et al. (1988)
	liver	male Sprague-Dawley rats	increased	calorie	reduced	Merry et al. (1987)
	hepatocytes	male Fischer 344 rats	no change	calorie	no change	Birchenall-Sparks et al. (1985)

Continued

TABLE I. Studies on the Effects of Food Restriction (*Continued*)

Physiological Parameters Studied	Tissues or Procedures Used	Species or Strain	Age Effect	Types of Restriction	Dietary Effect	Reference
RNA/DNA (capacity for protein synthesis)						
	liver, kidney	male Sprague-Dawley rats	prepubertal peaks	calorie	eliminated prepubertal peaks	Merry and Holehan (1985)
	heart, abdominal skin	male Sprague-Dawley rats	increased to approx. 200 days; decreased after 900 days	calorie	lower than control	Merry and Holehan (1985)
	small intestine	male Sprague-Dawley rats	increased till approx. 300 days	calorie	similar to control	Merry and Holehan (1985)
mRNA						
	liver, α-2μ globulin	male Fischer 344 rats	decreased	calorie	attenuated	Chatterjee et al. (1989)
	liver, senescence marker protein	male Fischer 344 rats	increased	calorie	prevented	Chatterjee et al. (1989)
	α-2μ globulin hepatocytes	male Fischer 344 rats	decreased	calorie	attenuated	Richardson et al. (1987)
	thyroid calcitonin	male Fischer 344 rats	increased	calorie	reduced	Kalu et al. (1988a)
11. Gene Expression:						
α 2μ-globulin						
	hepatocytes	male Fischer 344 rats	decreased	calorie	increased	Richardson et al. (1987)
	liver	male Fischer 344 rats	decreased	calorie	increased	Chatterjee et al. (1989)
senescence marker protein (SMP)						
	liver	male Fischer 344 rats	increased	calorie	decreased	Chatterjee et al. (1989)
calcitonin						
	thyroid C-cells	male Fischer 344 rats	increased	calorie	decreased	Kalu et al. (1988a)
catalase						
	liver	male Fischer 344 rats	decreased	calorie	increased	Rao et al. (1988)
superoxide dismutase						
	liver	male Fischer 344 rats	decreased	calorie	increased	Rao et al. (1988)
12. Locomotive Activity:						
behavioral test						
	rotor performance, running wheel	female C3B10RFI mice	decreased	calorie	improved	Ingram et al. (1987)
	motor coordination	female C3B10RFI mice	decreased	calorie	improved	Ingram et al. (1987)
neuromuscular performance						
	learning, wire clinging	male B6CBAFI mice	decreased	calorie	improved	Harrison and Archer (1987)
physical movement						
	spontaneous motion	male Fischer 344 rats	decreased	calorie	prevented	Yu et al. (1985)
	open field movement	male B6CBAFI mice	decreased	calorie	prevented	Harrison and Archer (1987)

Continued

TABLE I. Studies on the Effects of Food Restriction (*Continued*)

Physiological Parameters Studied	Tissues or Procedures Used	Species or Strain	Age Effect	Types of Restriction	Dietary Effect	Reference
13. Membranes:						
fatty acid composition	liver and kidney mitochondria, microsomes	male Fischer 344 rats	decreased in 18:2, 18:3; increased in 22:5, 22:6	calorie	increased in 18:2, 18:3; decreased in 22:5, 22:6	Yu et al. (1989); Laganiere and Yu (1989a); Choi and Yu (1988)
lipid peroxidizability	liver mitochondria, microsomes	male Fischer 344 rats	increased	calorie	decreased	Laganiere and Yu (1989a); Yu et al. (1989)
NADPH-cytochrome reductase	liver microsomes	male Fischer 344 rats	no change	calorie	no change	Laganiere and Yu (1989a)
vitamin E content	liver mitochondria, microsomes	male Fischer 344 rats	increased	calorie	lower	Laganiere and Yu (1989a)
phospholipid	liver mitochondria, microsomes	male Fischer 344 rats	decreased	calorie	increased	Laganiere and Yu (1989a)
thickness	glomerular basement membrane	male Fischer 344 rats	increased	calorie	decreased	Hayashida et al. (1986)
14. Peroxidation and anti-oxidant Activity:						
catalase	liver supernatant (600 xg)	female C3B10RF₁ mice	decreased	calorie	increased	Koizumi et al. (1987)
	liver cytosol	male Fischer 344 rats	decreased	calorie	increased	Yu et al. (1988)
	liver cytosol	male Lobund-Wistar rats	decreased	calorie	increased	Chen and Lowry (1989)
	kidney	male BAlb/c mice	decreased	protein	no change	Stoltzner (1977)
superoxide dismutase	liver supernatant (600 xg)	female C3B10RF₁ mice	no change	calorie	no change	Koizumi et al. (1987)
	brain, liver cytosol	female albino Swiss mice	slightly increased	calorie	decreased	Chipalkatti et al. (1983)
	liver cytosol	male Lobund-Wistar rats	decreased	calorie	increased	Chen and Lowry (1989)
lipofuscin	liver, electron microscopy	male Fischer 344 rats	increased	calorie	reduced	Iwasaki et al. (1988)
	brain, heart, fluorescence measurement	female Swiss albino mice	increased	protein	reduced	Enesco and Kurk (1981)
	brain, heart	mice	increased	calorie	decreased	Chipalkatti et al. (1983)
	liver	female Sprague-Dawley rats	increased	calorie	decreased	Ferland and Tuchweber (1988)
glutathione (GSH)	liver, cytosol	male Fischer 344 rats	decreased	calorie	prevented	Yu et al. (1989); Laganiere and Yu (1989b)
	blood	male Lobund-Wistar rats	decreased	calorie	prevented	Lang et al. (1989)

Continued

TABLE I. Studies on the Effects of Food Restriction (*Continued*)

Physiological Parameters Studied	Tissues or Procedures Used	Species or Strain	Age Effect	Types of Restriction	Dietary Effect	Reference
lipid peroxidation	liver, mitochondria, microsomes	male Fischer 344 rats	increased	calorie	decreased	Yu et al. (1989); Laganiere and Yu (1989a)
ascorbic acid	liver, cytosol	male Fischer 344 rats	no change	calorie	no effect	Laganiere and Yu (1989b)
	liver	mice	increased	calorie	decreased	Chipalkatti et al. (1983)
malondialdehyde oxidation	liver mitochondria	male Fischer 344 rats	decreased	calorie	attenuated	Yu et al. (1989)
15. Metabolic Activity:						
metabolic rate	whole animal, based lean body mass/24 hr	male Fischer 344 rats	decreased	calorie	same as control	McCarter et al. (1985)
minimal oxygen consumption	whole animal	male Fischer 344 rats	decreased	calorie	increased	McCarter et al. (1986)
sensitivity to T_3	T_3 injection	male Fischer 344 rats	increased	calorie	increased further	McCarter et al. (1986)
16. Muscle and Related Activity:						
relaxation	aorta, β-receptor	male Fischer 344 rats	decreased	calorie	further decrease	Herlihy and Yu (1980)
	aorta, catecholamine induced	male Fischer 344 rats	declined	calorie	lower relaxation than ad lib; no decline after 6 mos.	Herlihy and Yu (1982)
	aorta	male Fischer 344 rats	declined	protein	similar to control	Herlihy and Yu (1982)
contractility	aorta, potassium induced	male Fischer 344 rats	decreased	calorie	prevented	Herlihy and Yu (1980)
inotropicity	left atria	male Fischer 344 rats	not tested	calorie	no effect	Herlihy (1984)
chronotropicity	right atria	male Fischer 344 rats	not tested	calorie	no effect	Herlihy (1984)
actomyosin content	aorta	male Fischer 344 rats	slight increase	calorie	similar to control	Herlihy and Yu (1984)
actin/myosin ratio	aorta	male Fischer 344 rats	no change	calorie	similar to control	Herlihy and Yu (1984)
oxygen consumption	aorta, smooth muscle preparation	male Fischer 344 rats	increased to 12 months; declined after 24 months	calorie	similar to control	Herlihy and Yu (1982)
oxidative capacity	muscle, gastrocnemis	male Fischer 344 rats (retired breeders)	decreased	calorie (starting at 10 months)	increased (approx. 20%)	Rumsey et al. (1987)

Continued

TABLE I. Studies on the Effects of Food Restriction (*Continued*)

Physiological Parameters Studied	Tissues or Procedures Used	Species or Strain	Age Effect	Types of Restriction	Dietary Effect	Reference
17. Cellularity: number						
	epididymal and perirenal fat cells	male Fischer 344 rats	increased	calorie	prevented	Bertrand et al. (1980a, 1984)
	thyroid C-cells	male Fischer 344 rats	increased	calorie	reduced	Kalu et al. (1988)
	liver and kidney	male Swiss albino rats	increased	protein	decreased	Enesco and Samborsky (1986)
	brain and heart	male Swiss albino rats	increased	protein	no change	Enesco and Samborsky (1986)
size						
	epididymal and perirenal fat cells	male Fischer 344 rats	increased early in life; decreased in senescence	calorie	reduced	Bertrand et al. (1980a)
	epididymal fat cells	Long-Evans rats	no effect	calorie	reduced	Craig et al. (1987)
	thyroid C-cells, grid measurement	male Fischer 344 rats	increased	calorie	no change	Kalu et al. (1988a)
	hepatocytes, electron microscopy	male Fischer 344 rats	no change	calorie	reduced	Iwasaki et al. (1988c)
cell renewal						
	mitotic activity, small intestine	male ddy mice	not studied	calorie	slowed	Koga and Kimura (1979)
fat mass						
	depots; epididymal and perirenal	male Fischer 344 rats	increased	calorie	reduced mass	Bertrand et al. (1980a)
response to catecholamine	epididymal and perirenal fat cells	male Fischer 344 rats	decreased	calorie	improved	Yu et al. (1980)
response to insulin						
	epididymal fat cells	Long-Evans rats	not affected	calorie	improved	Craig et al. (1987)
lipolytic response to:						
glucagon						
	epididymal and/or perirenal fat cells	male Fischer 344 rats	lost	calorie	not lost	Voss et al. (1982)
	epididymal and/or perirenal fat cells	male Fischer 344 rats	lost	calorie (begun at 6 months)	recovered	Bertrand et al. (1980b)
glucagon binding						
	epidydmal and perirenal fat cells	male Fischer 344 rats	lost	calorie	not lost or recovered	Bertrand et al. (1987)
β-adrenergic binding						
	isolated epidydmal and perirenal fat cells	male Fischer 344 rats	declines early (4 to 12 mos)	calorie	no effect	Bertrand et al. (1987)
phosphodiesterase activity						
	homogenate of epidydmal and perirenal fat cells	male Fischer 344 rats	low Km = increase between 6 & 12 mos. hi Km = decrease between 6 & 12 mos.	calorie	no effect on K_D	Bertrand et al. (1987)

Continued

TABLE I. Studies on the Effects of Food Restriction (*Continued*)

Physiological Parameters Studied	Tissues or Procedures Used	Species or Strain	Age Effect	Types of Restriction	Dietary Effect	Reference
18. Brain						
cholinergic receptor binding	corpus striatum, cerebral cortex	male Wistar rats	no effect	calorie	higher	London et al. (1985)
choline acetyl transferase	hippocampus cerebellum	male Wistar rats	increased (6-24 mos), and decreased thereafter	calorie	higher	London et al. (1985)
	striatum, hippocampus, cerebellum	male Wistar rats	not tested	calorie	increased	Joseph et al. (1983)
glutamic acid decarboxylase	hippocampus, cerebellum	male Wistar rats	no change	calorie	no difference	London et al. (1985)
tyrosin hydroxylase	hippocampus, cerebellum	male Wistar rats	no effect	calorie	higher	London et al. (1985)
neuronal population						
19. Enzymes:						
ATPase	liver	male Charles River SD rats	decreased	calorie	delayed	Ross (1959)
	visual cortex (area 17)	Sprague-Dawley (retired breeder) rats	lost	calorie (restricted for 2 months)	delayed	Peters et al. (1987)
malic dehydrogenase	liver, kidney	female C57BL/6J mice	not changed	protein	reduced	Leto et al. (1976)
succinoxidase	liver, kidney	female C57BL/6J mice	not changed	low protein	lower	Leto et al. (1976)
cholinesterase	liver	female C57BL/6J mice	not changed	low protein	lower	Leto et al. (1976)
HMGA reductase	liver, microsomes	male Fischer 344 rats	decreased	calorie	attenuated	Yu et al. (1984)
acid phosphatase	liver homogenates	female Sprague-Dawley rats	no change	calorie	increased	Solomon (1984)
galactosidase	liver homogenates	female Sprague-Dawley rats	decreased	calorie	increased	Solomon (1984)
arylsulphatase B	liver homogenates	female Sprague-Dawley rats	decreased	calorie	increased	Solomon (1984)
cathepsin D	liver homogenates	female Sprague-Dawley rats	no change	calorie	no change	Solomon (1984)

Continued

TABLE I. Studies on the Effects of Food Restriction (*Continued*)

Physiological Parameters Studied	Tissues or Procedures Used	Species or Strain	Age Effect	Types of Restriction	Dietary Effect	Reference
20. Prostaglandin Synthesis:						
	spleen homogenates	Emory mice	increased	calorie	decreased	Meydani et al. (1988)
	kidney, microsomes	male Fischer 344 rats	decreased	calorie	increased	Choi and Yu (1988)
21. Others:						
body temperature						
	rectal probe	female B10C3F₁ mice	no change	protein or calorie	lower than control	Cheney et al. (1983)
	rectal probe	female C57BL/6J mice	no change	protein	lower than control	Leto et al. (1976)
	rectal probe, circadian	male Sprague-Dawley rats	young rats had higher initial body temp.	calorie	similar as control	Volicer et al. (1984)
cold tolerance						
	exposure to −27°C	male Sprague-Dawley rats	not tested	calorie	less effective	Campbell and Richardson (1988)
intestinal absorption						
	vitamin A	F₁-hybrid, male mice (for age study), female (for restriction study)	no change	calorie	increased	Hollander et al. (1986)
	calcium	male Fischer 344 rats	no change	calorie	no change	Kalu et al. (1988a)
wound healing						
	tail wound	male B6CBAF mice	slowed	calorie	slower	Harrison and Archer (1987)
urine concentration ability						
	urine osmolarity	male B6CBAF1 mice	reduced	calorie	prevented	Harrison and Archer (1987)
eye lens						
	gamma-crystallin	F₁-hybrid mice	decreased	calorie	retarded	Leveille et al. (1984)
cytochrome content						
	liver	male Fischer 344 rats	decreased	calorie	attenuated	Rumsey et al. (1987)
lean body mass						
	carcass analysis	male Long-Evans rats	decreased	calorie	minimally affected	Garthwaite et al. (1986)
	cyclopropane uptake method	male Fischer 344 rats	not changed	calorie	not affected	Yu et al. (1982)

RECENT WORK

Is It the Caloric Intake or a Specific Component of the Diet That Is Responsible for the Actions of Food Restriction?

In seeking possible mechanisms for this nutritional intervention, one of the more important and compelling questions asked is whether it is the reduction of calorie intake or of any specific dietary component that retards the aging processes. This question may have originated from early studies in which both calorie reduction and protein restriction produced qualitatively similar results (e.g., life span extension and reduction of age-associated diseases).

Although the available data did not permit a definitive conclusion, recent data do. For example, Yu et al. [1985] found that restriction of the protein component without caloric reduction produced only 10–15% increase in longevity as compared with 50% increase by calorie restriction of a similar magnitude. It appears, moreover, that the 10–15% increase in longevity is solely due to the effects on nephropathy [Maeda et al., 1985]. The level of protein restriction did not influence disease processes or physiological events not secondary to renal disease [Maeda et al., 1985]. A recent report [Masoro et al., 1989b] on the dietary modulation of the age-related progression of nephropathy shed further light on the relationship between protein and nephropathy. The study showed that restricting food intact by 40% without restricting protein intake was highly effective in reducing the severity of nephrotic lesions. The conclusion drawn from the study is that reducing the intake of protein is not the major reason for the retardation by food restriction of the age-related nephropathy in rats.

In pursuit of further refinement on the effect of various dietary components involved in nutrition manipulation, new pieces of information emerged. Iwasaki et al. [1988a] launched an experiment in which the fat and mineral component of the diet was restricted by 40% without restricting calories. The results indicate that neither the restriction of fat nor the mineral restriction affected the median or maximum life spans of rats. However, the pathological analyses interestingly revealed that fat but not mineral restriction did show some ability to retard the progression of nephropathy. The findings from these studies are significant from three points of view: 1) the broad spectrum of beneficial actions by food restriction does not seem to involve reduction in intake of fat or minerals; 2) the high incidence of chronic nephropathy in male Fischer 344 rats is not influenced by mineral restriction; 3) the high load of mineral consumption may not be the major factor involved in age-related nephropathy seen in male Fischer 344 rats.

Another interesting aspect of dietary study is the question of whether the quality of specific dietary components has some influence on the longevity and progression of age-related nephropathy. Iwasaki et al. [1988b] examined this question by studying the effects of the quality of protein on longevity and age-related disease in Fischer rats. Replacing dietary casein with soy protein without changing any calorie intake produced an improved median life span (844 days vs. 730 days for casein-fed rats). The results seemed to indicate that the extended life span may be related to the extent of attenuation of the progression of chronic nephropathy as shown by the data that only 7% of the rats on soy protein showed end-stage lesion of nephropathy compared with about 40% of the rats on the casein diet. In addition to the effect on nephropathy, soy protein showed some modulatory effect on age-related hyperparathyoidism and senile bone loss [Kalu et al., 1988b]. The age-related increase in serum PTH was effectively suppressed by soy protein feeding. In control group rats, there was an age-associated loss in serum 25-hydroxy vitamin D and bone, but this loss was retarded by a soy protein diet. On the basis of their findings, the authors concluded that a partial prevention of renal deterioration is responsible for the retardation of bone loss, and age-related progressive hyperparathyroidism [Kalu et al, 1988b].

Modulation of Lipid Peroxidation by Food Restriction

Since the time Harman proposed the free radical theory of aging [Harman 1956, 1983], free radicals and the related oxidative reactions have been strongly suggested as the possible factors underlying functional decrements and disease processes associated with aging [Hegner, 1980; Vladimirov et al., 1980]. Early attempts to intervene in the progression of free radical-induced damage and to prolong the life span of laboratory rodents by feeding antioxidants met with only limited success. Thus experimental data supporting the free radical hypothesis have been insufficient. However, recent studies from several laboratories on food restriction have generated experimental support for the hypothesis [Chipalkatti et al., 1983; Koizumi et al., 1987; Yu et al., 1989].

One of the early indications that dietary manipulation modifies the extent of free radical damage was the work of Enesco and Kruk [1981], who studied the effect of dietary protein restriction on fluorescent lipofuscin accumulation in brain and heart of male Swiss albino mice at 3, 5, 7, and 12 months of age. The level of fluorescent products was significantly reduced by diets containing 4% protein compared with the control diet with 26% protein. The authors suggested that the reduced lipofuscin accumulation was related to the decreased free radical-induced lipid peroxidation. Chipalkatti et al. [1983] reported that restricting food intake of mice to about one-half of that of ad libitum feeding resulted in reduction in lipofuscin content of brain and in lipid peroxidation as measured by malondialdehyde content in liver homogenates. Additionally, the age-related increase in activities of lysosomal enzymes in brain and heart was lowered by food restriction.

Further support for the modulating effect of food restriction on free radicals and lipid peroxidation was reported by Koizumi et al. [1987]. They investigated the age-related lipid peroxidation in liver homogenates from 12- and 24-month-old mice. In addition to lipid peroxidation measurements, the activities of three important liver enzymes, catalase, superoxide dismutase, and glutathione peroxidase, were measured. The strongest dietary effect was increased catalase activity and suppressed lipid peroxidation.

Recently, in-depth investigations on the antioxidative status in food-restricted rats were reported by Yu et al. [1989] and Laganiere and Yu, [1989b], using mitochondria, microsomes, and cytosols from rat liver. These studies included several liver cytosolic antioxidant systems: reduced glutathione (GSH), GSH reductase, GSH transferase, GSH peroxidase, catalase, ascorbic acid, and vitamin E. Food restriction modified age-related deteriorations of these cytosolic antioxidants. One of the major byproducts in the lipid peroxidation process is malondialdehyde (MDA), which is a well-known cytotoxic substance, but the metabolic fate of this potentially harmful MDA is not well characterized in aging animals. A recent work [Yu et al., 1989] showed that MDA is converted to CO_2 by mitochondrial fraction through aldehyde dehydrogenase reaction. The data indicate that MDA oxidation was modulated by both age and food restriction. With increasing age, the MDA oxidation was diminished, and by 22 months the activity decreased to about one-third the level of 6 months. In food-restricted rats, the MDA oxidation was significantly maintained at higher levels.

Modulation of Membrane Damages by Food Restriction

The notion that membrane structures and membrane components might have undergone age-related changes similar to other cellular constituents has attracted early attention of many gerontologists [Vorbeck et al., 1982; Harman, 1983; Naeim and Walford, 1985; von Zgliniki, 1987]. Hochshild [1971], for instance, recognizing the importance of membrane instability during age, proposed a lysosomal hypothesis of aging by predicting age-related increase in the lytic enzyme activity leading to the eventual cellular autolytic degeneration. He suggested that membrane stabilizers including antioxidants might be involved in life extension of cultured cells and fruit flies secondary to lysosomal membrane stabilization. Zs-Nagy [1978] proposed that cellular membrane deterioration during aging is the main event leading to cellular aging [Zs-Nagy, 1979]. His proposal was based on the results of X-ray microanalysis of abnormally elevated potassium concentrations in the nucleus and cytoplasm of brain and liver cells of old rats.

Membrane alterations during aging were well reviewed in an article by Grinna [1977], who highlighted age-associated mitochondrial and microsomal membrane deteriorations in relation to structural and compositional alterations. Recently, the destruction of membrane lipid asymmetry during aging was suggested as the basis for the mechanism underlying age-related membrane deterioration by Schroeder [1984]. His hypothesis formulated the idea that the idea that the peroxidative disruption of lipid asymmetric distribution of phospholipids and cholesterol results in membrane rigidity and eventually in dysfunctioning of membranes. As reviewed by Hansford [1983], an increased free radical-induced peroxidation of membrane lipids during the course of aging could be a major contributing factor. Such a possibility was substantiated by reports in which the hydroperoxide contents in mitochondria and microsomes from both liver [Laganiere and Yu, 1989a] and kidney [Choi and Yu, 1988] were shown to be increased with age.

Evidence indicating age-related changes in membrane fatty acid composition was first reported by Hegner [1980], but the data on the membrane constituents modulated by food restriction were not available until recently. Laganiere and Yu [1989a] and Yu et al. [1989] showed that restricting the intake of calories by 40% modified the age-dependent changes in fatty acid pattern in mitochondrial and microsomal membranes from rat liver at different ages ranging 6 to 24 months. With aging, the major changes taking place were in a decreased amount of 18:2 and 18:3 fatty acids and increased highly peroxidizable long polyunsaturated acids (20:4, 22:5, and 22:6) in ad libitum fed rats. A most interesting point is the fact that food restriction causes an age change in the opposite direction: increased 18:2 and 18:3 and decreased 20:4, 22:5, and 22:6 fatty acids. It was also revealed that the unsaturation/saturation ratio of membrane fatty acids was maintained by food restriction at high levels throughout the life span. Therefore, it seems that the two major membrane age-modulated properties—peroxidizability and unsaturation/saturation ratio—were influenced by food restriction to provide the organism with a well-maintained membrane integrity.

One of the important functions of biological membranes is that of providing the optimal environment for maximizing receptor–ligand interactions at the cell surface. Since biological membranes undergo a deteriorative process in the course of aging, it is plausible that many observed age-related changes in receptor activity may be influenced by membrane deterioration with aging rather than the deterioration of receptors per se [Dax, 1985]. Some examples of this idea are the age-related diminished catecholamine responsiveness in myocardial tissue [Herlihy and Yu, 1982], and changes in number of beta-adrenergic receptors in rat lung [Scarpace and Yu, 1987] and adipocytes [Bertrand et al., 1987]. Therefore, it is important to realize that how the age-dependent changes in receptor and other membrane associated activities are modified by the membrane-modulating food restriction.

Roth et al. [1984] first reported the dietary manipulation of dopaminergic receptor activity. The age-related decrease in dopamine receptors was about 40% in control rats between 3 and 12 months of age, and food restriction blunted the age-related decrease. Other recent evidence indicating modulation of food restriction at the receptor level was that of Scarpace and Yu [1987]. They investigated the effects of food restriction on the beta-adrenergic receptors and adenylate cyclase activity in lung isolated from Fischer 344 rats. The rats were either ad libitum fed or restricted to 60% of the ad libitum food intake starting at 6 weeks of age. By 27 months beta-adrenergic receptors decreased to 260 fmol/mg protein from 417 fmol/mg protein (at 6 months) in ad libitum fed rats, while the number was 360 fmol/mg protein in restricted rats. Isoproterenol-stimulated adenylate cyclase activity declined with age in the ad libitum rats, but not in restricted rats.

Modulation of Genetic Activity by Food Restriction

One of the new and exciting findings emerging from food restriction concerns gene expression utilizing molecular probes [Richardson and Cheung, 1982]. The significance of the new evidence is that the genomic activity of cells and their components can be readily modifiable by calorie restriction. This may not be surprising if one assumes that nuclear genetic constituents are as vulnerable as any other cellular constituent and thus as modifiable by nutritional factors. However, newly emerging evidence on genetic modification by an extrinsic factor such as food restriction should cast new light on the exploration of the mechanism of the antiaging action of food restriction.

There is sound evidence that the activity of transcription process in protein synthesis declines considerably with aging [Richardson and Semsei, 1987b]. The earliest indication for such age-related changes at the molecular level was reported by Chatterjee et al. [1981], who showed that changes in major senescence marker protein synthesis are due to age-related changes in mRNA. Richardson et al. [1987], examining the gene expression of $\alpha 2\mu$ globulin in isolated hepatoyctes from male Fischer 344 rats, showed a decrease of 90% in the activity between 6 and 22 months of age. They also found that the levels of $\alpha 2\mu$ globulin mRNA and the transcription genes showed a corresponding decrease of 80–85% between the ages of 5 and 24 months. In their study, the globulin synthesis, mRNA, and transcription activities were about two to three times higher for restricted rats compared with the ad libitum fed group. Their findings are the first evidence that calorie restriction regulates transcriptional activity at the gene level. A functionally important hormone and its regulation by food restriction at the gene level was recently revealed [Kalu et al., 1988a]. The report showed clearly that food restriction reduced the age-related increase in calcitonin levels by suppressing the calcitonin gene expression activity.

Further evidence on the action of food restriction on gene modification has been reported in the work of Chatterjee et al. [1989], who found that dietary restriction retarded age-associated loss in androgen responsiveness and age-related suppression of the $\alpha 2\mu$-globulin gene in rat liver and concomitantly, repression of the androgen-repressible senescence marker protein (SMP-2) gene. It was shown that the transcriptional action of the $\alpha 2\mu$-globulin gene occurs at the onset of puberty, resulting in the appearance of the corresponding mRNA in the postpubertal stage, peaking in young adulthood. The progressive loss in transcription of this gene was completely age-dependent. The steady-state levels of $\alpha 2\mu$-globulin gene mRNA in the livers of 24- and 27-month-old ad libitum fed rats were undetectable, while mRNA in food-restricted rats showed a detectable level even at 27 months of age. As expected, the steady-state level of the SMP-2 mRNA was increased and maintained in older ad libitum fed rats.

There is yet another aspect of genetic constituents that is known to modulate food restriction related to DNA molecules altered with age. There is considerable evidence suggesting that DNA damage by oxidative process may play a major role in causing many age-related dysfunctions. The oxidative insult responsible for DNA damage results from increased free radicals and other oxidative reactions. With the realization of the great consequences of DNA damage on cellular integrity, Gensler and her colleagues proposed "the DNA damage hypothesis of aging" [1989]. Many studies on the contribution of DNA damage to aging processes have also been reported; for example, single-strand breakage in the livers of female Swiss–Webster mice up to 24 months of age was found to increase with age [Lawson and Stohs, 1985], and this damage was significantly reduced by feeding antioxidants. Recent reports by Fraga and Tappel [1988] showed DNA damage concurrent with lipid peroxidation in rat liver slices. A study on the effect of dietary restriction on DNA repair during aging [Weraarchakul and Richardson, 1988] indicated that the age-related decline in DNA repair capacity, as measured by unscheduled DNA synthesis with the incorporation of [^{3}H]thymidine in both hepatocytes and kidney cells, was clearly modulated by food restriction. More recently Licastro et al. [1988] measured the effect of food restriction on DNA repair by [^{3}H]TdR incorporation in UV-irradiated splenocytes and showed that splenocytes from 45-month-old mice on ~ 50% restricted diet maintained DNA repair activity comparable to that of younger animals at 21–28 months of age with minimal restriction.

REFERENCES

Bale CW, Davis TA, Beauchene RE (1988): Long-term protein and calorie restriction: Alteration in nucleic acid levels of organs of male rats. Exp Gerontol 23:189–196.

Barrows CH, Kokkonen GC (1978): The effect of various dietary restricted regimes on biochemical variables in mouse. Growth 42:71–85.

Bertrand HA, Lynd FT, Masoro EJ, Yu BP (1980a): Changes in adipose mass and cellularity through the adult life of rats fed ad libitum or a life-prolonging restricted diet. J Gerontol 35:827–835.

Bertrand HA, Masoro EJ, Yu BP (1980b): Maintenance of glucagon-promoted lipolysis in adipocytes by food restriction. Endocrinology 107:591–595.

Bertrand HA, Stacy C, Masoro EJ, Yu BP, Murata I, Maeda H (1984): Plasticity of fat cell number. J Nutr 114:127–131.

Bertrand HA, Anderson WR, Masoro EJ, Yu BP (1987): Action of food restriction on age change in adipocyte lipolysis. J Gerontol 42:666–673.

Birchenall-Sparks MC, Roberts MS, Staecker J, Hardwick JP, Richarson A (1985): Effect of dietary restriction on liver protein synthesis in rats. J Nutr 115:944–950.
Campbell BA, Richardson R (1988): Effect of chronic undernutrition on susceptibility to cold stress in young adult and aged rats. Mech Aging Dev 44:193–202.
Chatterjee B, Nat TS, Roy AK (1981): Differential regulation of the messenger RNA for three major senescence marker proteins in male rat liver. J Biol Chem 256:5939–5941.
Chatterjee B, Fernandes G, Yu BP, Song C, Kim JM, Demyan W, Roy AK (1989): Calorie restriction delays age-dependent loss in androgen responsiveness of the rat liver. FASEB J 3:169–173.
Chen LH, Lowry SR (1989): Cellular antioxidant defense system. In Snyder DL (ed): "Dietary Restriction and Aging." New York: Alan R. Liss, Inc., pp 247–256.
Cheney KE, Liu RK, Smith GS, Meredith PJ, Walford RL (1983): The effect of dietary restriction of varying duration on survival, tumor patterns, immune function, and body temperature in B10C3F$_1$ female mice. J Gerontol 38:420–430.
Chipalkatti S, De AK, Aiyar AS (1983): Effect of diet restriction on some biochemical parameters related to aging in mice. J Nutr 113:944–950.
Choi JH, Yu BP (1988): Anti-oxidant action of food restriction on membrane alteration of aged kidney. FASEB J 2:A1208.
Christadoss P, Talal N, Lindstrom J, Fernandes G (1984): Suppression of cellular and humoral immunity to T-dependent antigens by calorie restriction. Cell Immunol 88:1–8.
Craig BW, Garthwaite SM, Holloszy JO (1987): Adipocyte insulin resistance: Effects of aging, obesity, exercise, and food restriction. J Appl Physiol 62:95–100.
Dax EM (1985): Receptors and associated membrane events in aging. In Rothstein M (ed): "Review of Biological Research in Aging," Vol. 2. New York: Alan R. Liss, Inc., pp 315–336.
Deyl Z, Juricova M, Rosmuss J, Adam M (1971): The effect of food deprivation on collagen accumulation. Exp Gerontol 6:383–390.
Enesco H, Kurk P (1981): Dietary restriction reduces fluorescent age-pigment accumulation in mice. Exp Gerontol 16:357–361.
Enesco HE, Samborsky J (1986): Influence of dietary protein restriction on cell number, cell size and growth of mouse organs during the course of aging. Arch Gerontol Geriatr 5:221–233.
Ferland G, Tuchweber B (1988): Lipofuscin accumulation in the liver of rats during aging and dietary restriction. FASEB J 2:A1208.
Fraga CG, Tappel AT (1988): Damage to DNA concurrent with lipid peroxidation in rat liver slices. Biochem J 252:893–896.
Garthwaite SM, Cheng H, Bryan JE, Craig BW, Holloszy JO (1986): Aging, exercise and food restriction: Effects on body composition. Mech Aging Dev 36:187–196.
Gensler HL, Hall JD, Berstein H (1989): The DNA damage hypothesis of aging: Importance of oxidative damage. In Rothstein M (ed): "Review of Biological Research in Aging," Vol. 3. New York: Alan R. Liss, Inc., pp 451–465.
Grinna LS (1977): Changes in cell membranes during aging. Gerontology 23:452–264.
Hansford RG (1983): Bioenergetics in aging. Biochim Biophy Acta 726:41–80.
Harman D (1956): Aging: A theory based on free radical and radiation chemistry. J Gerontol 11:298–300.
Harman D (1983): Free radical theory of aging: Consequences of mitochondrial aging. AGE 6:86–94.
Harrison DE, Archer JR (1987): Genetic differences in effects of food restriction on aging in mice. J Nutr 117:376–382.
Hayashida M, Yu BP, Masoro EJ, Iwasaki K, Ikeda T (1986): An electron microscopic examination of age-related changes in the rat kidney: The influence of diet. Exp Gerontol 21:535–553.
Hegner D (1980): Age-dependence of molecular and functional changes in biological membrane properties. Mech Aging Dev 14:101–118.

Herlihy JT (1984): Dietary manipulation of cardiac and aortic smooth muscle reactivity to isoprotererol. Am J Physiol 246:H369–H373.
Herlihy J, Yu BP (1980): Dietary manipulation of age-relaed decline in vascular smooth muscle function. Am J Physiol 238:H652–H655.
Herlihy JT, Yu BP (1982): Effects of age and diet on the catecholamine-induced changes in rat aortic smooth muscle contractility. Gerontology 22:49.
Herlihy JT, Yu BP (1984): Effects of food restriction on physical and biochemical properties of aging rat aorta. Physiologist 24:52.
Herlihy JT, Stacy C, Bertrand H (1989): Effects of long-term food restriction on the plasma thyroid hormones of Fischer 344 rats. FASEB J 3:A461.
Hochschild R (1971): Lysosomes, membranes and aging. Exp Gerontol 6:153–166.
Holehan AM, Merry BJ (1985): The control of puberty in the dietary restricted female rat. Mech Aging Dev 32:179–191.
Hollander D, Dadufalza V, Weindruch R, Walford RL (1986): Influence of life-prolonging dietary restriction on intestinal vitamin A absorption in mice. Age 9:57–60.
Ingram DK, Weindruch R, Spangler EL, Freeman JR, Walford RL (1987): Dietary restriction benefits learning and motor performance of aged mice. J Gerontol 42:78–81.
Iwasaki K, Gleiser CA, Masoro EJ, McMahan CA, Seo EJ, Yu BP (1988a): Influence of the restriction of individual dietary components on longevity and age-related disease of Fischer rats: The fat component and the mineral component. J Gerontol 43:B13–B21.
Iwasaki K, Gleiser CA, Masoro EJ, McMahan CA, Seo EJ, Yu BP (1988b): The influence of dietary protein source on longevity and age-related disease of Fischer rats. J Gerontol 43:B5–B12.
Iwasaki K, Maeda H, Shimokawa I, Hayshida M, Yu BP, Masoro EJ, Ikeda T (1988c): An electron microscopic examination of age-related changes in the rat liver. Acta Pathol Jpn 38:1119–1130.
Joseph JA, Whitaker G, Roth S, Ingram K (1983): Lifelong dietary restriction affects striatally mediated behavioral responses in aged rats. Neurobiol Aging 4:191–196.
Jung LKL, Palladino MA, Calvano S, Mark DA, Good RA, Fernandes G (1982): Effect of calorie restriction on the production and responsiveness to interleukin 2 in (NZBXNZW) F_1 mice. Clin Immunol Immunopathol 25:295–301.
Kalu DK, Cockerham R, Yu BP, Roos BA (1983): Lifelong dietary modulation of calcitonin levels in rats. Endocrinology 113:2010–2016.
Kalu DK, Hardin RR, Cockerham R, Yu BP, Norling BK, Egan JW (1984): Lifelong food restriction prevents senile osteopenia and hyperparathyroidism in F344 rats. Mech Age Dev 26:103–112.
Kalu DK, Herbert DC, Hardin RR, Yu BP, Kaplan G, Jacob JW (1988a): Mechanism of dietary modulation of calcitonin levels in Fischer rats. J Gerontol 43:B125–131.
Kalu DK, Masoro EJ, Yu BP, Hardin RR, Hollis BW (1988b): Modulation of age-related hyperparathyroidism and senile bone loss in Fischer rats by soy protein and food restriction. Endocrinology 122:1847–1854.
Koga A, Kimura S (1979): Influence of restricted diet on the cell renewal of the mouse small intestine. J Nutr Sci Vitaminol 25:265–267.
Koizumi A, Weindruch R, Walford RL (1987): Influences of dietary restriction and age on liver enzyme activities and lipid peroxidation in mice. J Nutr 117:361–367.
Laganiere S, Yu BP (1987): Anti-lipoperoxidation action of food restriction. Biochem Biophys Res Commun 145:1185–1191.
Laganiere S, Yu BP (1989a): Effect of chronic food restriction in aging rats. I. Liver subcellular membranes. Mech Aging Dev 48:207–219.
Laganiere S, Yu BP (1989b): Effect of chronic food restriction in aging rats. II. Liver cytosolic antioxidants and related enzymes. Mech Aging Dev 48:221–230.

Lang CA, Wu W, Chen T, Mill BJ (1989): Blood glutathione: A biochemical index of life span enhancement in the diet restricted Lobund–Wistar rat. In Snyder DS (ed): "Dietary Restriction and Aging." New York: Alan R. Liss, Inc., pp 241–246.

Lawson T, Stohs S (1985): Changes in endogenous DNA damage in aging mice in response to butylated hydroxyanisole and oltipraz. Mech Aging Dev 30:179–185.

Lee CJ, Panemangalore M, Wilson K (1986): Effect of dietary energy restriction on bone mineral content of mature rats. Nutr Res 6:51–59.

Leto S, Kokkonen GC, Barrows CH (1976): Dietary protein, life-span, and physiological variables in female mice. J Gerontol 31:149–154.

Leveille PPJ, Weindruch R, Walford RL, Bok D, Horwitz J (1984): Dietary restriction retards age-related loss of gamma crystallins in the mouse lens. Science 224:1247–1249.

Lewis SEM, Goldspink DF, Phillips JG, Merry BJ, Holehan AM (1985): The effects of aging and chronic dietary restriction on whole body growth and protein turnover in the rat. Exp Gerontol 20:253–263.

Licastro F, Weindruch R, Davis LJ, Walford RL (1988): Effect of dietary restriction upon the age-associated decline of lymphocyte DNA repair activity in mice. Age 11:48–52.

Liepa GU, Masoro EJ, Bertrand HA, Yu BP (1980): Food restriction as a modulator of age-related changes in serum lipids. Am J Physiol 238:E253–E257.

London ED, Waller SB, Ellis AT, Ingram DK (1985): Effects of intermittent feeding on neurochemical markers in aging rat brain. Neurobiol Aging 6:199–204.

Maeda H, Gleiser CA, Masoro EJ, Murata I, McMahan CA, Yu BP (1985): Nutritional influences on aging of Fischer 344 rats. II. Pathology. J Gerontol 40:671–688.

Masoro EJ, Compton C, Yu BP, Bertrand HA (1983): Temporal and compositional dietary restrictions modulate age-related changes in serum lipids. J Nutr 113:880–892.

Masoro EJ, Katz MS, McMahan A (1989a): Evidence for the glycation hypothesis of aging from the food-restricted rodent model. J Gerontol 44:B20–23.

Masoro EJ, Iwasaki K, Gleiser CA, McMahan CA, Seo EJ, Yu BP (1989b): Dietary modulation of the progression of nephropathy in aging rats: An evaluation of the importance of protein. Am J Clin Nutr 49:1217–1227.

McCarter R, McGee J (1987): Influence of nutrition and aging on the composition and function of rat skeletal muscle. J Gerontol 42:432–441.

McCarter R, Masoro EJ, Yu BP (1985): Does food restriction retard aging by reducing the metabolic rate? Am J Physiol 248:E488–E490.

McCarter R, McGee J, Herlihy JT (1986): Minimal oxygen consumption in rats: Effects of age, food restriction and thyroid hormone. Proc Int Congr Physiol Sci 16:273.

Merry BJ, Holehan AM (1984): The endocrine response to dietary restriction in the rat. In Woodhead AD, Blackett AD, Hollander A (eds): "Molecular Biology of Aging." New York: Plenum Press, pp 117–141.

Merry BJ, Holehan AM (1985): In vivo DNA synthesis in the dietary restricted long-lived rat. Exp Gerontol 20:15–28.

Merry BJ, Holehan AM, Lewis SEM, Goldspink DF (1987): The effects of aging and chronic dietary restriction on in vivo hepatic protein synthesis in the rat. Mech Aging Dev 39:189–199.

Meydani SN, Lipman R, Blumberg JB, Taylor A (1988): Calorie restriction decreases spleen prostaglanding (PG) E_2 synthesis in Emory mice. FASEB J 2:A1208.

Naeim F, Walford RL (1985): Aging and cell membrane complex: The lipid bilayer integral proteins, and cytoskeleton. In Finch CE, Schneider EL (eds): "Handbook of the Biology of Aging." New York: Van Nostrand Reinhold, pp 272–289.

Nelson J, Gosden RG, Felicio LA (1985): Effect of dietary restriction on estrous cyclicity and follicular reserves in aging C57BL/6J mice. Biol Reprod 32:515–522.

Peters A, Harriman KM, West CD (1987): The effect of increased longevity, produced by dietary

restriction, on the neuronal population of area 17 in rat cerebral cortex. Neurobiol Aging 8:7–20.
Rao G, Hedari A, Gu MZ, Waggoner S, Marquardt L, Richardson A (1988): Effect of dietary restriction on gene expression in rodents. FASEB J 2:A1209.
Reaven E, Reaven GM (1981): Structure and function changes in the endocrine pancreas of aging rats with reference to the modulating effects of exercise and calorie restriction. J Clin Invest 68:75–84.
Richardson A, Cheung HT (1982): Th relationship between age-related changes in gene expression, protein turnover, and the responsiveness of an organism to stimuli. Life Sci 31:605–613.
Richardson A, Semsei I (1987): Effect of aging on translation and transcription. In Rothstein M (ed): "Review of Biological Research in Aging," Vol. 3. New York: Alan R. Liss, Inc., pp 467–483.
Richardson A, Butler JA, Rutherford MS, Semsei I, Gu MZ, Fernandes G, Chiang WH (1987): Effect of age and dietary restriction on the expression of α 2μ-globulin. J Biol Chem 262:12821–12825.
Ricketts WG, Birchenall-Sparks MC, Hardwick JP, Richardson A (1985): Effect of age and dietary restriction on protein synthesis by isolated kidney cell. J Cell Physiol 125:492–498.
Ross M (1969): Aging, nutrition, and hepatic enzyme activity patterns in the rat. J Nutr 97:565–602.
Roth GS, Ingram DK, Joseph JA (1984): Delayed loss of striatal dopamine receptors during aging of dietarily restricted rats. Brain Res 300:27–32.
Rumsey WL, Kendrick ZV, Starnes JW (1987): Bioenergetics in the aging Fischer 344 rats: Effects of exercise and food restriction. Exp Gerontol 22:271–287.
Scarpace PJ, Yu BP (1987): Diet restriction retards the age-related loss of beta-adrenergic receptors and adenylate cyclase activity in rat lung. J Gerontol 42:442–446.
Schroeder F (1984): Role of membrane lipid asymmetry in aging. Neurobiol Aging 5:323–333.
Snyder DL, Towne B (1989): The effect of dietary restriction on serum hormone and blood chemistry changes in aging Lobund–Wistar rats. In Snyder DL (ed): "Dietary Restriction and Aging." New York: Alan R. Liss, Inc., pp 135–146.
Stoltzner G (1977): Effects of life-long dietary protein restriction on mortality, growth, organ weights, blood counts, liver aldolase and kidney catalase in ABLB/c mice. Growth 41:337–348.
Vladimirov YA, Olenev VI, Suslova TB, Cheremisina ZP (1980): Lipid peroxidation in mitochondrial membrane. Adv Lipid Res 17:173–249.
Volicer L, West CD, Chase AR, Greene L (1983): Beta-adrenergic receptor sensitivity in cultured vascular smooth muscle cells: Effect of age and of dietary restriction. Mech Aging Dev 21:283–293.
Volicer L, West C, Greene L (1984): Effect of dietary restriction and stress on body temperature in rats. J Gerontol 39:178–182.
von Zgliniki T (1987): A mitochondrial membrane hypothesis of aging. J Theor Biol 127:4127–132.
Vorbeck ML, Martin AP, Long JW Jr, Smith JM, Orr RR (1982): Aging-dependent modification of lipid composition and lipid structural order parameter of hepatic mitochondria. Arch Biochem Biophys 217:351–361.
Voss KH, Masoro EJ, Anderson W (1982): Modulation of age-related loss of glucagon-promoted lipolysis by food restriction. Mech Aging Dev 18:135–149.
Ward WF (1988a): Enhancement of food restriction of liver protein synthesis in the aging Fischer 344 rats. J Gerontol 43:B50–53.
Ward WF (1988b): Food restriction enhances the proteolytic capacity of the aging rat liver. J Gerontol 43:B121–124.
Weindruch RH, Kristie JA, Cheney KE, Walford RL (1979): Influence of controlled dietary restriction on immunological function and aging. Fed Proc 38:2007–2016.

Weindruch RH, Cheung MK, Verity MA, Walford RL (1980): Modification of mitochondrial rsepiration by aging and dietary restriction. Mech Aging Dev 12:375–392.

Weindruch R, Naylor PH, Goldstein AL, Walford RL (1988): Influences of aging and dietary restriction on serum thymosin$_{\alpha 1}$ levels in mice. J Gerontol 43:B40–42.

Weraarchakul N, Richardson A (1988): Effect of age and dietary restriction on DNA repair. FASEB J 2:A1209.

Yu BP, Bertrand HA, Masoro EJ (1980): Nutritional-aging influence of catecholamine-promoted lipolysis. Metabolism 29:438–444.

Yu BP, Masoro EJ, Murata I, Bertrand HA, Lynd FT (1982): Life span study of SPF Fischer 344 male rats fed ad libitum or restricted diets: Longevity, growth, lean body mass and disease. J Gerontol 37:130–141.

Yu BP, Wong G, Lee HC, Bertrand H, Masoro EJ (1984): Age change in hepatic metabolic characteristics and their modulation by dietary manipulation. Mech Aging Dev 24:67–81.

Yu BP, Masoro EJ, McMahan CA (1985): Nutritional influences on aging of Fischer 344 rats: I. Physical, metabolic, and longevity characteristics. J Gerontol 40:657–670.

Yu BP, Laganiere S, Kim JW (1989): Influence of life-prolonging food restriction on membrane lipoperoxidation and antioxidant status. In Simic GM, Taylor KA, Ward JF, von Sonntag (eds): "Oxygen Radicals in Biology and Medicine." New York: Plenum Pub. Corp., pp 1067–1073.

Zs-Nagy I (1978): A membrane hypothesis of aging. J Theor Biol 75:189–195.

Zs-Nagy I (1979): The role of membrane structure and function in cellular aging: A review. Mech Aging Dev 9:237–246.

Index

AAUAAA polyadenylation sites, 165
Abnormal filamentous structures, 139
Acetylcholine (ACh) synthesis, 186
   AChE, 187–189
   functioning, age-related changes, 192
   synthesis and release, 189
   systems, 190
ACh. *See* Acetylcholine
ACTH, 233
Actin, polymeric to monomeric ratio, 113
Acyclic females, 205
AD. *See* Alzheimer's disease
Aerobic metabolism, 58
Agalactosyl
   IgG
      increase, 113
      rheumatoid arthritis, 106
      N-linked oligosaccharides, 106
Age effect, food restriction, 350–360
Age-1, 24
   and fer-15 crosses, 24
   genetic locus example, 25
   mean life spans, 22
   mutant, 22
      C. elegans, 11
   mutation, 24
Age-associated changes, T cells, minor, 110
Age-related
   changes, receptor activity, 364
   dysfunctions, 366
Age-synchrony, 16
Aging
   diversity, 306
   genes, 8
   major factor, 58
   protozoan, 41
   rate, genetic control, 293

   relationships, 29
   and spontaneous locomotor activity, 63
   study models, multicellular, 57
   universality, 305
Agonist stimulation, 189, 190
Alcohol
   exposure chronic effects, 77
   life extension, 68
   tolerance, 68
Aldolase activity, 322
Aldosterone, 237, 238
Alpha$_1$-antichrymotrypsin, 338
Alpha crystallin formation, 323
Alpha-MHC
   isozyme, 237
   /beta-MHC mRNA ratio, 237
Alpha-sen DNA, 31, 32
   senescence, causal factor, 30
Alternative demographic phenotypes, 6
Alzheimer's disease (AD), 341
   and CTL, 167
   etiology, 123, 141
   cortex or hippocampus, 167
   families, onset age, 125
   hippocampus, 166
   impariments, 139
   and normal senescence, 337
   pathology generation, 151
Alzheimer's Disease and Related Disorders A
   (ADRDA), 134
   neurofibrillary tangle (NFT), 171
   symptomology, 163
Amicronucleates, 43
Aminoacyl tRNA synthetases, 317, 318
Amoeba cultures, 301
Amplifier T cells, 111
   activity, 111

Amygdala, lateral, 205
Amyloid
 angiopathy, 171
 beta-peptide (AP), 163
 classification, 171
 common properties, 163
 defined, 171
 deposition, 174
  etiology, 168
 deposits, AD, 175
  human brain, 173
 fibrils, 141, 339
 filament, 338
 NFT exclusion, 171
 peptide (AP) sequence, 164
 plaque core protein, 172
 precursor protein (APP), 140
  cDNA clones, 146
  coding sequences, 165
  gene, 140
  gene 5', 147, 164
  gene locus, 149
  gene maps, 144, 166
  gene products, mutant, 153
  gene promoter, 148
  gene structure, 143
  gene triplication in AD, 166
  mini-gene constructions, 152
  mRNAs, 164–168
  precursor molecules, 145
  protein, 147
  role, 152
  transcripts, 145, 146
 precursor protein (APP)-751 transcript, 164
 proteins, 171, 338, 340
  fibers, 139
  in plaques, 140
 relationship to aging, 174
Amyloidosis, 171
 AD and Down's syndrome factors, 141
 process, 151
Ancestral marcronucleus, 44
Androgens, 239
 levels, 238
 responsiveness, 365
Aneuploidies, 302
Animal housing effects, 234
Animal models, nervous system disorders, 150
Antagonist dissociation occurence rates, 76
Antagonistic pleiotropy, 4, 7, 8, 10, 64
 genetic locus, 25

 hypothesis, 9
 mechanism, 4, 5
 theory, 63
Antiaging action, 365
Anti-beta-protein antiserum, 175
Anti-CD3 (OKT3)
 antibody, 93, 94
Anti-human gamma globulin antibody response, 109
Anti-IL2R, 91
Anti-inflammatory agents, 61
Antioxidants
 decline, 58
 defense loss, 58
 effect, longevity, 60
 levels, hydrogen peroxide effect, 59
 life span prolonged, 47
 status effects, 363
Antioxienzymes, 35
AP. See Amyloid peptide
APP. See Amyloid precursor protein
Apparent immortality, 41
Arcuate nucleus, 207, 205, 258
^{3}H-arginine, 233
Artificial chromosome vectors (YACs), 16
Aryl hydrocarbon hydroxylase, 267
Asexual vegetative cycle, 41
Asn-Gly sequence, 324
Asparaginyl deamidation, 324
Astrocytes, 77
Asx residues, 324
AT-rich decamer, 147
ATP, 32
 -dependent calcium, accumulation, 320
ATPase activity, 236, 320
Auto-anti-idiotypic antibody, 109
 levels, 114
 regulation, 105, 110
Autofluorescence
 coincidence, 17
 granules, 16
Autoimmune
 disorder, 77
 NZB strains, 111
Autonomous aging, 295
Autosomal dominant mode, AD genetic
  transmission, 125
B6, 77
 function results, 74
BALB/c defective dopaminergic mechanisms, 77

Index

Barrier-protected animals, 231
Basement membrane (BM), 287
B cell
    activation, 108, 109
    bone marrow generated, 108
    and CD5 B cells compared, 110
    differentiation (BCDF), 112
        age increase, 115
    down-regulation, 109
    function, 75, 107
    growth factors, 93
    7m8oGuo restored, 109
    population, peripheral, 109
    proliferation
        antigen-stimulated, 111
        T cell presence, 108
    subpopulation, CD5 positive, 106
BCDF. *See* B-cell differentiation
Behavior ordering and sequencing, 181
Benign condition, 106
Beta-adrenergic
    receptors decrease, 364
    -stiumulated, 245
Beta-amyloid, 340
    gene, 125
    precursor, 338
        protein, 164, 338–340
Beta-andrenoceptor, density, 76
Beta-cells isolated perfused, 235
Beta-MHC cardiac activity, 236
Beta-protein, 174
    amyloids, 171, 176
        deposits CNS, 175
        detection, 172
    immunostaining, 175, 176
    precursors, 172
Beta-sen 31, 32
Binary fission, 41
Bioavailability IL2, 86
Biological life span (RC), 34
Biomarker aging, 7, 76
Birefringence exhibition, 171
B lymphocytes, 105, 113
Bombyx mori 68
Bone marrow
    cells, 109
    chimeras, 76
BOS. *See* L-buthionine-(R,S)-sulfoximine
Bovine serum albumin (BSA), 16
Brain amyloid, aging, 172
Brain regions, hippocampus, 77

Brisk c-myc induction, 286
BSA. *See* Bovine serum albumin
B suppressor and helper T cell effects, 115
Bud abscission and scars, 33

Ca^{2+}
    channels, 95
    cytosolic free, 94, 95
    concentration, 30
    ion, 95
    mobilization, 192
    stimulated ATPase activity, 320
Caenorhabditis elegans, 10, 15, 294
    mutant allele, 7
    self-fertilizing, 22
Calcitonin levels increase, 365
Calcium
    binding, protein, 151
    concentrations, 245
    effects, 30
    ion movement, 244
    mobilization, 245–247
    movement nonstimulated changes, 248
    transport rate differences, 321
    uptake decrease, 319
Caloric restriction, 306, 365
cAMP
    phosphodiesterase activity, 7
    production vasopressin-dependent, 233
    -stimulated ^{32}P incorporation, 185
Carbohydrate moieties PHA binding, 83
Carboxymethyl-transferase, 325
Carcinogen treatment, 270
Cardiac
    calcium pump, 320
    muscle contraction, 245
    myosin ATPase activity, 237
    sarcoplasmic fractions, 319
    /skeletal muscle system, 236
CA3 region, 216
CAs. *See* Catecholamines
Casein, dietary, 361
Catalase, 35, 58, 59, 362
Catalog data food restriction, 349
Cataract formation relation, 323
Catecholamines (CAs), 193
    estrogen regulation, 204
    hypothalamic, 255
    metabolism, 259
Cathepsin B, 319
    cel, Ce2, CeX, 18
    D, 18

CD5 B cells, 107
  clones, 110
  disregulation, 110
  further research need, 115
CD3 complex, 93
Cell cycles
  genes, senescent cells, 274
  listed, 274
  regulation, 46
Cellular aging, 106
  process, 267
  differentiation, 29
Central nervous system, 181
Central processing area, 181
Cerebellum deposits, AD early onset, 176
Cerebrovascular amyloid, 141
    protein material, 139
CF3 foreskin fibroblasts, 274
C-fos, 274
CG. *See* Chorionic gonadotropin
Channels, transient, long-lasting negative, 244
ChAT. *See* Choline acetyl-transferase activity
C_3H female mice, 256
Chinese hamster fibroblast line V79-8, 270
Chitin major scar component, 33
Choline acetyl-transferase activity (ChAT), 186
  immunocytochemical staining, 188
Cholinergic function, 189, 191
Chorionic gonadotropin (CG), 238
C-H-ras
  DNA, 271
  proto-oncogene, 303
Chromatid gaps, 302
Chromatin structure, 49
Chromosomal
  mapping experiments, 144
  region loss, 301
Chromosome
  21, 133, 144, 164
    linkage, Alzheimer's, 124
  breakage 302, 305
Chronic CNS demyelinating disease, 150
Ciliates, 45, 48
  characteristics, 42
  model eukaryotes, 42
  protozoa, 301
Circulating sIL2R, elderly presence, 93
CJD. *See* Creutzfeldt-Jakob disease
Clastogens, 302
Clonal
  life span, Paramecium tetraurelia, 44
  senescence, 301

C-myc
  and ODC mRNA, 274
  proto-oncogene expression, 286
CNS. *See* Central nervous system
Collagen cross-linking, 306
Concanavalin A (Con A), 83, 97
Conformational isomerization, 331
Congo red staining, 172
Conjugation results, 302
Contigs, 15–16
Control-CM, 284
Control (CTL) tissues, 166
Controversy cell type survival, 295
Corpora lutea absence, 254
Corticosterone levels, 219, 234
  secretion maximum, 233
Cortisol, 233
  rhythms, 219
  secretion, 2341
Cosmid clones, 15
C-ras, 274
Crawlers, 58
Creutzfeldt-Jakob disease (CJD), 171
Crosslinking studies, 90
C-sis cDNA, 271
CTL. *See* Control tissues
Cu-An superoxide dismutase glycation, 325
Cumene hydroperoxide, 267
Cyanide
  -insensitive SOD, 59
  -resistant respiration, 29, 59
Cyanogen bromide fragments chromatography, 329
Cyclase stimulation loss, 185
Cyclic nucleotide and IP_3 formation, 194
Cystatin C, 163, 171
Cysteine residues, reactive, 331
Cytochrome oxidase, 32
Cytofluorograph, 107
Cytokines, 98
Cytological changes, 33
Cytosolic free calcium, 30, 96
Cytotoxins, 193

DA. *See* Dopamine
Daf mutants, 19
D_2 autoreceptors, 190
D_2 concentration decrements, 185
D_1 D_2 ratio increases, 183
Deactivation kinetics, 322
Deamidation asparagines, 329
  residues, 323

Index 377

Dementia risk, 132
Dexamethasone (DEX), 268
DHT. *See* Dihydrotestosterone, 238
Diabetes, 212
   age-correlated, 209
   prevalence age levels, 210
   type II, 219
Diabetogenic mutations, 212
Diet restriction (DR), 74
   biological systems, 74
   and increased longevity, 75
Dietary effects, 362
   food restriction, 350–360, 365
      studies, 295
      manipulation, dopaminergic receptor, 364
Diethylstilbesterol, 207
Differentiation, 29
Dihydrotestosterone (DHT), 238, 239
Diuretics, zinc depletion, 114
D-loop primer, 304
DNA
   cytosines, 298
   damage, 305
      agents, 18
      oxidative process, 366
   fragments, 15
   loss, 301
   methylation, 298, 299
   polymerase 36, 48
   repair 18, 35, 265, 366
      food restriction effect, 366
      theory, 73, 74
   replication, 46
   strand
      breaks, single-, 266
      rejoining, 266
   synthesis, 273
      inhibitors, 269
      PBL, 89
   -replicating enzymes, 85
   transfection experiments, 148
Dopamine (DA)
   decline
      concentration, 255
      hypothalamus, 257
   K+ evoked release, 190, 191
   receptors, neuronal, 243
   striatal, 182
      functioning striatal age-related changes, 192
   synthesis, DOPA accumulation assessment, 182

Dopaminergic
   ergot drugs, 256
   neurons, 258
Down's syndrome (DS), 171, 338
DR. *See* Diet restriction
D_2 receptor binding, 184
Drosophila melanogaster, 57, 294
   life span, 61
   lines, 7
   species, 5
   stocks, 8
   strains predictions, 65–66
DS. *See* Down's syndrome
DsRNA, 37
D_1 system, modulatory capability, 186
Dunce allele effects, 7
Dutch disease, hereditary, 338
Dystrophic neurites observed, 339

Ecdysterone-deficient mutant, 68
Ectopic expression, derepressed genes, 299
EF1-alpha, 300
EGF. *See* Epidermal growth factor
Egg-laying rate, 5
Egg production, female survival decrease, 63
EGTA chelating effects, 94
Elongation factor, 273
EMS. *See* Ethyl methanesulfonate
Endoplasmic reticulum, 244
Endotetraploidy, 302
Enzyme
   aging, 321
      importance, 316
   changes, 317
      DR results, 74
   conformational isomerization, 328
   levels, biological effects, 315
   oxidation, 326
      model shortcomings, 327–328
EPI. *See* Epidermal proliferation index
Epidermal growth factor (EGF), 267, 283
   -receptor complex, 269
   response to, 283
Epidermal keratinocytes, 288
   and melanocytes, 281
Epidermal melanocytes, 286
Epidermal proliferation index (EPI), 285
Epidermal thymocyte activating factor (ETAF), 284
Epinephrine, 238
Epistasis, 10

Erpobdella octoculata, 9
Error catastrophe theory, 321
Erythrocyte
    accumulation, 324
    function, 316
Estradiol, 206
    valerate (EV), 207
Estrogen, 201
    catecholamine regulation, 204
    elevated levels, 206
    hyposecretion, 205
    hypothalamic neuron damage, 258
    secretion decrease and delay, 204
    stimulates PRL, 259
    toxic effects, 203, 219
Estrogenic hysteresis, 203, 204, 208
Estrous cycles, 258
ETAF. *See* Epidermal thymocyte activating factor
Ethyl methanesulfonate (EMS), 69
Etiology
    AD, 155
    AP deposition, 163
Eukaryotes, higher characteristics, 48
EV. *See* Estradiol valerate 207
Evolutionary
    equilibrium, 5
    theory, 3
        reliability, 6
        senescence, 4
Exogenous IL2, 97
Exon coding, 86
Extracellular amyloic plaque cores, 141
Eye-lens, 322
    accummulations, 324
Familial AD (FAD)
    locus, 140, 155
        role, 141
    gene linked to AD, 141
Family history studies
    Ad, 127
    studies, error sources, 128
Fat restriction, 361
Fecundity
    allele promoted, 63
    and metabolic rate, 65
Feedback inhibition, 201
    loss, 205
Female
    fecundity collapse, 7
    longevity, male exposure, 63
    survival, mated and unmated, 67

Fer-15 gene, 23
Fertilization
    cell cycle, 44
    occurence, 41
FGF. *See* Fibroblast growth factor
Fibrillar proteinaceous substance properties, 171
Fibroblasts
    cultures, 265
    differentiated functions, 281
    growth factor (FGF), 268
    -like cells, 281
    normal human, 266
        treatment, 270
    replicative life span, 297
    studies limitations, 281
Filamentous fungi
    senescence, 36
    vegetative propagation, 304
Filter elution technique, 266
Fission age cell, 41
Fliers, 58
Flies, dewinged, 62
Fluorodeoxyuridine (FUdR), 16
Follicle-stimulating hormone (FSH), 238
Follicular depletion acceleration, 205
Food restriction, 208, 365
    caloric intake, 360–361
    effects, 209, 215, 350–360
    free radicals, 362
    information accumulation, 349
    intervention, 349
    modulation, 363
    and neuroendocrine impairment, 202
    popularity, 349
    role, 204
    specific diet component, 360–361
Foreskins, 283
    newborn, 287
Free radicals, 29, 35, 58, 259, 265
    -mediated oxidative stress, 38
    role, 267
    theory, 75
        aging, 362
FUdR. *See* Fluorodeoxyuridine

Galactosylation IgG fluctuates, 106
Gamma
    IFN production, 112
    -irradiation, 266
    radiation, 18

GAP 43 and actin, 342
Gene
  expression
    defect IL2, 87
    molecular probes, 365
    promoter minigene, 152
    regulation, 73
      catalase activity, 74
Generation time, 33
Genetic
  dissection, aging processes, 20
  heterogeneity AD, 126
  marker defined, 123
Germ line, senescence exemption, 294
G_1 event, 273
G_0G_1
  arrest point, 272
  levels, 267
  phase, cell cycle, 96
  progression, 273
G6PD. *See* Glucose-6-phosphate dehydrogenase
G_0 phase, 97
G-protein
  activation, 192
  interactions, 193
G_1S
  boundary, 273
  phase, cell cycle, 97
G_0 T cells, 95
GH. *See* Growth hormone
GHRH. *See* Growth hormone releasing hormone
Glial hyperactivity, 206, 207
Gliosis, 206, 258
Glucagon, 211
  biosynthesis, 235
Glucocorticoids
  feedback inhibition, 216
  hysteresis, 203, 215, 216, 219, 233, 296
  regulation, levels, 201
  toxic effects, 219
Glucose, 201
  -induced damage, 214
  intolerance, 211
    age-correlated, 210
  metabolism, 236
  oral, 210
  polymers, 77
  reduction, 209
  regulation, 208–210
  tolerance, 208
  toxic effects, 219

Glucose-6-phosphate dehydrogenase (G6PD), 318
  and GPDH activity, 322
  postsynthetic, 318
Glucose-6-phosphate isomerase (GPI), 323
Glutathion peroxidase, 316
Glutathione (GSH), 29
  decrease, 59
  levels, 267
  reduced, 363
  reductase, 316
Glx residue, 324
Glyceraldehyde-3-phosphate-dehydrogenase (GPDH), 318
Glycolysis, 106
Glycoproteins, 85
  S-28, 16
Glycosylation sites identified, 325
GnRH. *See* Gonadotropin releasing hormone
Gonadotropin releasing hormone (GnRH), 255, 258
  secretion rise, 255
  -suppressive effects, 239
GPDH. *See* Glyceraldehyde-3-phosphate-dehydrogenase
GPI. *See* Glucose-6-phosphate isomerase
Growth hormone (GH), 254
  decline, 260
  secretion, 256, 257
Growth hormone releasing hormone (GHRH), 256, 257
GSH. *See* Glutathione
GTP binding proteins intermediaries, 85

HACU. *See* High-affinity choline uptake
Haloperidol, 257
Hemizygosity, 301
Heterosis, 35
Heuristic value, new animal models, 76
High-affinity choline uptake (HACU), 186
  assessed, 187
Hippocampal neurons, 201
  CA3, 216
Hippocampus, 190
Histone
  composition, 43
  H_1^0, 300
  H3, 273
Holoclones, 283
Holometabolous insects, 68
Homeostasis, changes, 63

Index

Homeostatic
  mechanism deterioration, 65
  pathways, 296
Homogeneous
  antibody, 106
  immunoglobulin, 107
Homozygous populations, constructing, 23
Hormesis, 19
  defined, 47
Hostone proteins, 45
HPLC peptide mapping, 330
HPRT deficiency, 154
^{3}H thymidine modulation, 366
Human genes
  beta-globin, 149
  chromosome 21, 151
  lambda-crystalin, 149
Human epidermis, 281
Human fibroblast
  aging, 305
  foreskin, 271
Human precentral motor cortex, 189
Humoral immunosenescence, 105
Hydrogen peroxide
  effect, 59
  production, 259
4-hydroxynonenal, 267
Hyperglycemia, 208, 210, 213
  episodes, 209
Hyperglycemic hysteresis, 203, 219
Hyperinsulinemia, 214
Hyperparathyroidism progressive, 362
Hyperprolactinemia, 204, 205, 207
Hypersecretion, 204
  pancreatic, 211
  phases, 203
Hypodiploidies, 302
Hypomethylation, 298
Hypomorphy, negative effect, 69
Hypothalamic
  catecholamines (CAs), 255, 257, 258
  DA, 255
  dopamine impairments, 207
  dopaminergic neurons, 207
  involvement, 260
  neuron damage, 259
Hypothalamus, 255
  importance, 253
  role, 259
Hysteresis, 201, 203

Idiopathic paraproteinemia, 106, 107, 114
IFN. See Interferon
IgD receptors, CD4-positive T cells, 111
IGF-I. See Insulin-like growth factor
IgM
  anti-DNA antibody production, 111
  production, 108
IL1. See Interleukin 1
IL2. See Interleukin 2
IL2R. See Interleukin 2 receptor
Immortal
  human cell lines, 272
  strains, 41
Immunoglobulin production, 106
  depressed, 114
Immunosenesence, 105
  expression, 114
Incoloris mutants, 30
Indomethacin, 107
  supplementation, 114
Inheritance
  patterns, AD family history, 129–132
  trait onset, 132
Inositol
  phosphate (IP), 95
  triphosphate (IP$_3$), 244
INS. See Insulin
Insects, aging, 57
Insertional (IS) element, 36
Instar larvae, 68
Insulin (INS), 268
  biosynthesis, 235
  output, 210
  resistance, 210, 214
Insulin-like growth factor-I (IGF-I), 268
  transmembrane receptor system, 268
Interferon (IFN), 284
  -alpha-like protein, 286
Interleukin 1 (IL1)
  production, 113
  secretion monocyte, 86
Interleukin 2 (IL2), 83
  and B-cell stimulatory factors, 112
  biosynthesis, 86
  expression, 96
  gene, normal expression, 88
  and IL3 production age decline, 115
  levels, 96
  mRNA, 88, 97
    expression, age effects, 87

Index 381

product, 86
production, 87, 112
question, 88
secretion, 96
   cellular effects, 87
   decrease, 86
   -specific mRNA, 88
   T-cell function improved, 111
Interleukin 2 receptor (IL2R), 91, 97
   expression, 83, 85, 89–92
   lower density, 90
   subunits, alpha and beta, 90
Intermeiotic span, 305
Intracellular messengers, transduction, 93
Ionophores, PMA and Ca, 96
IP. *See* Inositol phosphate
Irradiation, cosmic gamma ultraviolet, 47
IS. *See* Insertional
Isomerization asparagines, 329
Isoproterenol-stimulated muscle, 245
Isozyme, type I, skeletal, 236

Kal DNA, 36
Kalilo
   cytoplasms, 36
   senescence factor, 36
   strains, 36
Karyotypic abnormalities, 302
Keratinocytes
   adult, 283
   cultivation, human, 282
   cultures, 284, 288
   generations, 282
   immune modultor production, 281
Kidney maltase, 329
Kunitz protease inhibitor, 164
Kuru plaques, 171

L-buthionine-(R,S)-sulfoximine (BOS), 267
L-dopa, 256, 257
L-type
   channels, 245
   current, 245
Lambda clones, 15
Late reproduction
   lines, 64
   selection, 65
Learning ability, age-related changes, 76
Lens enzymes, 322
Leucine transport, aged worms, 18
LH. *See* Luteinizing hormone

Life cycles, 254
Life extension effects, 306
Life span, 254
   and calorie restriction, 295
   differences, genetic basis, 66
   extension, 64
      high-density cultures, 65
      hermaphrodites RI strains, 21
   reduction, 74
Life-long food restriction, 208
Light microscopic immunohistochemistry, 340
Linear plasmids, kalilo and maranhar, 36
Linkage analyses and studies, AD, 133
Lipid peroxidation, 151, 320
   decrease, 75
   food restriction, 362
Lipofuscin
   accumulation, 16, 258
   cell aging indicator, 37
   granules, autofluorescent, 296
   -like compounds, 58
Lipofuscinoses, 338
Lipopolysaccharide responses, 109
Liver, catalase activity increase, 75
Locus coeruleus, 258
Long- and short-lived mice, 76
Long-lived strain chromosomes, 66
Longevity, 5
   DR mice increased, 75
   heritability, 293
   houseflies, 58
   increase, 74
   tool, diet restriction, 74
Lordosis, 207
Luteinizing hormone (LH), 204, 238, 239
   impairments, 204, 205
   levels, 255
   surge, 205
Lymphokines, 83
   age-associated changes, 112
   production, 106
Lysates, APP-770 transfected COS cells, 164
Lysis, 34
Lysomal
   hydrolases, 18
   proteases decline, 18
Lysosomotropic agents, 18

mAChR. *See* Muscarinic receptors
Macronuclear development, Tetrahymena, 46

Index

Macronucleus, 49
  longevity role, 43
  protozoan life span reasons, 43
mADhR, 193
Major histocompatibility complex (MHC), 107
Malondialydehyde polymers, 37
Mammalian senescence, 305
MAP 1b and 2, 342
Maranhar, 36
Marker protein (S-28), 16
Mathematical theory, 6
MBP. *See* Myelin basic protein
MDA metabolic fate, 363
Mean population doublings (MPD), 296
Mechanism of aging theories, 265
Median life span, improved, 361
Mediolateral and dorsoventral striatal axes, 184
Melanocytes, 287–288
  pigment production, 281
Membrane
  alterations, 363
  bound proteins, 319
  deterioration, 364
  fatty acid composition changes, 364
Memory
  age-related changes, 76
  and learning impairment, 77
Meningeal amyloid, AD and Down's brains, 141
Meninges and cerebral cortex, 139
Metabolic
  efficiency increased, 75
  rate, 29, 61, 65
Metazoa senescence, 293
Methylase activity nuclear extracts, 299
Methyl methanesulfonate (MMS), 18
7-methyl-8-oxoguanosine (7m8oGuo), 109
MFO. *See* Mixed-function oxidation
MHC. *See* Major histocompatibility complex;
    Myosin heavy chain
Mice, age span differences, 258
Microinjection technique, 271
Microisolator cages, 231
Micro-/macronuclei differentiation, 47
Micronucleus, 43, 49
  transcriptionally inactive, 43
  versus macronucleus, 48
Mineral restriction, 361
Mitochondrial
  DNA

    loss, 61
    structural alterations, 304
  (mt) plasmid DNA (pl DNA), 30
  mutations, 37
Mitogenic potential decline, 106
Mitotic cells, 296
Mixed biological function results, 74
Mixed lymphocyte response (MLR) compared, 110
Mixed-function oxidation (MFO) systems, 326
  systems, 327
MLR. *See* Mixed lymphocyte response
MMS. *See* Methyl methanesulfonate
Molting failure, 68
Monoclonal
  antibodies receptor complex, modified, 85
  gammopathy, 106
Monosodium glutamate, 206, 212
Motor performance retarded, 182
Mouse studies, 73
MPD. *See* Mean population doublings, 296
mRNA, 165, 270
  x c-myc, 273
mt-IS element, 36
mtDNA, 36, 304
Multifocal leukoencephalopathy (PML), 150
Multiple endoreduplication, 44
Multiplex families, 124
Multivariate analyses, comparing, 337
Murine enzyme, 155
Mus musculus, 78, 258
  aging, 73
Muscarinic receptors (mAChR), 188
  /agonist stimulation sensitivity, 191
  binding, 187
  blunt, 191
  deficits responsivity, 190
  ligand activation, 192
  responsiveness, 189, 190
  subtypes, pattern loss, 188
Mutation accumulation, 3, 4, 64
  hypothesis, 5
  mechanism 5, 8–10
  theory, 63, 68
Mycelia, 31
Myelin basic protein (MBP), 150
  transgenic mice, 151
Myoplasmic calcium trnsient, 320
Myosin heavy chain (MHC), 236–237

Index 383

N-acetylcysteine (NAC) 267
NADH, 61
NADP$^+$ coenzyme effect, 318
NADPH, 267
  labeled, 339
National Institute of Neurological and Communcative Disorders and Stroke (NINCDS), 134
Natural selection, 3
  force, 6
N-CM, 284
NE. See Norepinephrine
Necropsied brains, 77
Nematode histones, 16
Nephropathy progression, 361
Nephrotic lesion reduction, 361
N-ethylmaleimide, 319
Neural regeneration, 341
Neurites, 340
Neuritic
  degeneration, 340
  dystrophy neural epithelium nose, 340
  plaque, 141
    core content, 139
    formation, 163
  response, 340
Neuroendocrine
  control, 201
  deterioration, 201
  impairments, 203–205
Neurofibrillary tangles (NFT), 151, 338, 339, 343
  AD brains, 139
Neurohormonal hysteresis, 217–218
Neurohumoral hysteresis, 201, 203
  features, 202
Neurohyphyses, superfused, 234
Neuroleptic drugs, listed, 257
Neuronal
  cell loss, 151
  degeneration, 206
  feedback inhibition, 203
Neuropathological definition Alzheimer's, 337
  marked, 172
Neurophyophysis, 233
Neurospora, 29, 32
  intermedia, 361
NFT. See Neurofibrillary tangles
Nigrostriatal system sensitivity, 193
NIH 3T3 cells, 271

N-linked glycosylation sites, 142
Nonfibroblastic cell aging models, 281
Nonlysosomal pathway, 18
Norepinephrine (NE), 255
  decline, hypothalamus, 257
  secretion, 238
Normal aging
  correlation, 339
  process, 172
Normal human fibroblasts, 272
Northern blot studies, 166, 167
N-terminal peptide sequences, 141
Nuclear protein, 300
Nucleic acid processing, 45
NZB mouse model, autoimmunity, 75

Obesity, 212
O-CM, 284
ODC. See Ornithine decarboxylase
OKT3. See Anti-CD3
Oncogenes, 269
Optimal body weight departures, 74
Ornithine decarboxylase (ODC), 273
  mRNA, 274
Osteoporosis, senescence-prone lines, 76
Ovariectomy, 205, 207, 254
  estrogen reduction, 204
Ovaries
  exhaustion, 205
  impairments, 204
  transplanted, 254
Oxidative damage, 321
Oxide dismutases (SOD), 58
Oxygen
  consumption, winged flies, 62
  free radicals, 29, 295, 316

Pacemaker organ, 295
Paired helical filaments (PHF), 140
Pancreas, 210 hypertrophy, 213
Pancreatic
  glucagon plasma levels, 211
  hypersecretion, 212
Paramecium, 44
PCR. See Polymerase chain reaction
PD. See Population doublings
PDGF proteins, 271
P21EJ-ras, 271
Pentobarbital metabolism, 75
Periperal tissue expression, 146

Index

Peromyscus leucopus, 258
Peroxidase, 58
Peroxidative challenges, skin fibroblasts, 267
PGK. See Phosphoglycerate kinase
PHA. See Phytohemagglutinin
Pharmacologic agents, D_1 and D_2 specific, 186
Phase III proliferative senescence, 286
Phenathiazines, 257
PHF. See Paired helical filaments
Phorbol esters, 287
  tumor-promoting, 95
Phosphatidylinositol (PI), 192
Phosphoglycerate kinase (PGK), 318, 329
  aging reversibility, 330
  cardiac muscle, 329
  initial oxidation, 331
Phosphoinositol bis-phosphate (PIP_2), 85
  hydrolysis, 95
Phosphorylcholine-specific splenic B cells, 109
Photoreactivation and longevity, 48
Physarum, 29
  polycephalum microplasmodia, 30
Physiological parameters studied, food restriction, 350–360
Phytohemagglutinin (PHA), 97
  binding, 83
  receptors, 86
  -stimulated, 94
    lymphocyte, 106
    cells, 89, 112
PIP_2. See Phosphoinositol bis-phosphate
PKC. See Protein kinase C
Plasma
  membrane channels, 244
  norepinephrine and sleep, 238
Plasmid-derived insertion sequences, 37
Platelet-derived growth factor (PDGF), 268
Platelet membrane fluidity, AD, 133
Pleiotropic
  effects, fitness characters, 68
  patterns, 10
PMA/ionomycin, 97
  stimulation, 96
PML. See Multifocal leukoencoephalopathy
PO_2 reduction 297
Podospora, 29, 31, 32
Pokilotherms temperature effect, 65
Poly A + RNA, 270
Polygenic, 66
  etiology, 155

Polymerase chain reation (PCR) technology, 146
Polypeptides, 163
  chains T3 complex, 83
  degraded, 323
Polyploidy, 302
Population doublings (PD), 282
Posttranslation protein modification, 322
Premature aging syndromes, 319
Prion PrP, 163
PRL. See Prolactin
Procedures used, food restriction, 350–360
Processed antigen, recognized, 84
Progesterone, 254
  implants, 208
  role, 204
  treatments, 208
Prokaryotes characteristics, 48
Prolactin (PRL), 205
  elevated, 205
  hypersecretion, 205
  secretion, 255, 256
Proliferative capacity in vitro, 83
Propyl gallate, 60
  life-extending property, 61
Prostaglandins, 107
Protease inhibitor domain, 172
Protein
  damage, oxygen radicals, 326
  degradation, 322
  expression, quiescence and senescence, 300
  glycation, 325
  kinase C (PKC), 85, 94
    activated, 96 functions, 95
    reduction, 361
    and membrane component, 319
    and nephropathy relationship, 361
  sequence information
    amyloid protein, 142
    revealed, 139
  synthesis, 254, 256, 260, 322
    changes, 62–63
    decrease, 106
    enzyme role, 317
Proteolysis
  introcellular, 327
  lens crystallins, 322
Proteolytic activity, 18
Proto-oncognes, normal cellular, 269
Protozoa, 41
Pseudopregnancy vaginal pattern, 205

Quiescence genetics, 30

Racemization Asp residue, 325
Range research, 76
RC. *See* Reproductive capacity; Biological life span
RDNA Paramecium, 47
Receptor
    /ligand interface, 193
    synthesis, impaired regulation, 185
Recombinanat, inbred strains, 20
Redox
    balance, 29,
    status, 316
Rejuvenation process, 302
Reporter gene, 152
Reproductive capacity (RC)
    and bud scars, 34
    phenotype, 35
    typically, 20–30
    see also Biological life span, 33
Reproductive neuroendocrine impairments, 204
Reproductive schedule, population, 4
Reproductive senescence, 204
    phase, 204–205
Rescue hybridization, 35
Reserpine, 257
Resistance starvation decline, 67
Restriction, food, 350–360
Reverse transcriptase activity, 31
Reversed selection experiment, 67
Rhodamine-B-isothiocyanate (RITC), 16
    -BSA intestinal granules, 17
RNA
    gel blot analysis, 164
    polymerase, 36

S-28, 16
SAC. *See Staphylococcus aureus*
SAF (scrapie-associated fibrils), 171
Sarcolemma preparations mulation, 320
Sarcoplasmic reticulum vesicles, 320–321
Scavengers, dietary, free radical, 37
SCE. *See* Sister chromatid exchange
Scrapie and Creutzfeldt-Jacob disease, 164
Self-fertilization, 45
Senescence
    accelerated, 4, 76
    delay, 37

events dependency order, 20
evolution, 3
    theories, 63
    postponed, 7
female reproductive, 203
filamentous fungi, 37
inducement, 36
life span factors, 305
phase III, 286
plasmids, 37
stochastic process, 65
strain characterization, 304
theories
    contrasted, 64
    tested, 67
Senescent
    cells, 269
    striatum indices, 186
Senile plaques, 151, 339
    appearance, 172
    cerebral cortex, 174
        distribution, 176
    composition, 338
    relationship, neurofibrillary, 338
Sensorineural hearing loss, progressive B6, 78
Serial transplantation experiments, 295
Serine protease inhibitors, 164
Serotonin increase hippocampus, 77
Serum
    25-hydroxy vitamin D, 362
    PTH, 362
    titers, IgA, IgG, IgM, 105–106
Sex evolution, 3051
Sexual activity and longevity, 7, 67
Sexual cycle, 41
Short-lived strain chromosomes, 66
sIL2r, 93
Single-celled organisms, 41
Sister chromatid exchange (SCE), 73, 303
Skin biopsies, 281
    paired, 282
Sleep
    quality decline, 238
    /wake pattern, 238
SOD. *See* Superoxide dismutase
Sohal group study results, 59
Somatomedin secretion, 254, 260
    decline, 256
Somatostatin, 256
    levels increase, 77
    pancreas, 235

Soy protein, 361
    diet retarded, 362
Species or strain food restriction, 350–360
Specific pathogen-free (SPF), 231
    animals, 233
        barrier-maintained, 237
        barrier-protected, 234
        status, 231
Sperm production decrease, 255
SPF. See Specific pathogen-free
S phase, 273
    cells, 296
        young, 274
Spiperone
    ^{3}H, 183
    -labeled sites, 184
Spleen
    cells, PMA-costimulated, 96
    function, 75
Splenic precursor frequencies, age decline, 109
Splenocyte cell surface, MHC antigens, 113
Splenocytes, 107
    stimulated, 96
Sprague-Dawley rat, 183, 185
S100 protein, 151
Staphylococcus aureus (SAC)
    B-cells
        assess growth, 107
        -responding, 108
        Cowan I, 108
Statin synthesis, 300
Steroidogenic enzymes, 233
Stochastic loss, 65
Streptoxotocin, 213
Striatal
    D_1 and D_2, 189
    function interference, 182
    membrane fluidity, 185
    receptors, 183, 185
Striatum
    aged organism alteration, 181
    homogeneous structure, 184
    senescence, 185
Study references, food restriction, 350–360
Succinate dehydrogenase activity, 319
Succinimide formation, 324
Superfused neurohypopheses, 233
Superoxide dismutase (SOD), 35, 267, 317
    activity, 30, 58, 59, 322
        longevity, 60

    cyanide-insensitive 29, 1
Superoxide radical generation, 29
Suppressed lipid peroxidation, 362
Suppressor cell function, age affected, 110
Suppressor T cells evidence, 86
SUSM-1 immortal cell line, 269
SV40 T-antigen expression, 150
Sympathetic nervous sytem, 238
Synapses number decrease, 189
Synthetic peptide, beta-protein, 172

Takeda group, 76
TATA
    box, 147
    and CAAT box absence, 147
Tau
    Alzheimer neurites, 343
    mRNA, 342
    protein, 343
        immunoreactivity, 174, 342
3T3 cells, 282
    feeder, 283
T cells
    accumulation nonresponding, 88
    activation, 94, 98
    activity, 111
    antigen
        recognition, 85
        response and mitogen, 83
    bone marrow removal, 109
    CD3+, 93
    defect, 94, 106
    defective signal transduction, 98
    deficient, 109
    development and function, 98
    function, age-dependent decline, 75
    PMA stimulated, 90, 92
        -costimulated, 96
    purified 88, 1
    population defect, 86
    receptor (TCR), 83, 98
        antigen recognition role, 93
        -CD3 complex, 84
T3 complex, 83
TCR. See T-cell receptor
Telomere terminal transferase, 301
Telomeric repeats, 301
Test-retest reliability, AD informants, 129
Testosterone, 238
    decline, 255
    secretion, 254

Tetrahymena, 42, 45
Tetraploidy increase, 302
Thermoregulation, age-related decline, 78
Thymic lymphocytes subtypes, 249
Thymidine kinase, 272–274
Thymidine triphosphate (TTP) synthesis, 273
Thymus gland, 254
Thyroid
  deficiency, 236
  hormone, 236, 237
  status, 236
Thyroxine (T_4), 236
T-independent and dependent antigens, 108
Tissues used, food restriction, 350–360
Titers serum, increase and decline, 105
TJ143 RI strain longest-lived, 24
T lymphocytes, 83
  cultures, 97
TNP. *See* Trinitrophenyl
TNP-ABC. *See* Trinitrophenyl antigen binding cells
Tolerance induction, 109
TPE. *See* Triose phosphate isomerase, 323
Transcription, 49
  process, protein synthesis, 365
Transfection experiments, 149
Transgenic animals
  disease model
    AD mouse, 152
    limitation, 154
    human, 153
  models development, 150
  viral gene products, 150
Translocation, 302
Transmembrane signaling, 96–98
Trinitrophenyl (TNP), 108
  antigen binding cells (TNP-ABC), 108
Triodothyronine (T_3), 236
Triose phosphate isomerase (TPI), 323
Triplicated DS, 166
Trisomy chromosome 21, 1441
Trypsin
  activity cell lystates, 145
  inhibitor activity, 164

TTP. *See* Thymidine triphosphate
Tumorigenesis DNA transfection, 271
Tumors mammary and pituitary, 254
Turbatrix aceti, 18
Twin studies, AD, 126–127

Ubiquitin antibodies, 339
Ultra violet (UV), 18
  -induced dimers, 303
  and longevity, 48
Unified theory, 30

Vaginal cornification, 205, 208
Vasopressin
  levels, 233
  secretion, 233, 234
Vegetative cell cycle, 46
Ventromedial hypothalamus (VHM), 201, 213, 214
  damage, 204, 219
  estrogen receptors, 206
  lesions, 211, 212, 216
  neurons, 205, 208
Ventromedial nucleus, 207
Verapamil, chelating effects, 94
VHM. *See* Ventromedial hypothalamus
Vitamin C, 47
Vitamin E, life span effect, 19
Volga German families, 125

Wear and tear hypotheses, 295
WI-38 cells, 267, 273
Wound repair rate, 78

X-ray microanalysis, 363
Y-CM, 284
YACs. See Artificial chromosome vectors
Yeast senescence factor, 35

Zinc
  depletion, 114
  supplementation, 115